33 Springer Series in Solid-State Sciences
Edited by Manuel Cardona

Springer Series in Solid-State Sciences

Editors: M. Cardona   P. Fulde   H.-J. Queisser

Volume 40 **Semiconductor Physics** – An Introduction   By K. Seeger

Volume 41 **The LMTO Method**   Muffin-Tin Orbitals and Electronic Structure
By H. L. Skriver

Volume 42 **Crystal Optics with Spatial Dispersion, and Excitons**
By V. M. Agranovich and V. L. Ginzburg

Volume 43 **Resonant Nonlinear Interactions of Light with Matter**
By V. S. Butylkin, A. E. Kaplan, Yu. G. Khronopulo, and E. I. Yakubovich

Volume 44 **Elastic Media with Microstructure II**   Three-Dimensional Models
By I. A. Kunin

Volume 45 **Electronic Properties of Doped Semiconductors**
By B. I. Shklovskii and A. L. Efros

Volume 46 **Topological Disorder in Condensed Matter**
Editors: F. Yonezawa and T. Ninomiya

Volume 47 **Statics and Dynamics of Nonlinear Systems**
Editors: G. Benedek, H. Bilz, and R. Zeyher

Volume 48 **Magnetic Phase Transitions**
Editors: M. Ausloos and R. J. Elliott

Volume 49 **Organic Molecular Aggregates,** Electronic Excitation
and Interaction Processes
Editors: P. Reineker, H. Haken, and H. C. Wolf

Volume 50 **Multiple Diffraction of X-Rays in Crystals**
By Shih-Lin Chang

Volume 51 **Phonon Scattering in Condensed Matter**
Editor: W. Eisenmenger, K. Laßmann, and S. Döttinger

Volume 52 **Magnetic Superconductors and Their Related Problems**
Editors: T. Matsubara and A. Kotani

Volume 53 **Two-Dimensional Systems, Heterostructures, and Superlattices**
Editors: G. Bauer, F. Kuchar, and H. Heinrich

Volumes 1 – 39 are listed on the back inside cover

A. I. Kitaigorodsky

Mixed Crystals

With 134 Figures

Springer-Verlag
Berlin Heidelberg New York Tokyo 1984

Professor Dr. *Alexander I. Kitaigorodsky*

Institute of Elemento-Organic Compounds, Vavilova Street 28
SU-117334 Moscow B-312, USSR

Series Editors:
Professor Dr. Manuel Cardona
Professor Dr. Peter Fulde
Professor Dr. Hans-Joachim Queisser

Max-Planck-Institut für Festkörperforschung, Heisenbergstrasse 1
D-7000 Stuttgart 80, Fed. Rep. of Germany

ISBN 3-540-10922-6 Springer-Verlag Berlin Heidelberg New York Tokyo
ISBN 0-387-10922-6 Springer-Verlag New York Heidelberg Berlin Tokyo

This work is subject to copyright. All rights are reserved, whether the whole or part of the material is concerned, specifically those of translation, reprinting, reuse of illustrations, broadcasting, reproduction by photocopying machine or similar means, and storage in data banks. Under § 54 of the German Copyright Law, where copies are made for other than private use, a fee is payable to "Verwertungsgesellschaft Wort", Munich.

© by Springer-Verlag Berlin Heidelberg 1984
Printed in Germany

The use of registered names, trademarks, etc. in this publication does not imply, even in the absence of a specific statement, that such names are exempt from the relevant protective laws and regulations and therefore free for general use.

Offset printing: Beltz Offsetdruck, 6944 Hemsbach/Bergstr. Bookbinding: J. Schäffer OHG, 6718 Grünstadt.
2153/3130-543210

Preface

The two-word title of this book can only give an indication about its content and approach to the subject it deals with. In the course of time, the term has gradually become somewhat blurred. The reason is easy to see: similar problems are now more and more frequently studied by different branches of natural science. The term "mixed crystals" has acquired specific connotations in physics, chemistry, biology, and geology. One and the same term can now serve as a name for things which are either not quite the same or sometimes quite different.

And this is precisely what happened to the two words in the title of the book. One of them, the term "crystal", for which crystallography had an unambiguous definition, is now employed by biologists to describe the structure of cell membranes and by chemists who use it to denote degrees of polymer crystallinity. "Crystal" has thus become a broad term that can help describe any solid, or just a condensed state of a substance, if the solid has a sufficient degree of order in the arrangement of its components.

But the book is called "Mixed Crystals". The other word in its title, the adjective "mixed", has also developed several meanings. It is now thought applicable to both homogeneous and heterogeneous systems, that is, to crystals composed of different molecules and also to solids that are a mixture of crystals with different structures.

What is the subject matter of this book then? By and large, it deals with ordered mixed systems, which are studied by crystallographers, physicists, organic and inorganic chemists, and biologists, and which arouse the interest of engineers, technologists, and metallographers.

Was it worthwhile to examine such diverse material in one book, the size of which does not allow going into great detail? The author believes it was: technological progress has split natural science into thousands of clans; in contrast books like this are helpful in providing a kind of unifying cross-sectioning through its body. It has long become clear that fascinating discoveries take place where several sciences overlap, in other words, when the ideas and methods of one branch of the natural sciences are borrowed by another. It seems to me important to stimulate such processes: to get a metallographer whose subject is lattice transformations in alloys interested in conformational changes in organic molecules, and to acquaint a biologist who studies muscle contraction with decomposition of solutions of atomic crystals.

One can use different approaches to tackle such a task. It is possible to confine the study to a group of properties, to become involved with a general theory, or to take a purely pragmatic attitude, that is, to analyze the industrial benefits of composite materials. The author has, naturally, chosen the sectioning of the problem which best agrees with his knowledge and experience. So the book deals with the structure of mixed crystals - "structure" is used here in the broadest sense of the word implying atomic, molecular, cluster, or microcrystalline structures.

The field of research and the taste of the author account for two more features that distinguish this book. First, and of particular importance, is the physics of molecular crystals, specifically organic compounds. The second is that the book establishes the principle of close packing as the most important one in examining the structure of solids. The book is thus addressed to a broad audience of physicists, chemists, and technologists who are interested in the atomic and molecular structures of substances. Although it offers no new facts to those who specialize in a particular field, they might find it both interesting and useful to learn more about their "neighbors" in science and might appreciate the "close-packing approach" to the problems of the structure of solids.

Moscow, December 1983   A.I. Kitaigorodsky

Acknowledgement

This book could not have been written without the assistance of colleagues and collaborators.

Many thanks are due to Professor N.B.Chanh, Talence, France, for sending me reprints of his articles on inorganic and organic solutions. The results of his research group have been widely used. It is a pleasure to express my deep gratitude for his permission to reproduce some of the figures.

I would like to thank Professor R. Fouret and Dr. F. Baert, Lille, France, for sending me the material necessary for writing the section on solid solutions of enantiomers. I am very grateful to Professor David O. Mac Nicol, Glasgow, United Kingdom, for sending me reprints of an extremely interesting work on packing complexes.

I also express my gratitude to Professor V.S. Stubican, Pennsylvania, USA, for his permission to reproduce a figure from his papers.

My Russian colleagues have contributed greatly to my work with their constructive criticism and large amount of material. Professor Ya.S. Umansky, Moscow, helped me in writing the chapters dealing with metallic compounds, Dr. Yu.A. Dyadin, Novosibirsk, greatly facilitated the writing of the sections on clathrates, and Dr. W.B. Varshal, Moscow, supplied me with perfect photographs of ceramic materials. I would like to thank Dr. L.Z. Rogovina, who read attentively the chapter on polymers and made valuable remarks.

I wish to acknowledge the valuable collaboration of Dr. N.E. Esipova in writing the chapter on biopolymers.

I acknowledge with gratitude and affection the generous help I have received from the people from my laboratory.

The investigations performed by Professor D.Ya. Tsvankin and his pupils, and his interesting work in collaboration with Dr. S.G. Tarasov proved to be very helpful to me when I wrote Chapter 14.

About half of an important chapter of the book, devoted to the investigation of the structure of organic solutions, is a short account of the work carried out of Dr. R.M. Myasnikova and her pupils.

I should like to thank Dr. S.V. Semenovskaya for her generous help. She has looked through all the parts of the book dealing with the theory of shortrange order and the theory of order-disorder transformations and has made valuable remarks.

My thanks are due to Dr. O.F. Polivtsev for his assistance in writing Section 15.2.

I would like to warmly acknowledge the help rendered me by Mrs. I.E. Kozlova. She helped me to select the literature, to calculate the values of the increase in density and to prepare the manuscript.

The author appreciates the invaluable help of Professor Robert M. Metzger, University of Mississippi (USA), in clarifying details in the manuscript which had been obscured in the translation. He is also thankful to Mrs. Ingrid Blass for undertaking the job of preparing the master manuscript on the basis of a substantially copy-edited manuscirpt.

In this brief list of names I could not mention all the people to whom I owe a debt of gratitude. Sometimes even a little remark or suggestion was helpful to me. Therefore my thanks are due to all my friends, colleagues, collaborators, and pupils.

Contents

1. Introduction 1
1.1 The Early History 1
1.2 Developments of the Twentieth Century 8
1.3 Outline of the Book 14

2. Phase Diagrams 17
2.1 Fundamentals 17
2.2 Conditions of Stability in Mixed Crystals 20
2.3 Types of Phase Diagrams 26
    2.3.1 Crystallization of Two Pure Components from the Liquid Phase 27
    2.3.2 Unlimited Solubility 28
    2.3.3 Limited Solubility 31
    2.3.4 Component Forming a Complex 33
    2.3.5 Intermediate Phases 34
    2.3.6 Transformations in the Solid State 36
    2.3.7 Phase Diagrams of Optical Enantiomers 39
2.4 Determination of Phase Diagrams 40
    2.4.1 Differential Thermal Analysis 41
    2.4.2 X-Ray Phase Analysis of Polycrystalline Specimens 42
    2.4.3 Growth of Mixed Single Crystals 43
    2.4.4 Method of Contacting Specimens 46

3. Particle Packing in a Crystal 49
3.1 Elements of the Theory of Space Lattices 50
3.2 Geometric Model 60
    3.2.1 Close-Packed Spheres 60
    3.2.2 Representation of Atoms by Spheres 61
    3.2.3 Representation of Atoms by Truncated Spheres 63

	3.2.4 Hexagonal Packing	65
	3.2.5 Body-Centered Cubic Structure	67
	3.2.6 Ionic Crystals	68
	3.2.7 Molecular Crystals	71
3.3	Structure of Single-Phase Mixed Crystals	75
	3.3.1 Ordered Crystals	75
	3.3.2 Disordered Crystals	76
	3.3.3 Degree of Long-Range Order	77
	3.3.4 Static Concentration Waves	78
	3.3.5 Short-Range Order	80
	3.3.6 Size and Shape Factors	81

4. Free Energy of a Solid Solution ... 85

4.1	General Formula for the Free Energy	86
4.2	Solid Solution Characterized by One Interaction Parameter	91
	4.2.1 Free Energy of Mixing	92
	4.2.2 The Ordering Model	94
4.3	Limited Solubility	96
	4.3.1 Relationship Between the Energy of Mixing and Solubility	97
	4.3.2 The Calculation of Solubility by a Given Substitution Scheme	98
4.4	Permutation of Particles in a Rigid Lattice	101
4.5	Lattice Distortions	104
4.6	Thermal Vibrations	109

5. Heterophase Systems ... 111

5.1	Eutectic Crystallization	112
	5.1.1 Mechanism of Uncontrolled Crystallization	112
	5.1.2 Unidirectional Crystallization	119
5.2	Decomposition of Solid Solutions	123

6. X-Ray Scattering ... 126

6.1	X-Ray Scattering by a Single Crystal	127
6.2	Scattering by One- and Two-Dimensional Systems	129
6.3	Thermal Vibrations	130
6.4	X-Ray Diffraction by Solid Solutions	132
	6.4.1 Lattice Distortions	132
	6.4.2 Structure of the Average Unit Cell	133
6.5	X-Ray Diffuse Scattering by Solid Solutions	136

	6.5.1 Short-Range Order	136
	6.5.2 Energy of Mixing	138
6.6	X-Ray Structure Analysis	139

7. Intermetallic Compounds ... 142

7.1	Classification Schemes for Intermetallic Compounds	144
7.2	Geometrical Analysis	146
7.3	AB Compounds	147
	7.3.1 AuCu Type	147
	7.3.2 CsCl Type	148
	7.3.3 NaTl Type	149
	7.3.4 Zinc Blende (ZnS) Type	149
	7.3.5 NiAs Type	149
	7.3.6 NaCl Type	150
7.4	Compounds AB_2	151
	7.4.1 $MgCu_2$ Type	151
	7.4.2 AlB_2 Type	151
	7.4.3 $CuAl_2$ Type	152
	7.4.4 $MgZn_2$ Type	152
7.5	Compounds AB_3	153
	7.5.1 $SiCr_3$ Type	153
	7.5.2 Type BiF_3	154
	7.5.3 $AuCu_3$ Type	154
7.6	Number 13	155
	7.6.1 Compounds AB_{13}	155
	7.6.2 Compounds A_mB_n with $m = 1$ and $n = 12$	158
	7.6.3 Compounds A_mB_n with $m = 2$ and $n = 11$	159
	7.6.4 Compounds A_mB_n with $m = 3$ and $n = 10$	159
	7.6.5 Compounds A_mB_n with $m = 4$ and $n = 9$	160
	7.6.6 Compounds A_mB_n with $m = 5$ and $n = 8$	161
	7.6.7 Compounds A_mB_n with $m = 7$ and $n = 6$	161
7.7	Close-Packing Principle for Intermetallic Compounds	162

8. Solid Solutions of Metals ... 164

8.1	Unlimited Solubility	164
	8.1.1 Elements with the fcc Lattice	165
	8.1.2 Elements with the HCP Lattice	166
	8.1.3 Elements with the bcc Lattice	166
8.2	Terminal Solid Solutions	167

8.3 Solid Solutions on the Basis of Intermetallic Compounds 168
8.4 Ordering 171
     8.4.1 Investigation of the Order-Disorder Transition by X-Ray
            Diffuse Scattering 172
     8.4.2 Disorder in Cu_3Au 173
     8.4.3 Fe-Al Phase Diagram 174
     8.4.4 Determination of the Degree of Long-Range Order by X-Ray
            Diffraction 176
     8.4.5 Short-Range Order 177

9. **Inorganic Solid Solutions** 181
9.1 Isomorphous Substitutions in Alkali Halide Salts 181
9.2 Effect of the Difference in Ionic Size on the Solubility Limits
     in Systems of Inorganic Compounds 188
9.3 Theory of Ionic Substitutional Solid Solutions 190
     9.3.1 General Remarks 190
     9.3.2 Review of Suggested Theories 192
9.4 Interstitial Solid Solutions 195

10. **Conditions of Formation of Substitutional Organic Solid Solutions** 200
10.1 Geometrical Analysis of Substitution 201
      10.1.1 Major Rule of Solubility 201
      10.1.2 Exceptions to the Basic Rule of Solubility 203
10.2 Conditions of Continuous Solubility 204
      10.2.1 Crystals of the Components Belonging to the Same
             Structure Type 205
      10.2.2 Continuous Loss of Symmetry 206
10.3 Energy Calculations 208
      10.3.1 Approximate Estimation of Mixing Energy 208
      10.3.2 Possibility of Calculating Phase Diagrams 209
10.4 Unit-Cell Dimensions of Solid Solutions 210
      10.4.1 Lattice Loosening 211
      10.4.2 Interblock Solubility 214

11. **Ordering in Organic Solid Solutions** 217
11.1 One-Component Crystals with Orientational Disorder 219
      11.1.1 Disorder with Increase in Crystal Symmetry 219
      11.1.2 Disorder with No Increase in Crystal Symmetry 222
11.2 Orientational Disorder in Binary Systems 223
      11.2.1 Disorder Due to Admixture Molecules 223
      11.2.2 Lattices with Crystallographically Inequivalent Positions 225

12.	**Structures of Organic Solid Solutions**	227
12.1	Diphenyl-2,2'-Dipyridyl [12.2,3]	228
12.2	Para-Dibromobenzene — Para-Chloronitrobenzene [12.4]	232
12.3	Acenaphthene — α-Nitronaphthalene [12.8]	235
12.4	Naphthalene — Coumarine [12.11]	238
12.5	Dicarboxylic Acids [12.21,22]	243
12.6	Naphthalene and Its β Derivatives [10.14,16,11.25]	243
12.7	Solid Solutions of Optically Active Substances or Enantiomers	248
	12.7.1 Carvoxime	249
	12.7.2 Carvoximebenzene [12.34]	251
12.8	Durene-Para-Dibromobenzene [12.36]	251
	12.8.1 X-Ray Diffraction Analysis	251
	12.8.2 Substitution Energy Computation	255
	12.8.3 Probability Calculations of Different Guest Molecule Orientations	257
	12.8.4 Calculation for Oriented Short-Range Order Correlation	260
12.9	Tolane Diphenylmercury [12.39]	262
	12.9.1 Diphenylmercury Phase	264
	12.9.2 Tolane Phase	265
	12.9.3 Diffuse X-Ray Scattering Studies	267
13.	**Complexes**	275
13.1	Quasi-Valence Bonds Between Molecules of the Components	275
13.2	Complexes of Aromatic and Nitro Compounds	280
13.3	Packing Complexes	283
	13.3.1 Hydroquinone-Based Clathrates	287
	13.3.2 Clathrate Hydrates	289
	13.3.3 Urea and Thiourea Complexes	298
	13.3.4 Inclusion of Molecules into Intramolecular Cavities: Tri-o-Thymotide and Cyclodextrin Complexes	301
	13.3.5 Inclusion of Molecules into Intermolecular Voids	305
	13.3.6 Non-Inclusion Complexes	314
13.4	Layered Complexes	314
14.	**Polymers**	319
14.1	The Structure of Polymeric Materials	320
	14.1.1 A Polymer Molecule	320

	14.1.2 Packing of Polymer Molecules	321
14.2	The Structure of Polymer Blends	324
	14.2.1 Both Polymers are Amorphous	324
	14.2.2 One Component Crystallizes	325
	14.2.3 Both Components Crystallize	326
	14.2.4 Some Remarks on the Technology of Polymer Blends	327
14.3	Block Copolymers	328
	14.3.1 Packing of Molecules	328
	a) Layered Structures	328
	b) Hexagonal Structures	330
	c) Cubic Structure	331
	d) Segment Copolymers	331
	14.3.2 Deformation of Block Copolymers	332
14.4	Solid Solutions	339
	14.4.1 Solid Solutions of a Polymer in a Monomer	340
	14.4.2 Limited Solubility of Low-Molecular-Weight Substances in a Solid Polymer	343
14.5	Deformation of an Eutectic Mixture	349
14.6	Crystalline Complexes of a Polymer and a Low-Molecular-Weight Compound	350
15.	**Biopolymers**	354
15.1	Proteins	355
	15.1.1 Globular Proteins	356
	15.1.2 Protein Crystals	358
15.2	Systems Based on DNA Molecules	362
	15.2.1 DNA Crystals	362
	15.2.2 Chromosomes	365
	15.2.3 Viruses	366
References		371
Subject Index		383

1. Introduction

1.1 The Early History

The use of mixed substances for various purposes has been known since ancient times. Such names as "the Bronze Age", the Age of Iron and Steel" remind us of that. Using these names emphasizes the fact that it has been known since ancient times that variations in the composition of solid solutions of elements can lead to change in their material properties.

Production engineering of materials preceded the science concerned with their structure. However, the first scientific research began in the third century B.C. when Archimedes, in the well-known story, measured the specific weight of the gold-silver system in order to determine the composition of the crown of Hieron the Younger, the Tyrant of Syracuse. One may say that Archimedes foresaw a linear relationship between the specific volumes and the composition of a continuous series of solid solutions, such as gold-silver alloys.

When considering the development of any field of science, one has to "jump" from ancient Greece to the seventeenth or eighteenth century. In fact, the study of matter was begun by French scientists at the end of the eighteenth and the beginning of the nineteenth century. Chemists, with a balance in their hands, took to investigating products of chemical reactions. The law of constant proportions of the elements in chemical compounds, which is the fundamental law of chemistry, was proved by the famous French scientist PROUST. He was the first to say clearly: "Compounds, both natural and artificial, produced in laboratory, are characterized by exactly definite proportions of the constituents" [1.1]. Experiments led to the same conclusion, and arguments based on the atomistic theory of Dalton gave the same result.

Another talented French researcher, Berthollet, who can be regarded as the founder of the concept of equilibrium of matter, challenged Proust's ideas. The discussion began in 1801, lasted about ten years, and ended in the victory of Proust.

BERTHOLLET [1.2] asserted that ratios of the elements constituting compounds are not constant and elements can combine in any proportions. As evidence, he mentioned metallic alloys, glasses, and solutions. He took part in Napoleon's campaign in Egypt, where he was concerned with mining saltpeter and the formation of soda in lakes. Though his aims were rather practical, he did formulate some profound ideas on the essence of the thermodynamic equilibrium between the components of complex systems.

His ideas were undoubtedly influenced by the famous geometer Monge, a close friend. Berthollet completely accepted the ideas on the continuity of functions and also those on the continuity of motion, developed by Laplace. He considered it possible to extend these ideas to the transformation of matter. Having realized that variations in temperature can result in equilibrium systems of various compositions, Berthollet came to the conclusion that matter followed this behavior in all cases without exception. Berthollet's theoretical views, contradicting the atomistic theory, were unanimously rejected, and his ideas on the essence of the equilibrium of matter were long forgotten.

Only in the 1870's, after GIBBS [1.3], an eminent American physicist, published the fundamental papers on phase equilibrium did it become clear that the discussion between Proust and Berthollet had been, in fact, a discussion about words. All that Proust had said is valid for stoichiometric chemical compounds ("daltonides"), and the assertions by Berthollet hold for multicomponent systems ("berthollides").

Gibbs' papers defined for the first time the concepts of phases and components, derived the conditions of thermodynamic stability of mixtures, and outlined the ways for developing a rational classification of phase diagrams. VAN DER WAALS [1.4] and, to a much greater extent, ROOZEBOOM [1.5] demonstrated the exceptional importance of Gibbs' ideas, and, at the end of the nineteenth century, derived all possible types of phase diagrams. Thus, the theory for mixed substances was formulated only at the very end of the nineteenth century, and the experimental determination of phase diagrams began only in the twentieth century. Yet, extensive studies of the dependence of the properties, the melting point first of all, on the composition of the mixture were already performed much earlier.

Solutions of salts and melts of metals were investigated quite independently for a long time.

Quantitative data on the solubility of salts in water were available as early as the seventeenth century, but the concept of dissolution was vague at that time. The transition from a solid state to a liquid one was generally meant by this term. A distinction was made, however, in LOMONOSOV's thesis

[1.6] between dissolution of salt in water and that of metal in acid. LOMONOSOV [1.6] and BLAGDEN [1.7] established that the freezing temperature of aqueous solutions is lower than that of pure water.

A systematic study of dissolution of salts in water was carried out by Gay-Lussac. He was the first to formulate the exact definition of a saturated solution. In his papers we find the first graphical representation of the change of solubility with temperature.

Metallurgists of that time often asked university professors questions concerning the behavior of metallic alloys. To provide answers, researchers followed the way of Berthollet (leaving aside his theoretical arguments), who as early as 1786 published a treatise proving that differences in the properties of steel, wrought iron, and cast iron could be explained by variations in their carbon contents.

Faraday became engaged in finding a steel which would be in no way inferior to the metal of sword blades that were made in ancient Damascus [1.8]. In his investigations he noticed that even small amounts of impurities could influence the properties of steel. It was found that chromium made steel fragile, and the best kinds of steel were obtained with the addition of noble metals. FARADAY and STODART in 1822 [1.9] described the production of alloys of platinum with iron.

At the beginning of the nineteenth century the technical advisers of metallurgical factories in all European countries were, as a rule, acquainted with the use of a microscope. It became clear that not only the chemical composition, but also the graininess of the microstructure affected the properties of an alloy. As a result one can say, without exaggeration, that the modern knowledge of metallic compounds was born in steel-making factories.

Experimental investigations again preceded theoretical studies. In 1831 the Russian engineer Anosov, the managing director of the largest factory in the Urals, employed for the first time a consistent combination of techniques, which is now called the micrographic method and forms the basis of metallography [1.10]. He used a microscope to study the structure of polished and etched steel. Later on, similar investigations were performed in Sheffield (U.K.) by SORBY [1.11], who in 1864 reported on the structure of steel and on "microscopic photos" of various kinds of iron and steel.

The first investigations of melting such two-component alloys as lead-tin, bismuth-tin, lead-bismuth, and zinc-tin were made by RUDBERG in Upsala [1.12] and by Kupfer in Petersburg [1.13]. Not wishing, probably, to enter into conflict with Berthollet's opponents, Kupfer presented the results of his measurements as if it were a matter of discrete chemical compounds, rather than of a phase of variable composition: five parts of tin, one part of lead; four parts of tin, one part of lead; etc.

In the nineteenth century many scientists were studying the melting point of inorganic, metallic, and organic systems. HEINTZ [1.14] was the first to plot a melting curve. He marked off temperature as ordinate and composition as abscissa, opposite to the way Gay-Lussac plotted solubility curves. The first system with high melting points, for which the melting curve was constructed, probably was the silver-copper system, studied by ROBERTS-AUSTEN [1.15]. Thermocouples had not yet been invented, and the solidification point of mixtures was determined by means of an iron cylinder of known mass that was removed from the melt at the moment crystallization began and was placed into a calorimeter. The amount of heat acquired by the cylinder was used to calculate its temperature and hence the temperature of the alloy. The errors of this peculiar technique were as large as 200°C.

The analogy in the behavior of aqueous solutions of salts and metallic alloys became clear after the work by GUTHRIE [1.16]. He was the first to investigate the complete melting curve of the water-potassium nitrate system. Guthrie introduced the term "cryohydrate", though at that time he could not, of course, answer the question of whether a cryohydrate was a chemical compound or not. It was Guthrie who introduced the concept of eutectic [1.17] as follows:

> The content of this communication is associated with the existence of complex bodies, which are characterized, mainly, by a low melting point. This property of bodies can be called the eutectic state, and bodies which possess this state are called eutectics, from the Greek word ευτήκειν meaning "melts well or easily". The word was used in this meaning even by Aristotle. It is quite obvious that cryohydrates are, in essence, eutectics.... . Let me invite the reader to draw a parallel between water and any other molten substance, considering the former as molten ice.... .

Only after these papers were published did it become clear that melting of metallic alloys and dissolution of salts in water are phenomena of the same nature. One of the reasons why the complete analogy between all these phenomena had not been noticed until 1884 is that those who studied salts plotted the composition as the ordinate and the temperature as the abscissa, while those who studied alloys did it the other way around!

In the second part of the nineteenth century researchers began to study transformations of matter in the solid state. The concept of polymorphic modifications of a solid was introduced by Berzelius in 1841. The polymorphism of iron was discovered by CHERNOV [1.18] and studied in detail by OSMOND [1.19]. The phenomenon of co-crystallization, i.e., the formation of solid solutions, was discovered in the early 1900's. Even in 1819 Mitcherlich generalized what was known about the formation of mixed crystals of substances similar in composition, such as KH_2PO_4, KH_2AsO_4, and $NH_4H_2PO_4$, and called this phenomenon isomorphism. Mendeleev in 1856 pointed out that substances in which halogens can be interchanged are isomorphous; sulphur-selenium-tellurium and calcium-strontium-barium are examples of isomorphous series. Later, when compiling his Periodic Table, Mendeleev used extensively the phenomenon of isomorphism to classify the elements into groups. At the end of the last century some researchers discovered the similarity in molecular volumes of isomorphic substances.

Thus ended the first era in the study of mixed systems. The information about the dependence of the properties of mixed systems on their composition was scanty. There was no clear understanding of the distinctions between solutions, melts, and chemical compounds. Experimental techniques were far from being perfect.

The second period began at the end of the nineteenth century, when Gibbs' phase theory gained widespread acceptance by the scientific world. Even the first experimental investigations carried out to verify and illustrate Gibbs' theory showed clearly how mixed systems were to be described. Proceeding from the theory researchers began to look deliberately for new phase combinations that previously had not received attention.

The paper by GIBBS, "Equilibria in Heterogeneous Systems", was translated into German by W. Ostwald in 1892 and into French by Le Chatelier in 1899 [1.20]. Reviews dealing with the phase theory were published in Russia in 1898 and 1902. Le Chatelier explained the possibility of determining the phase diagram by a thermal technique, i.e., by measuring melting points of mixtures with the aid of a thermocouple, and invented a thermoelectric pyrometer. He also explained that alloys of iron with carbon were in essence solid solutions, and pointed out that pure elements or compounds could segregate from molten steel with a decrease in temperature. Gibbs' theory and the experimental technique developed by Le Chatelier led to a rapid progress in the study of metallic alloys.

In his memoirs Roozeboom gave an example of a phase diagram of steel. Extensive investigations of aluminum-magnesium and antimony-tin were conducted. Various amalgams (i.e., alloys of mercury with metals), brass, and bronze were studied thoroughly.

In 1903, the German scientist TAMMANN began to publish a series of papers in the field of metallurgy [1.21-23]. We present here a quotation from his article "On Determining the Composition of Chemical Compounds Without Analysis" [1.23]:

5

> The theory of equilibrium of heterogeneous systems yields a number of techniques for answering the question of whether two substances form chemical compounds with each other and what is the quantitative composition of the compound formed. The answers to these questions can be obtained from the melting curve and from the dependence of properties of crystallized alloys on their composition. These techniques should be used when the crystals cannot be separated mechanically from the alloy in which they formed, for instance, in studying compounds of various metals and silicates precipitated from their melts. Only cases where mixed crystals are not formed are taken into consideration.... . If one takes into account the temperature of eutectic crystallization when constructing the diagram (and especially the duration of eutectic halts), one can obtain a sufficient number of fundamental points to determine the composition of compounds formed even in the most complicated cases.

Thereafter, Tammann gave detailed instructions concerning the analysis of the cooling curves and the construction of phase diagrams of the system on the basis of these curves. Equal weight amounts of the solution or the alloy should be taken in experiments, then the time of segregation of the eutectic mixture, which is different for alloys of various compositions, will be directly proportional to the amount of the eutectic solution and maximum for the alloy with eutectic composition. In an alloy, constituting a chemical compound, the eutectic halt is absent; the melting point of such a compound is the highest melting point of the alloys, its composition being determined by the composition of the alloy. If the compound decomposes before its melting point is reached, its composition can be established by extrapolating the maximum of the melting curve, by determining the position of the eutectic horizontal line, and from the duration of eutectic crystallization.

In [1.24] "On the Application of Thermal Analysis to Anomalous Cases" TAMMANN wrote:

> .... it is known that the thermal analysis, which permits determining the composition of crystalline substances without their mechanical separation from the cooling curves, is the only technique which has practical application to the study of refractory crystalline conglomerates.

In this paper Tammann emphasized more clearly than in his first one the characteristic feature of the thermal analysis: not only the beginning of crystallization (liquidus, the term introduced by Roozeboom) and the end of crystallization (solidus), but also all the smallest halts and changes in the rate of cooling should be fixed thoroughly on cooling curves.

One should not think that all the studies of phase diagrams were dictated by practical needs. The interest in the verification of the laws discovered by Gibbs was enormous. Theory suggested problems for experimental investigations. That is why even alloys with no practical importance were also studied.

For instance, the school of KURNAKOV thoroughly studied the gold-copper system [1.25]. The solid phase of this system was found to decompose. Later, this decomposition was discovered to result from the transition of a disordered solid solution to an ordered one. However, in 1914 when this work was done, one could only say that two intermetallic compounds, AuCu and Au_3Cu, were formed from the solid solution. Phase diagrams were constructed for mixtures of an ever-increasing number and variety of substances and were much more complicated than those studied at the beginning of our century.

One of the first triple melting diagrams constructed from experimental data was that of a system of the three most abundant substances in nature: lime, alumina, and silica. It was a result of long research carried out by the American scientists SHEPHERD et al. [1.26]. The knowledge of the phase rules enabled them to construct phase diagrams without errors. They used Gibbs' triangle to plot the diagram of a triple system.

It was of great practical importance to find compositions with minimal melting points; these are systems which constitute most of the slags obtained in producing cast iron from ores in a blast furnace. Finding the most easily melted slags is very important for saving energy in metallurgy.

In the first quarter of our century great attention was paid to the study of complex salt equilibria. Pioneering work in this direction was done by VAN'T HOFF (in the period 1898-1910) [1.27]. A similar systematic investigation of solid salts formed by the decomposition of the system sodium chloride-magnesium sulfate was performed by Kurnakov (see [1.9]). Such studies were necessary to develop the most appropriate technique for extracting salts from lakes and oceans. Crystallization conditions for various salts were found. Phase diagrams became a guideline for techniques of separating pure substances.

Researchers began to study *organic systems* as well. The first system, oleic acid-margaric acid, was studied by CHEVREUL as early as 1823 [1.28]. He prepared 99 mixtures, in which the content of margaric acid increased successively by 1%, and determined the melting curve. Some other practically important systems (aliphatic acids, cholic acids, systems containing urea) were studied from the same viewpoint, i.e., to establish the relationship between properties and composition.

But the study of organic systems became really widespread after the contours of structural organic chemistry were well defined. Since at that time organic molecules could already be considered as atomic systems possessing a certain configuration, attempts were made to find a relationship between formulations of chemical structure and the form of the phase diagram. Many investigations were carred out by TIMMERMANS [1.29] and also by EFREMOV [1.30]. There was little progress in this direction because the concept of a molecular shape did not exist at that time.

Yet the failures did not hold back the researchers, for they believed, and quite justly, that the study of crystallization and phase transformations of organic multicomponent systems, which were of no practical interest, was important in itself and as a model for other systems. The results obtained were successfully applied to metallic systems, the behavior of which was quite similar, but whose direct study was much more complicated because of opacity and high melting points.

1.2 Developments of the Twentieth Century

A new stage in the development of the theory of mixed crystals began with the advent of X-ray diffraction analysis. After the discovery of X-ray diffraction in crystals in 1912, the possibility of determining the arrangement of atoms in a crystal was demonstrated. The first investigations were performed on single crystals, hence they did not attract the attention of researchers who were studying the phase diagrams of multicomponent systems. But soon Debye showed that the X-ray diffraction technique, as applied to micro-crystalline powders, permitted one to obtain extensive information on the atomic structure of high-symmetry crystals.

It was found at the beginning of the 1920s that in a large number of cases, metallic alloys had cubic lattices. If the lattice parameters are not very large (as is the case with most metallic alloys), each X-ray reflection can easily be assigned a certain set of indices. This assignment can also be made easily if crystals have hexagonal or tetragonal lattices.

The invention of the high-temperature X-ray cameras enabled investigations at high temperatures. As a result many researchers began systematic studies of the structure of alloys.

X-ray patterns obtained even in the early days made it possible to determine the number of phases of the system, and parameters of the lattice cell for each phase. If the structure was not very complicated, one could determine the location of atoms relative to one another and calculate interatomic distances. Even such a relatively complex structure as that of γ-brass was analyzed quite correctly as early as 1926.

In the early 1920s information was obtained on the dependence of the lattice parameters of a solid solution on its composition. With great interest physicists accepted the discovery of the superstructure, which appears when a disordered solid solution is cooled; and the first theories were formulated to describe this phenomenon.

The survey of a large number of samples of various compositions at several temperatures permitted a detailed determination of the phase diagram of an alloy. Unfortunately, the X-ray diffraction technique had low sensitivity. Moreover, possible overcooling and overheating did not permit the determination of a reliable equilibrium phase diagram. Small contamination of the specimen under investigation could affect the phase transition temperature. That is why the data in the literature were conflicting. If a system was of practical interest, it was, as a rule, studied rather extensively.

The X-ray phase analysis was improved as years passed. The accuracy and sensitivity of the technique increased. New methods for preparing objects of high purity came into general use. And finally, not only the Debye-Scherrer powder method, but also the method for determining the structure of single crystals was applied to metallic alloys. True, such a labor-intensive investigation was carried out only when the structure of a low-symmetry intermetallic compound had to be determined. Parameters of solid solutions with a high-symmetry crystalline lattice are determined even today by the Debye method with the aid of a large-radius camera or diffractometer, recording scattered X-rays by means of ionization detectors or photomultipliers.

With the advent of neutron diffraction techniques, the potential of diffraction analysis increased considerably, because it became applicable to alloys consisting of atoms with almost the same number of electrons.

The possibilities of determining complex structures of intermetallic compounds increased significantly with the advent of computers, which are now used to calculate X-ray patterns. X-ray phase analysis has reached a high degree of perfection in the last decade. Information on phase diagrams yielded by X-ray analysis is much more profound than that given by other techniques, because here comprehensive data on the crystal lattice structure, in addition to melting curves, are obtained.

The X-ray phase analysis has been supplemented by electron microscopy. In the case of multiphase systems, electron microscopy permits one to observe individual crystallites. It has also been applied successfully to the study of modulated structures appearing upon the decomposition of alloys.

More than two thousand binary metallic alloys have been studied by X-ray diffraction techniques. But a comprehensive structural description of phase diagrams was obtained only for about 200 systems.

Today, metallic systems are investigated rather intensively. This is primarily due to the great interest aroused by substances with extraordinary electrical and magnetic properties. But the search for new refractory materials still remains an important problem. Finally, since metallic systems are frequently employed as construction materials, even the slightest increase in strength or an improvement of other mechanical properties is very important. That is why there has been a persistent search for new metallic systems. Hence, the study of the structure of all possible phases formed by various components is still of the utmost interest.

The investigation of salt solutions in the manner initiated by Van't Hoff has probably come to an end. Researchers are not interested in the structure of mixed crystals of salts. Most of the phase diagrams of salt systems have recently been studied by thermal analysis, without using modern physical techniques. This may be due to the rather limited practical applicability of solid salt systems.

However, mineralogists pay great attention to structural investigations. They seek in the isomorphism of structures an explanation of many regularities which are important for geochemistry. We should bear in mind that most minerals are solid solutions.

The interest in studying ceramic materials has not diminished, either. There is a vast amount of literature dealing with the systematic study of the dependence of various properties of ceramic materials (mechanical, thermal, electrical, optical) on the composition. Most ceramic materials employed in industry are mixtures of oxides or mixtures with glass. Like metals, they are important as structural materials and electrical insulators. But in recent years their applicability has extended due to their potential use in engineering. The piezoelectricity of quartz, the magnetic properties of crystals akin to iron oxide that permit their use in high-frequency technology and as memory elements for computers, the semiconducting properties of certain ceramic materials which make them competitive with germanium and silicon — these examples are sufficient to point out the great interest in oxides, selenides, and tellurides, which are the main types of ceramic salts.

The phase diagrams of multicomponent systems are important, but the properties of a material could also be changed drastically by varying the polycrystalline structure. The size of a grain, the density of grain packing, the possibility of "wrapping" a crystallite of an oxide into a thin glass — these are features that can affect material properties more significantly than a variation of the chemical and phase compositions.

Yet, there is still a certain interest in the arrangement of atoms in a crystallite, too. It has long been known that the impact resistance of various oxides and similar materials as a function of the interatomic distance can be represented by smooth curves. For instance, the extraordinary impact resistance of beryllium oxide can be explained by the extremely close packing of the atoms.

The theory of the structure and properties of mixed organic crystals developed as follows. Hundred of phase diagrams were constructed with the use of the thermal method. Since organic substances are transparent and melt easily, a number of interesting problems were solved by postulating an analogy between the behavior of organic substances and opaque refractory materials, the direct investigation of which is practically impossible. For instance, the regularities in eutectic crystallization were discovered, as will be shown later, by observing the behavior of mixtures of organic crystals, whose melting points range from room temperature to $100^{\circ}C$.

Attempts to interconnect the configuration of molecules and the shape of phase diagrams failed until enough information on the structure of organic crystals was obtained by the X-ray diffraction technique. This did not occur until the middle of the 1940s. Organic crystals are of low symmetry, i.e., a unit cell usually contains only between one and four molecules, but many atoms in general positions. Consequently, the structure can be determined only for single-crystal samples, and sophisticated methods are required to analyze X-ray patterns.

After the establishment of the main principle of organic crystallochemistry (the *close-packing* or *key-to-the-lock principle* that molecules in a crystal should be packed so closely that the projecting end of one molecule fits into the hollow space left by the adjacent one), it became possible to begin a systematic study of the relationship between the structure of component molecules and the form of the phase diagram. Experimental investigations of single crystals of solid solutions of organic substances brought to light solubility rules which possess predictive power. In many cases these rules are of great practical value. For instance, one must know the rules of solubility in the solid state when an impurity molecule is to be introduced into a crystal (scintillators) or a pure substance is to be obtained (e.g., to remove the carbazole impurity from anthracene).

Applications of complexes of organic molecules in industry can be extremely varied. Packing complexes are used as catalysts, for removing undesirable impurities from the mixture, and even for peculiar "storage" of such molecules which if uncomplexed or in direct contact with air may result in an explosion. The principle of packing permits the prediction of cases where such complexes may be formed.

The investigation of *mixed molecular crystals* is also important because it helped, as will be shown below, to extend and refine the concept of complex (molecular compound) and solid solution.

The regularities established in studying mixed molecular crystals can be used also for synthetic polymer materials, which are of great practical significance, and also for biopolymers in molecular biology.

In the 1920s several materials in frequent everyday use were polymeric: wood, fur, wool, silk - all of these substances are made of enormous molecules, usually chains of similar links, the so-called monomers. The idea that synthetic polymers can be of practical use arose in that period. Even the first steps in this direction were successful, and such materials as polyethylene and nylon appeared on the market. The industry of synthetic polymers developed very rapidly; at present there is a large-scale production of synthetic fibers, synthetic rubber, and a great variety of plastics. The annual production of these materials amounts to several tens of billions of dollars. Every year new materials are produced. It is not surprising, therefore, that in a short period the physics of polymers became a vast branch of science involving thousands of researchers.

There are practically no polymers with a completely disordered structure. This applies especially to their most important class, that of materials constructed of chain molecules. At the same time, the order of the so-called highly crystalline polymers is much lower than the order which exists in crystals. This fact, naturally, makes it difficult to study the structure of polymers and to find correlations between their structure and properties.

The main laws determining the formation of ordered domains in polymers proved to be completely similar to those discovered for low-molecular-weight substances. A tendency towards the most compact filling of space, the arrangement of adjacent molecules according to the "*key-to-the-lock*" principle - these are the main ideas for understanding the structure and properties of these materials. Organic crystallochemistry becomes, naturally, a basis on which the theory of polymer structure is constructed.

We consider it legitimate to treat polymer materials in this book. Even if they are constructed of molecules of one kind, they are mixed systems. Regular polymers are the mixture of crystalline and amorphous regions. Block copolymers are also heterogeneous systems. Mixtures of polymorphous substances, consisting of different molecules, are of great practical importance. Thus, mixed organic crystals - solid solutions and complexes - are of considerable interest for polymer physics. These objects, whose structure can be studied rather accurately, may serve in a large number of cases as model systems for materials formed from macromolecules.

A very large number of physicists work in the field of molecular biology due to the enormous potential of *X-ray diffraction analysis*. The study of very complex biological systems was begun by Perutz, who interpreted the structure of the myoglobin molecule, and by the discovery of a double helix, which provides the structure of desoxyribonucleic acid (DNA). Only the structural investigations and the discovery of the close-packing principle made it possible to explain the biological mechanism of replication, a partial understanding of the mechanism of cross-linking of amino acids into a chain of protein molecules, the action of viruses and bacteriophages, the action of enzymes - phenomena which underlie all of molecular biology.

Thus, the study of complex mixed-ordered systems with the aid of X-ray diffraction analysis and electron microscopy has reached a high degree of development. The interpretation of experimental results would involve significant difficulties, were it not for the possibility of applying the laws established for one-component molecular crystals and especially the laws established for mixed organic crystals.

As mentioned in the Preface, the *"packing approach"* will be predominant in the discussion of the structure of various mixed systems. Qualitative predictions can be made with confidence using this simple model, which rules out the existence of structures in which adjacent molecules do not tend to be packed as close as possible.

There is a way to improve the model, thereby predicting the structure and properties of solids not only qualitatively, but quantitatively as well. The idea of close packing leads to pairwise interaction of atoms, the interaction potential being dependent only on the distance between atoms that are not covalently bonded ("non-bonded atoms"). The *atom-atom potential* method has been widely used to solve various problems of structural organic chemistry, the crystallography of organic substances, and the physics of polymers and biopolymers. Naturally, this model cannot be universal. If the problem cannot be solved without taking electron interactions into account, the situation still remains very difficult.

The *free energy* can be calculated accurately only for the simplest systems. It is practically impossible to take into account the effect of electron distributions on the structure of a solid. At best, the electronic theory of solids helps in explaining the structure, but very seldom can it actually predict the structure. However, the optical and especially electrical properties of a body can be understood approximately only within the framework of the electronic theory of solids. But this large part of solid-state physics is not considered in our book.

The physics of *polymers* and *biopolymers* is not the only field where the theory is difficult to apply. The same situation is observed in all fields of materials technology. There is no doubt that theory permits technologists to perform a deliberate search for new materials. But at present theory can only offer a guess in the search for materials with desirable properties. Rigorous theory also encounters great difficulties in the case of simple atomic crystals: even the calculation of the structure and properties of an ideal defectless crystal, whose atoms are bound by ionic, covalent, and metallic bonds, is an almost unsolvable problem. Therefore, the intuition of a technologist, who is guided by qualitative arguments borrowed from the physics and chemistry of solids, but primarily by experiment, still remains the most important factor in the solution of technical problems.

In the field this book deals with, the gain of experimental data and the search for empirical laws will play the leading role for some time to come. The main problems, which in our opinion nowadays face researchers engaged in the study of mixed crystals, may be formulated very briefly: namely, the study of the structure of mixed crystals of systems that are of direct practical importance, the study of systems that simulate complicated objects inaccessible for direct investigation, and the development of approximate, semi-empirical, computer-based calculation techniques.

1.3 Outline of the Book

The book is divided into a general part (chapters 1 through 6) and parts concerned with metallic, inorganic, organic, and polymer systems (chapters 7 through 15).

Chapter 2 is a concise presentation of the theory of phase equilibria as applied to binary mixtures. All the main types of phase diagrams are considered.

Chapter 3 presents the fundamentals of crystallochemistry that may prove necessary for the reader. The treatment of any crystal structure as the packing of spherical atoms is shown to be unreasonable. It is pointed out that characterization of an atom by the atomic radius is feasible only in the simplest cases.

Chapter 4 considers the limited possibilities that we have at our disposal to calculate the free energy of a system of particles. Theoretical derivations are omitted. Only the final equations needed for interpreting experimental results are presented. Attention is given to assumptions which have to be made to complete the calculations. The reader will understand from this chapter that the simplest models for the structure of a solid in general, and mixed systems in particular, turn out to be rather successful in a number of cases.

It is essential to note that the potential energy of the interaction of atoms and molecules is the leading term in the equation for the free energy, and is crucial in determining the structure of a crystal and its thermodynamic properties. Moreover, the most important part in this term is that which corresponds to the interaction between a particle and its nearest neighbors.

Max Born and his students showed some time ago that the model of the pairwise central interaction of atoms provides an excellent description for the simplest crystals constructed of single-atom particles (frozen noble gases).

The attempts to use this approximation for ionic crystals constructed of single-atom ions also proved rather successful, though the electrostatic, i.e., long-range part of the interaction energy had to be taken into account. The representation of the energy as the sum of pairwise interactions of metallic atoms also leads to correct results.

The model of pairwise interactions seemed, for a long time, inapplicable to crystals constructed of polyatomic particles. At the end of the 1940s, the present author noted that one could apparently apply the principle of pairwise interaction to molecular systems, introducing the assumption of additivity of the interaction of atoms of which a molecule is constructed. This method of calculation, called the *atom-atom potential method*, was tested only after computers had been invented, and proved to be correct.

The *packing approach*, used in this book, is in essence the same atom-atom potential method. The fact that molecules are considered to be rigid bodies of certain shapes only implies that we implicitly assume the rectangular atom-atom potential to be applicable. The principle of close packing, i.e., the principle of crystal architecture according to which molecules are arranged so that the "projection" of one molecule fits the "hollow" of another one (*key-to-the-lock* principle), is the manifestation of the additivity of non-bonded atoms' interactions.

The principles of the atom-atom potential method are considered elsewhere. Here we only present the final results of some calculations performed by this method. We recall that the packing approach cannot lead to quantitative results.

Chapter 5 considers very briefly the general laws of the crystallite structure of heterophase systems. Chapter 6, which deals with the main concepts and formulas of the theory of X-ray scattering, concludes the general part.

Metallic systems are considered in two chapters, one of which (Chap.7) is devoted to intermetallic compounds and the other (Chap.8) to solid solutions.

All the problems are considered from one viewpoint. We want to know how simple "packing" arguments can be used to understand the formation of an intermetallic compound and its structure, and to what extent the mutual solubility of metals can be explained. The electron theory of metals is beyond the scope of the book.

The next short chapter (Chap.9) deals with mixed crystals of the simplest salts and oxides. Substitutional and interstitial solid solutions are considered. The author found it impossible to present in this book the extensive and involved information concerning silicates and other inorganic systems. A brief survey of this topic would hardly be of any use, and the author is not an expert in the field of inorganic physical chemistry.

The main part of the book is that dealing with organic substances. It presents the results of investigations performed for about 40 years in author's laboratory.

Chapter 10 treats the conditions for the formation of substitutional solid solutions of organic substances. The behavior of molecular crystals is so specific that it deserves special consideration. The same is true for Chap.11, which shows the necessity of extending the concept of ordering to the case of mixed molecular crystals.

Chapter 12 deals in great detail with a large portion of the available experimental data on binary organic systems.

There is a vast amount of literature dealing with molecular complexes. Chapter 13 treats the main types of the complexes, yet the principal attention is given to molecular compounds called packing complexes. These interesting compounds have been studied in detail only in recent years. They have a wide practical applicability, so it is important for technologists to know the principles of their construction.

Polymer systems are considered in Chap.14. The main principles of the structure of polymer solids are investigated, and it is shown how essential the packing approach is for this field.

The book ends (Chap.15) with a discussion of protein crystals which are, in essence, ordered liquid solutions. The author found it necessary to dwell upon the structure of some bioorganic complexes in order to illustrate the universality of the *key-to-the-lock* principle. This principle explains a number of specific features of these complexes so important for molecular biology.

2. Phase Diagrams

We shall start our discussions with a brief review of the classical theory of phase equilibrium, developed by Gibbs and his successors. This theory, an inalienable part of physics and physical chemistry, is treated in many books (see, e.g., [2.1-4]). For the convenience of the reader, we give here a brief account of the basic principles that will be needed later.

2.1 Fundamentals

This book is devoted to mixed systems. Therefore we consider here single-component systems only when they are two- or three-phase mixtures or multi-component systems in any state. Since our aim is to give the reader a general outline of the subject and to emphasize, first and foremost, the similarities and distinctive features of mixed systems, it is not necessary to dwell here on all the details. Consequently, the presentation will be limited to two-component or binary systems. The following cases will be considered: single-component mixtures of polymorphous modifications, ordered crystals, binary single-phase solid solutions, and binary heterophase systems.

Much attention will be given to two-component ordered crystals. These objects are usually called *complexes* by organic chemists, molecular compounds by inorganic chemists, while metal researchers prefer the term intermetallic compounds. The term "chemical compound" often used in statistical physics is obviously a misnomer because any single-component organic compound is a chemical compound. To prove to the reader that we really must distinguish between these terms, we should like to recall the definition of a component as it is given in Gibbs' theory. A *component* is a constituent part of the system such that its composition, at least in one state of aggregation, does not depend on the concentration of the other parts. Of course, sugar consisting of carbon, oxygen, and hydrogen atoms, and naphthalene consisting of carbon and hydrogen atoms, are single-component systems because the fraction of atoms of each kind is strictly fixed in the solid as well as in the liquid and gaseous states. Sodium chloride is also a one-component system, despite the fact that it decomposes into chloride ions and sodium ions during melting or dissolution. Though the decomposition of the substance has occurred, the numbers of anions and cations cannot be varied independently - they must be the same. Hence sugar, naphtalene, and sodium chloride are *chemical compounds*.

Crystals of γ-brass of the composition Cu_5Zn_8 are fully ordered. To find out if these crystals are two-component, we must melt them. They decompose into Zn and Cu atoms whose relative concentration can be varied arbitrarily. The term "chemical compound" cannot be used here; γ-brass is an *intermetallic compound*.

Many organic molecules in the solid state form complexes rather tightly bonded with iodine. They are named complexes and not chemical compounds because upon dissolution they decompose into the constituents and the quantity of iodine in the solution can be varied arbitrarily. Thousands of complexes of this kind are known. Chemists write their formulas separating two parts of the complex by a dot. "Halves" of the complex are chemical compounds, but to use the term "chemical compounds" for the whole complex is unwise. If one does not like the word "complex", the term *molecular compound* may be used instead.

Thus, it is clear in what cases the system can be called the two-component system. In principle, we can prepare a mixture of two individual chemical compounds of any composition. Depending on the temperature and the relative content of the components, single-phase or two-phase solids are formed. For certain definite compositions, regularly ordered crystals can form, while for variable compositions we obtain solid solutions or heterophase mixtures.

Nevertheless, one has to realize how unambiguous the notion of a two-component system is. In organic chemistry we often deal with molecules which have the same composition and atom sequence but differ in shape. Such conformers can be converted into one another by simple rotation around one or several valence bonds. But the barriers of such a conformational transformation can be different. It is possible to prevent the transformation at low temperatures. Then we have two different molecules whose mixture is either a two-component or a three-component one in solution. At higher temperatures, conformational transformation may again become possible. We may also find a mixed binary crystal which consists of two molecules differing in shape. While they are melted, the conformers are in a state of dynamic equilibrium and their fractions cannot be changed. Thus, the system is a single-component one.

If the barrier of transformation is not very high or not too low, it is convenient, as a rough approximation, to speak of it as a two-component system in one case, and as a single-component system in another case, i.e., to use two different approximations at different temperatures. In any case, the rigorous notion has become somewhat conventional.

There are some other reasons which make the term "component" not applicable to most macromolecules and to practically all synthetic polymers. In these cases solids consist of molecules differing in weight. A material can be characterized by a certain distribution in weights. No matter how narrow such a distribution is, we have, strictly speaking, a multicomponent system, because we can always add some more molecules of a certain length, thus changing the distribution in molecular weights. This circumstance can affect the thermodynamic behavior of such substances. All equilibrium curves and all phase transitions are "blurred", i.e., they occur not at a precise temperature, but within some temperature range. Pure polymers, for example, polyethylene or polystyrene, cannot be considered as classical one-component systems; moreover, they are not single-phase systems.

In thermodynamics a *phase* is defined as follows. First of all, it is a part of the system which can, in principle, be separated from the other parts of the system. Consequently, sharp macroscopic boundaries between phases must exist in any solid or liquid body. In any two-phase system, in particular in a solid one, the boundary between two phases is not impenetrable. The phase equilibrium is dynamic. There are polymorphous solids in which two modifications of a single-component system are at equilibrium. The interface between the phases does not move, but the molecules penetrate through it. In the case of equilibrium, the number of molecules travelling, say, from right to left is the same as the number of molecules travelling in the opposite direction.

Now, what about the equilibrium solution which consists of a solvent and a multitude of precipitated crystals? If the crystals are built from the same molecules and belong to the same polymorphous modification, they and their solution are a one-component, two-phase system.

There is no such clear classification of most polymer systems. Each polymer is characterized by its own "*degree of crystallinity*". Long molecules in some parts of a solid are packed in order, forming a crystalline phase. The same molecules in other regions are disordered, forming an amorphous, or liquid phase (with low mobility of particles). Thus, solid polymers such as polyethylene can be considered as one-component single-phase systems only in a rough approximation.

Consider another interesting example, which proves that there are cases when classical concepts of the phase theory cannot be applied unconditionally. *Proteins* can crystallize and give excellently faceted crystals. Their X-ray patterns are also of a very good quality and consist of tens of thousands of reflections. But protein crystals contain a large number of water molecules, sometimes amounting to 70-80%. Water molecules are in a disordered state, which is confirmed by sharp peaks in the nuclear magnetic resonance spectrum of the protons in the water molecules. It is well known that all molecules in a protein structure have exactly the same molecular weight. Thus, the number of components in such a crystal is, no doubt, two. But protein molecules form a crystal lattice whereas water is in a liquid state. A protein crystal is a system which in principle is different from other crystals with water molecules distributed in an ordered way. Protein crystals can be regarded as a solution of protein in water. Then it is a single-phase system. But the orderliness of protein molecules allows this specific solid to be considered as a two-phase system. Anyhow, it is hardly possible to use the classical approach to describe such crystals.

Composite materials and artificially created heterophase systems such as epitaxial systems (which are widely used in modern technology) should not be treated in the framework of classical thermodynamics. Here molecular exchange between the different phase layers is practically impossible, and the concept of a phase diagram is not applicable to such non-equilibrating composites.

But nevertheless, a great number of solids obey the rules predicted by Gibbs' theory, the importance of which for solids and, in particular, for mixed crystals is immense.

2.2 Conditions of Stability in Mixed Crystals

Thermodynamic equilibrium conditions in any body, simple or mixed, consisting of one or more species of particles, are absolutely the same. The law of nature which determines these conditions is of universal character. In mechanics, the condition for stable equilibrium is the minimum of potential energy. In the case of energy transformations of various kinds occurring in a crystalline system, it is possible to find a function describing the state of the system which is analogous to the potential energy in mechanics. There are several functions of this kind in thermodynamics, depending on the conditions under which the system exists. It is especially important to know the condition of equilibrium when the system is under constant pressure and temperature. From the general laws of thermodynamics it follows that such a function, called the *thermodynamic potential* or Gibbs' *free energy*, is given by

$$G \equiv U - TS + pV , \qquad (2.1)$$

where U is the internal energy, S the entropy, p the pressure, V the volume, and T the temperature.

Any irreversible changes cause the thermodynamic potential to decrease. The *condition of equilibrium* can be written as that of the minimum of G, i.e.,

$$dG = 0 . \qquad (2.2)$$

In this book only a few cases will be encountered where the pressure is different from atmospheric pressure, so that the last term in (2.1) is negligibly small. In practice, solid solutions under either zero or atmospheric pressures behave in the same way. Instead of considering the thermodynamic potential we shall confine ourselves to analyzing and calculating the free energy (Helmholtz free energy) of the system

$$F \equiv U - TS . \qquad (2.3)$$

The equilibrium condition (2.2) is equivalent to the condition of a minimum in the free energy.

In some cases, it is convenient to relate the thermodynamic potential and the free energy to a mass unit, to a mole fraction, or to a molecule. The thermodynamic functions are additive. Therefore, for a complicated system, the thermodynamic potential and the free energy can be written in the form

$$G = \sum_i N_i \frac{\partial G}{\partial N_i} , \quad F = \sum_i N_i \frac{\partial F}{\partial N_i} , \qquad (2.4)$$

where N_i is the number of molecules in each phase.

Let us consider the equilibrium condition in a system consisting of r *phases* and n *components*. It has the form

$$\sum_{i=1}^{r} dG_i = 0 . \qquad (2.5)$$

Each differential related to an individual phase can be written in the form

$$dG_i = \frac{\partial G}{\partial N_1^i} dN_1^i + \frac{\partial G}{\partial N_2^i} dN_2^i + \ldots + \frac{\partial G}{\partial N_n^i} dN_n^i . \qquad (2.6)$$

When the equilibrium is reached, the phase ratio can change, but the number of particles of each species remains constant throughout the system. Taking this into account and performing simple transformations, we obtain the $n(r-1)$ conditions of stability for a multiphase system, originally formulated by Gibbs,

$$\frac{dG}{\partial N_1^1} = \frac{\partial G}{\partial N_1^2} = \ldots \frac{\partial G}{\partial N_1^r} \ldots \frac{\partial G}{\partial N_n^1} = \frac{\partial G}{\partial N_n^2} = \ldots = \frac{\partial G}{\partial N_n^r} \quad . \tag{2.7}$$

We should recall the two important consequences of these equations. The first is the *principle of polyphase equilibrium*. It states that two phases which are in equilibrium with the third one will be in equilibrium with each other when brought into contact. Consequently, the equilibrium does not depend on the number of molecules in each phase.

The second consequence is Gibbs' *phase rule*. It is evident that to characterize the relative phase composition, it is sufficient to indicate only $(n-1)$ of the n concentrations. The composition of all the phases is determined by $r(n-1)$ concentrations. Adding pressure and temperature, we obtain the total number of parameters which determine the system:

$$r(n-1) + 2 \quad . \tag{2.8}$$

Thus, the number of equations (2.7) which determine the equilibrium is $n(r-1)$, and the number of the unknowns is $r(n-1)+2$. The system of equations can be solved only if the number of constraint equations does not exceed the number of unknowns, i.e., $r \leq n+2$.

This is the *phase rule* which reads: if the number of independent components is n, then not more than $(n+2)$ phases can be in equilibrium. The number of variables which can be changed arbitrarily without upsetting the equilibrium is called the *number of thermodynamic degrees of freedom* and is equal to $(n+2-r)$.

What does the phase rule mean to the binary system $(n=2)$? To answer the question, we must consider a three-dimensional diagram. Let us plot temperature (T) versus concentration (x) and direct the third axis, pressure (p), towards the reader. The mixture of the two components can give a single-phase system. Molecules can be mixed in a liquid state (melt) or in a solid state (solid solution). Three parameters are available — pressure, temperature, and concentration. Consequently, a single-phase system is associated with a three-dimensional region in the diagram. If we restrict ourselves to studying the behavior of mixtures under fixed pressure, the diagram will be two dimensional and the single-phase system occupies a limited area of the TX plot. If an equilibrium system is a two-phase one $(r=2, n=2)$, there are two degrees of freedom. The two-phase equilibrium is attained on curved surfaces limiting single-phase regions. If the pressure is constant, and only one cross section (T-x) of the diagram is considered, the equilibrium state of the two phases (melt-component, melt-solid solution, two solid solutions, etc.) is characterized by curves limiting single-phase regions. Equilibrium between three phases is also attainable. In the general case (pTx surface), there is only one degree of freedom. But if one considers the cross section of the three-dimensional diagram at fixed pressure, then the equilibrium is characterized by a point on the two-dimensional diagram. At that point (zero degrees of freedom) three phases, or the melt and two solid phases, can be in equilibrium.

The equilibrium equations for an arbitrary binary system can be written in the form (for normal pressure, G is replaced by the Helmholtz free energy F)

$$\frac{\partial F_\alpha}{\partial N_1^\alpha} = \frac{\partial F_\beta}{\partial N_1^\beta} \quad , \quad \frac{\partial F_\alpha}{\partial N_2^\alpha} = \frac{\partial F_\beta}{\partial N_2^\beta} \quad . \tag{2.9}$$

When referring to one particle, the equations take the form

$$\frac{\partial f_\alpha}{\partial x_\alpha} = \frac{\partial f_\beta}{\partial x_\beta} \quad , \quad \frac{\partial f_\alpha}{\partial (1 - x_\alpha)} = \frac{\partial f_\beta}{\partial (1 - x_\beta)} \quad , \tag{2.10}$$

using particle fractions (or mole fractions) x of component A. Here components are denoted by capital Roman letters, and phases by Greek letters.

The second equation follows from the first one, hence only one independent equation is left:

$$\frac{\partial f_\alpha}{\partial x} = \frac{\partial f_\beta}{\partial x} \quad . \tag{2.11}$$

Subscripts indicate the phases in equilibrium.

In accordance with the phase rule, both concentrations have to be fixed by equilibrium conditions. Then, one more equation is necessary. Equation (2.11) shows that the tangents drawn to the f-curves at the points corresponding to the equilibrium concentrations of the two phases must be parallel: but such a requirement is necessary but not sufficient. For sufficiency, the tangents must coincide; it is easy to show that this condition can be written in the form

$$f_\alpha - x \frac{\partial f_\alpha}{\partial x} = f_\beta - x \frac{\partial f_\beta}{\partial x} \quad . \tag{2.12}$$

Two-component solids can be obtained from either a solution or a melt. In the former case we deal with a three-component system, and a theoretical consideration becomes rather involved. It is also easier to carry out an experimental study of the crystallization of mixed crystals from the melt. Therefore we shall consider the latter case, the crystallization of solid substances from a mixture of two liquids.

To begin, Fig.2.1 represents a plot of the free energy f per particle versus the concentration of the mixture. The quantities f_A and f_B are the specific or molar free energies of the two pure components that can form the mixture. Assume that the liquids do not intermix, i.e., the system is heterogeneous and consists of two phases. Then the free energy per particle in such a mechanical mixture can be represented by the line STU and by the equation:

$$f = (1 - x) f_A + x f_B \quad .$$

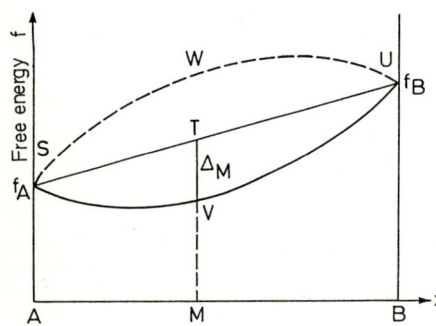

Fig.2.1. Free energy f of a particle versus concentration x of the mixture

If the melted substances can be mixed in any proportion, the mixture is homogeneous. The free energy of such a single-phase system is lower than the free energy of a mechanical mixture for all concentrations. The dependence of the free energy on concentration is depicted by a sagging curve SVU.

Suppose we have a two-phase system stable in a heterophase state. If, using some artificial method, we could bring the substances into a homogeneous molecular mixture, then the curve of this single-phase system, SWU, would lie above the straight line of (2.12) because of the greater stability of the heterophase state.

The free energy of the stable (homogeneous) mixture M is thus given by curve SVU in Fig.2.1 and can be written in the form

$$f = (1 - x) f_A + x f_B + \Delta_M \quad , \tag{2.13}$$

where Δ_M, a negative quantity, is the free energy of mixing (represented by the line TV in Fig.2.1).

All that was said above can be applied to any homogeneous mixture in the liquid, the gaseous, or the solid state.

Experience shows, in full accordance with the phase rule, that a stable equilibrium is attainable between a homogeneous melt and the crystals of one of the components. Figure 2.2 illustrates the curve PQ of the free energy f of the melt in equilibrium with the crystals K of the pure phase A. It is quite obvious that the point K must lie below the left end of the curve PQ: since the crystals of the pure phase A are stable, their free energy must be lower than that of the liquid (at P). It is important to note that the two-phase system of melt plus crystals K is stable only if the concentration x_M of the melt at the point M lies on the tangent to the curve drawn from the point K.

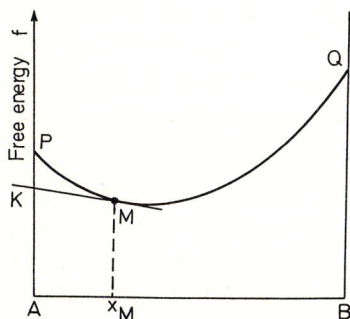

Fig.2.2. Free energy f of the mixture M being in equilibrium with crystals K of pure A

Figure 2.2 shows that the mechanical mixture (melt plus crystals K) is stable to the left of the concentration x_M, whereas the melt is stable to the right of x_M. Hence, the crystals K precipitate from the melt when the concentration is lower than x_M, up to the moment it reaches the value x_M. If the concentrations are higher than x_M, the crystals K will melt.

The line KM is called a *connode*. For a mixture of any composition between K and M, the fraction of the melt can be determined with the help of the "*lever rule*".

Under constant pressure, the free energy is a function of concentration and temperature. We have considered the fx cross section. But in practice it is essential to know another cross section of this surface. This means that Fig.2.2 should also contain the temperature axis constructed at right angles to the drawing plane. The surface representing the stable states of the system will then look like a groove. Of interest is the behavior of the projection of the point M onto the temperature-concentration plane. The resulting curve, which is called the solubility curve, must look like Fig.2.3. Above the curve the system is in the molten state. During cooling, the representative point will fall on the melting curve. At that point, the mixture of the requisite concentration becomes two-phase, and the crystals of the component A precipitate; they are in equilibrium with the melt. Upon further cooling of the

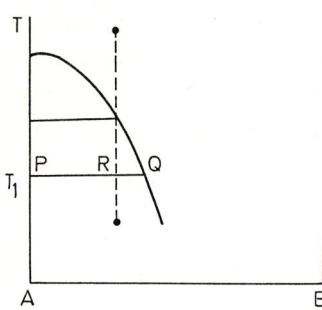

Fig.2.3. Solubility curve

system the point representing the melt, *"the representative point"*, remains on the curve and moves to the right because the melt keeps losing component A and becomes richer in component B. At any temperature T_1, the connode PQ connecting the curve with the left-hand ordinate is divided by the point R into two segments, PR and RQ, which permits determining the relative quantities of crystals (at P) and the melt (at Q) which add up to the total concentrations M of component A at the point R. The average concentration of the mixture is, of course, constant. It is possible to consider the movement of the representative point both upwards and downwards. Consequently, the solubility and crystallization curves are equivalent notions.

2.3 Types of Phase Diagrams

Phase diagrams are graphs that represent, in a simple form, phase states of complicated systems under different pressures and at different temperatures. Of course, multi-dimensional diagrams are difficult to construct and hence, as a rule, systems are described by two-dimensional graphs. As said, we shall confine ourselves in this book to the data on the structure and phase states of binary systems. We are not going to complicate graphical representation by considering systems under nonzero (by which we mean nonatmospheric) pressures.

Under these restrictions, complete data on the system can be obtained from the temperature versus mixture concentration (Tx) diagram. In accordance with the phase rule, we divide the diagram into regions, each line being a line of a two-phase equilibrium. The lines can intersect, thus giving triple points. Points of higher multiplicity are impossible. An area of the phase diagram corresponds either to a single-phase system or to a two-phase mixture. In the latter case the composition is determined with the help of a connode drawn through the representative point. The ratio of the intercepts of the connode on both sides of this point determines the percentage of the two phases.

There are a limited number of types of phase diagrams. Their form depends on the polymorphism of the phases and on their miscibility. Now, we proceed to the derivation of the different types of phase diagrams. Predicting the shape of the phase diagram for given components is a different and much more complicated problem. Possible solutions of some of these problems will be considered in Chap.4.

2.3.1 Crystallization of Two Pure Components from the Liquid Phase

A plot of the free energy f for a homogeneous liquid mixture of two components is shown as the sagging curve SCLV in Fig.2.4. It can be related both to a melt or to a solution, for example, to a solution of rock salt (sodium chloride) in water. If the curve is plotted for a temperature lower than the melting temperature of the pure components, then their free energies are depicted by the points K and P. Component A is in equilibrium with the melt C, component P with the melt L. Now imagine, as in the previous section, that the temperature axis is perpendicular to the plane of the sheet. The points C and L in such a three-dimensional diagram will generate certain curves, which are precisely the curves of the two-phase equilibrium. If we map these curves onto the plane Tx, we shall obtain the picture represented in Fig.2.5. As the temperature decreases, the points C and L of Fig. 2.4 get closer and then merge into a triple or eutectic point E of Fig. 2.5. In the case of freezing water solutions, the eutectic point is called cryohydrate.

The simple phase diagram thus obtained has four regions. Above the solubility curves there is a single-phase region 1 of a homogeneous liquid mixture. Below the horizontal passing through the eutectic point E there is a region (A+B) of a solid, consisting of two phases, crystals of the component A and crystals of the component B; the other two regions (1+A and 1+B) correspond to two-phase systems consisting of the melt (or solution) and crystals of A or B depending on whether the representative point is taken to

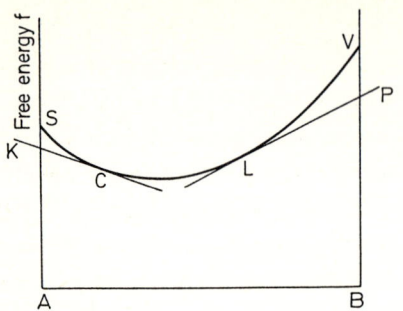
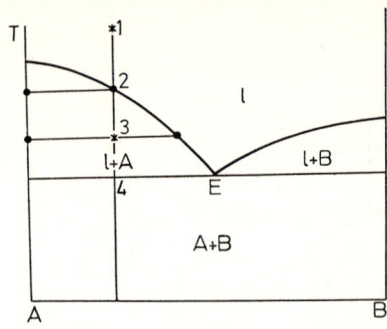

Fig.2.4. Free energy f of a homogeneous two-component liquid mixture

Fig.2.5. Diagram with eutectic point

the left or to the right of the eutectic point E. Figure 2.5 also shows a vertical line (12345) which describes the process of crystallization or melting of the mixture.

Crystallization proceeds in the following way. When the temperature is cooled from T_1 to T_2, and the representative point meets the curve of the two-phase equilibrium (point 2), crystals of component A begin precipitating and the composition of the melt approaches that of the eutectic. When the temperature becomes equal to that of the point E (T_4), the remaining melt (or solution) solidifies at constant temperature. The crystals of the pure component that separate are usually much larger than the crystals forming at the eutectic. Crystallization of alloys described by this simple phase diagram is of great practical importance. Depending on the melt and cooling conditions, various microstructures can be obtained, examples of which will be considered in Chap.5.

2.3.2 Unlimited Solubility

Let us consider the types of phase diagrams when molecules of two components have unlimited (i.e., continuous) solubility both in the liquid and in the solid state. Figure 2.6 represents two isotherms of free energy: for the solid state and for the liquid state. Since the latter isotherm is the lower, at this temperature the melt is the stable state. Again, as in the previous case, imagine the temperature axis to be perpendicular to the sheet. The curves will move along the temperature axis forming two grooves which must intersect somewhere. Figure 2.7 represents this intersection occurring at a lower temperature than that of Fig.2.6. It is seen that at this temperature a two-phase equilibrium is attained: crystals of composition K are in equilibrium with the melt of composition C, when the slopes df_s/dx and df_l/dx are equal.

28

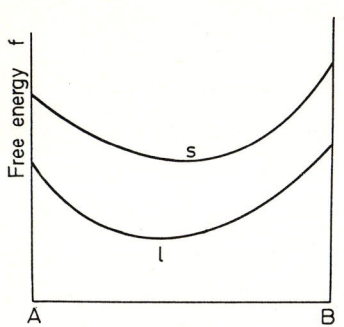

Fig.2.6. Free energy isotherms for the solid and the liquid state at high temperature

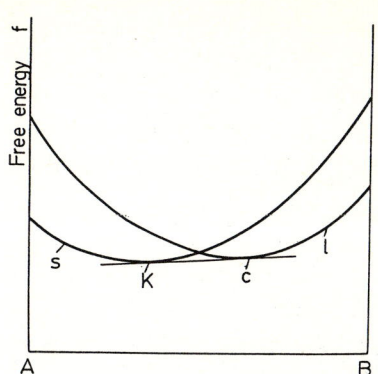

Fig.2.7. Free energy isotherms at a lower temperature: two-phase equilibrium for the case of unlimited solubility. The slopes df_s/dx and df_l/dx are equal at points K and C

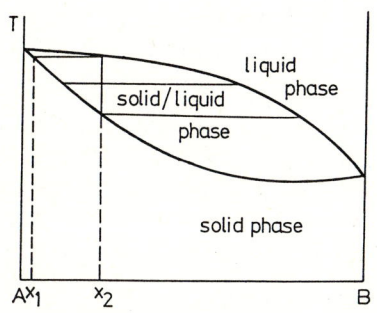

Fig.2.8. Cigar-shaped diagram of the state demonstrating unlimited solubility

If these two points, corresponding to solid phases in equilibrium, are mapped onto the plane Tx, a diagram will be obtained which has the shape of a cigar (Fig.2.8). It is divided into three parts: the upper part corresponds to the liquid phase, the lower one to the solid phase (solid solution), whereas the points inside the "cigar" correspond to a solid-solution - melt two-phase (s+l) system. The upper curve is called *liquidus* and the lower one *solidus*.

Crystallization and melting of such a system proceed differently from the cases described above. When, as the temperature decreases, the point depicting liquid reaches the liquidus curve at point 2, a solid solution of composition x_1 (at the same temperature) starts separating from the melt instead of the pure component. With a further temperature decrease the state of the system is characterized by the connodes shown in Fig.2.8. The liquid becomes A-deficient, while the crystals, separating from the melt, become A-enriched.

When cooled down slowly, solid solutions have a concentration varying within the range $x_1 - x_2$, indicated in Fig.2.8. This phenomenon, *liquation*, leads to the formation of inhomogeneous alloys.

Depending on the mutual arrangement of free-energy surfaces of liquid mixture and solid solution, other types of phase diagrams may occur. In the above-considered example, when the temperature is decreased (from Fig.2.6 to Fig.2.7), the surface of the free energy function for the solid mixture crosses that of the liquid mixture (Fig.2.7). But this is not necessarily the case. As the temperature is lowered, the two f versus x first meet at some intermediate point such that the surfaces of the liquid and the solid have one common tangent at the same composition. Thus the liquidus and solidus curves will merge at a certain *maximum* temperature. With a further temperature decrease the system behaves as the one described before. Thus the phase diagram will be of the type shown in Fig.2.9.

Finally, a third case is possible. The free-energy surfaces intersect at both ends before they intersect with equal slopes at some intermediate x at a lower temperature. It is quite obvious that the intersection will yield a certain *minimum* temperature at which the solid and liquid phases of the same composition reach the equilibrium. This is illustrated in Fig.2.10. It should be emphasized that the extremum points J (Fig.2.9) and K (Fig.2.10) have no unusual properties: they do not correspond to any simple component ratio, or to any special relationship between the components, or to the formation of new types of ordering.

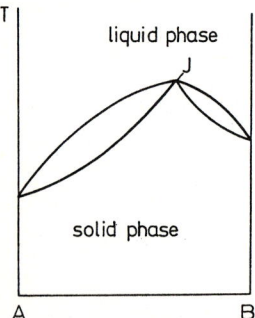
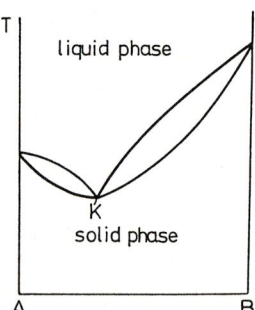

Fig.2.9. Tx Diagram (with a maximum) demonstrating unlimited solubility

Fig.2.10. The same as in Fig.2.9, but with a minimum

2.3.3 Limited Solubility

The free-energy surface is not always represented by a sagging curve. Assume that this surface of the solid mixture is depicted by a saddle-shaped curve (s in Fig.2.11), whereas that of the liquid mixture is given by a sagging curve (1 in Fig.2.11), as in the previous cases. It is the cross section for the temperature below the melting point of the two components. The crystals of composition x_1 are in equilibrium with the melt of the composition x_2, and the crystals of composition x_4 are in equilibrium with the melt of composition x_3. With a decrease in temperature the two tangents finally merge into a single horizontal tangent, which contacts the curve of the solid mixture at two points and the curve of the liquid mixture at some intermediate point. Thus, at this temperature an equilibrium is attained between two solid phases and one liquid phase. Hence we have an eutectic, but as distinct from the case considered in Sect.2.3.1, at this triple point the melt is in equilibrium with two solid solutions and not with pure components. The corresponding phase diagram is represented in Fig.2.12. This figure is divided into six parts. Above the liquidus curve there is a region of homogeneous melt. On the left and on the right there are regions of two different solid solutions α and β. In the middle part, below the horizontal line DEF passing through the eutectic point E, there is a region of a solid which is the mixture of crystals of the solid solutions α and β. Thus, different solids can be obtained depending on the composition of the cooling melt.

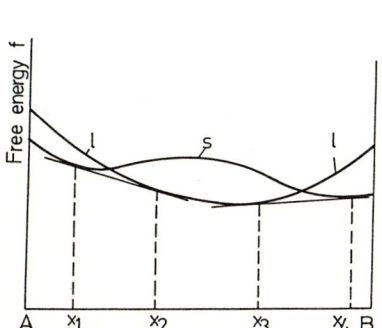

Fig.2.11. The crystals x_1 are in equilibrium with the melt x_2, and the crystals x_4 are in equilibrium with the melt x_3

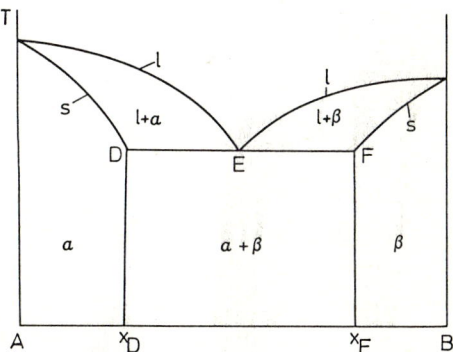

Fig.2.12. Diagram of the state for melt in equilibrium with two solid solutions

Crystallization of the melts to the left of point D and to the right of point F does not differ from crystallization of the melts with continuous solubility. In the intermediate case ($x_D < x < x_F$), the process occurs in the following way. First, crystals of the α-phase (or β) precipitate from the melt, and then, as a result of its deficiency of component A (or B), the point depicting the melt moves along the liquidus curve until the connode representing the state of the system merges with the line passing through the eutectic point. Then the eutectic crystals begin precipitating, i.e., mixed crystals of two solid solutions α and β begin to form.

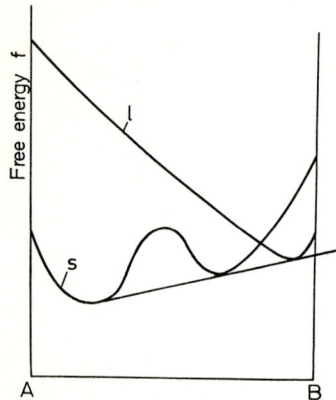

Fig.2.13. The triple point of two solid phases and a liquid being higher than the melting point of the more easily melting component

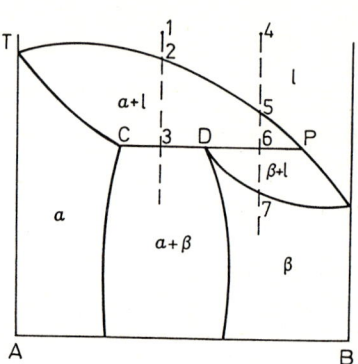

Fig.2.14. Phase diagram with a peritectic point (P)

Another type of limited solubility is observed when the difference between the melting temperatures of the components is very large. It may so happen that the temperature of the triple point of the two solid phases and the liquid is higher than the melting point of an easily melting component. The temperature cross section of the three-phase equilibrium (fx diagram, Fig.2.13) illustrates the distinction of this case from the previous one. We see that equilibrium is attained when one of the two solid phases has a concentration which is intermediate between the liquid and the other solid phase. This leads to the phase diagram shown in Fig.2.14. The point P in which a three-phase equilibrium is reached is called *peritectic*. The phase diagram is again divided into six regions: three solid, one liquid, and two liquid-solid regions.

How does the crystallization of such a melt proceed? For alloys whose concentration is lower than C or higher than E it proceeds as before. If the cooling liquid is represented by a point lying between C and D, then on cooling, crystals of an α-phase begin to separate (at point 2). When the peritectic temperature is reached at point 3, the two-phase system (α-solid)-liquid transforms into a two-phase solid-system, but this time it is a mixture of crystals of the α- and β-phases. The β-crystals of the composition D separate from the melt, and the process continues till complete solidification. If the representative point lies between D and P, the system (α-crystals)-melt (α+1) must first transform into the system (β-crystals)-melt (β+1) at point 6, which means that a solution of α-crystals begins with the simultaneous separation of β-crystals. After this transformation the representative point will move in the two-phase region β+1, and at last crystals of solid solution β will form at point 7.

It should be kept in mind that the phase diagram describes the behavior of the system on heating and cooling only if the conditions of thermodynamic equilibrium are fulfilled (to a reasonable approximation) at all times. This means, in other words, that the processes must be carried out slowly.

2.3.4 Component Forming a Complex

In binary metal alloys, organic systems, or systems in which one of the components is an element and the other is an organic compound, and systems composed of pairs of different salts, there are cases where at an exact, and often very simple stoichiometric ratio, one or several complexes of the composition $A_m B_n$ are formed in the solid state. A complex of any composition $A_m B_n$ may or may not form solid solutions with its components.

Let us consider a common case where the two components do not form solid solutions either with each other or with a compound. A temperature cross section of the free-energy curve of the mixture is depicted in Fig.2.15. We shall now indicate on this graph the free energy of a complex. Let us assume there is a compound AB. Then its free energy is represented by the point

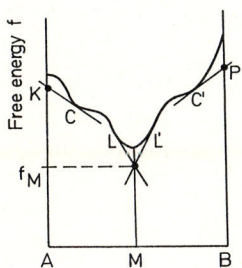

Fig.2.15. The two components that do not form solid solutions either with each other or with a compound

f_M at M, the midpoint of the abscissa. Suppose the section shown in Fig.2.15 refers to a temperature which is below the melting points of the components and the compound. If the temperature is higher than the melting point of the compound, the point f_M will be above the curve. We can draw two tangents from the points K and P corresponding to the free energies of the components and two from the point f_M representing the free energy of the compound.

To obtain the phase diagram we use the same geometric method. Again, assume that the temperature axis is perpendicular to this sheet and let us see how the points of contact C, C' and L, L' will move when going from one temperature cross section to another. We are interested in the projections of lines of contact of these points with the free energy surface. These projections onto the plane Tx are shown in Fig.2.16. It is easy to see that the melting diagram thus obtained is the sum of two diagrams like Fig.2.5. The abscissa of M (x_M) gives the composition of the complex. Note that in this case (contrary to Fig.2.15) point M is not the midpoint of the abscissa.

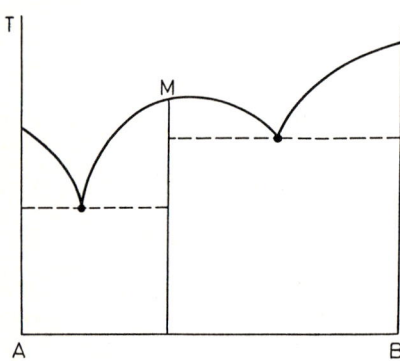

Fig.2.16. Simplest Tx diagram of a binary system forming molecular compounds. This diagram is the sum of two simpler diagrams similar to Fig.2.5

2.3.5 Intermediate Phases

A complex and a component cannot form a continuous series of solid solutions because it requires the particles to be completely similar in the arrangement of their unit cells (we shall return to this problem). Instead, they often form limited solid solutions. This is also true for two neighboring compounds in a phase diagram, especially for metal alloys.

Let us consider some examples. Figure 2.17 (and Fig.2.16) show phase diagrams that are the sum of two simpler diagrams similar to Fig.2.12 (and Fig.2.5). Here three types of solid solutions are realized. Crystals of solid solution α in the left part of the diagram are the solutions of the component B in the crystal A (strictly, α is the solution of the A_xB_y complex in A). The same holds for the crystals of the β-phase in the right-hand side

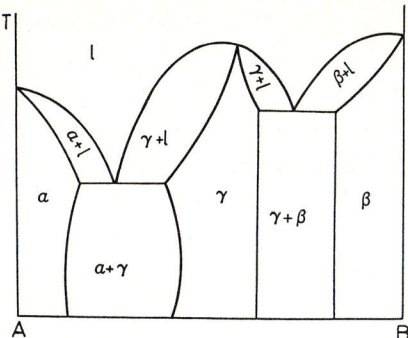

Fig.2.17. Phase diagram that is the sum of two simpler diagrams like Fig.2.12

of Fig.2.17. Solid solutions of crystals of the complex are somewhere in the middle part of the diagram: an intermediate solid γ-phase has appeared whose concentration range includes the complex (compound) of the components of exact stoichiometric composition.

Some latent intermediate phases can exist in a binary system which are solid solutions on the basis of the complex (compound) of the components A and B. The melting curve of such a phase diagram has no maximum corresponding to the melting point of the compound A_mB_n. In other words, the complex cannot be brought up to the melting point. It can decompose into the melt of the components and separate into crystals of another phase. One of the possible phase diagrams of such a type is represented in Fig.2.18. The shaded region is the domain of such an intermediate phase (γ). If the composition corresponds to the stoichiometric ratio m : n, the ordered crystals of the complex are stable in this temperature region (at the points on the dashed line). In other points of the shaded region, solid solutions are stable if they have the same structure type as the A_mB_n crystals.

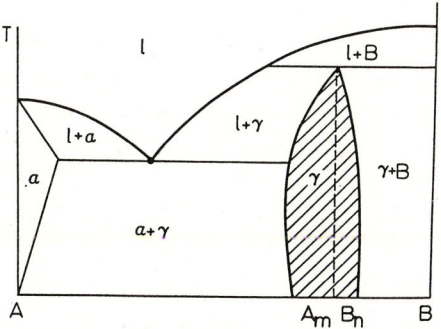

Fig.2.18. The complex (γ) cannot be brought up to the melting point

Thus, compounds can exist in systems whether or not their diagrams have a maximum for a certain composition.

The opposite statement, i.e., that there must exist a compound if there is a maximum in the phase diagram, is not valid.

We recall here what was said about Fig.2.9: phase diagrams are possible in which solidus and liquidus curves merge at a certain point and generate a maximum; this point does not differ from the neighboring points: definite stoichiometric ratio and an ordered crystal structure do not necessarily correspond to such a point.

The shapes of free-energy surfaces of solid and liquid phases can be such that the phase diagram resembles Fig.2.17 but that the highest point on the middle curve does not correspond to any stoichiometric ratio. Structure investigations have proved that for such a composition there exists only a disordered solid solution but no complex. If the region of the intermediate phase is rather wide, however, a complex can exist in the region of γ-solid solutions somewhere in the vicinity of the maximum point.

Finally, thermodynamics admits the existence of such intermediate phases which decompose just before melting; they give no maximum on the melting diagram. Such phases are solid solutions with a structure different from the structures of the components. Formally, the phase diagram can be like that shown in Fig.2.18, the only difference being that there are no ordered structures in the concentration of the γ-phase.

All this leads to a general conclusion: a maximum on the melting curve is not a necessary condition for the existence of a complex, and vice versa, the existence of a complex does not require the existence of a maximum on the melting curve.

We shall now consider another interesting problem which arises due to the polymorphism inherent in many chemical elements and compounds.

2.3.6 Transformations in the Solid State

Transformations in the solid state due to temperature or pressure changes play an important role in the technology of practically all materials.

Transformations in the solid state of simple and mixed crystals can be caused by *polymorphic transformations*. In the case of mixed crystals another important process occurs: the decomposition of solid solutions due to the dependence of solubility on the external conditions and, above all, on temperature.

The rearrangement of the crystal lattice requires the overcoming of energy barriers by the particles changing their positions in the lattice. Hence, the temperature at which a transformation should yield a more stable thermodynamic state can be easily overstepped. Then quenching can occur, i.e., fixation by fast cooling of structures which are thermodynamically stable only at higher temperatures.

Very often significant *supercooling* is observed when an attempt is made to obtain a phase state stable at low temperatures. This can explain the inconsistent data found in the literature for the temperatures and mechanisms of transformations in the solid state.

Transformations in the solid state are hindered by the elasticity of the medium in which the nuclei of new phases must grow. This leads to the extremely interesting phenomena of periodic growth in polymorphic transformations, and to the formation of *periodic macrostructures* as a result of the decomposition of solid solutions. These phenomena will be considered in Sect.5.2. Here we are interested only in the equilibrium phase diagrams typical of such transformations.

If the components do not form solid solutions, transformations in the solid state can occur when one or both components, as well as the molecular compounds (if they exist in the system under consideration), possess two or more polymorphic modifications.

In the absence of solid solutions, or in the eutectic zone in the case of limited solubility, a mixed crystal is a mixture of different crystals. With a change in temperature, crystals of one phase transform into another modification and the crystal grows from the crystalline medium.

If A and B form a simple phase diagram with one eutectic point and substance B has two polymorphic modifications, two types of phase diagrams are possible. In Fig.2.19 the temperature of a polymorphic transformation of B (B_α to B_β) is above the eutectic temperature. In this case, the equilibrium transformation in the solid state does not occur. Figure 2.20 shows a transformation in the solid state; here the polymorphic transformation occurs at a temperature below the eutectic temperature.

It is obvious that a phase diagram can also be divided into regions in which a polymorphic transformation occurs for both components and for the molecular compound of the 1 : 1 composition (Fig.2.21). When A and B form solid solutions and are subject to polymorphic transformations, different types of phase diagrams are possible. It is not expedient to demonstrate here the various schemes of such processes. The reader can easily imagine the phase diagrams which can be obtained using the following principle: plot any of the diagrams represented in Figs.2.5,8-10,12,14 below the diagrams shown in Figs.2.8-10.

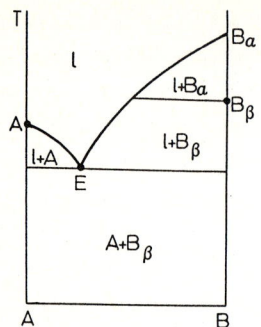
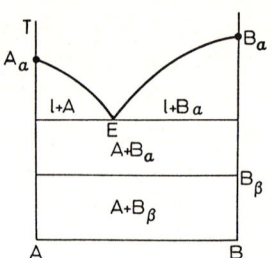

Fig.2.19. Phase diagram for the case in which the temperature of the polymorphic transformation of B (B_α to B_β) is above the eutectic temperature

Fig.2.20. Polymorphic transformation (B_α to B_β) occurring at a temperature below the eutectic temperature

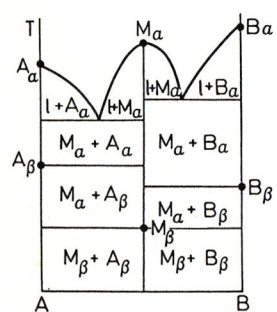

Fig.2.21. Phase diagram illustrating a polymorphic transformation occurring in both components and in the 1 : 1 molecular compound

Special attention should be given to the phase diagram that contains the "cigar" (Fig.2.8) of the continuous series of solid solutions, below which there is a region showing one or more compounds growing in the solid-state crystals that intermix only partially or not at all. In some cases it is difficult to obtain experimentally the part of the phase diagram in which transformations in solids take place. This is due to the fact that reconstruction of the crystal lattice occurs with difficulty. Hence investigators often plot only bits and pieces of equilibrium lines.

Anyhow, two situations have to be distinguished (the mechanism of the phenomenon will be discussed in Chap.4). First, with a temperature decrease the solid solution can become ordered. It is possible that a complex with a regular crystalline structure will form at a certain stoichiometric composition. If the composition differs from the composition of this compound, we can reach a two-phase region, i.e., a mixture of crystals of a new compound and of the solid solution from which it has formed.

Another situation is possible: the decomposition of a solid solution. Assume that decomposition occurs owing to the solubility change with temperature and not due to polymorphism. The corresponding phase diagram is presented in Fig. 2.22. The curve bka limits the region of heterogeneity. The vertex of the curve (k) is its critical point. A heterophase region can be narrow or can involve the whole concentration range. Crystals of the α-phase at all concentrations have, of course, a lattice of the same type, differing only by lattice parameters. Hence the decomposition of solid solutions, which occurs when the representative point intersects the curve bka at a certain temperature, consists in the appearance of slightly different crystals α_1 and α_2. Precise experiments proved that the system is really a two-phase system.

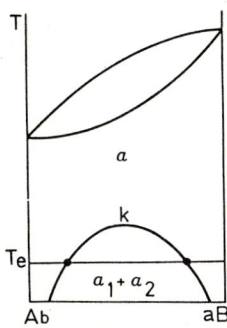

Fig.2.22. Decomposition occurring due to solubility changes with temperature

2.3.7 Phase Diagrams of Optical Enantiomers

Mixed systems of *optical enantiomers* have the same three solid-state forms as mixed crystals of two chemically different substances. Namely, we can have two-phase mixtures of crystals of the left- and right-handed molecules, and solid solutions. But the ordered complexes of the left and right molecules can exist only if the composition is 1:1: in this case the crystal possesses no optical activity and is called a *racemic compound*.

The derivation of all the possible phase diagrams for this case has been given by ROOZEBOOM [2.3]. He showed that mixed crystals of enantiomers cannot give phase diagrams with limited solubility. If there are no solid solutions, only two types of diagrams are possible, namely, those shown in Figs.2.5 and 2.16. But naturally, the diagrams must be symmetrical. Then both the eutectic point and the maximum correspond to the 1:1 composition, and the two parts of the diagram are quite similar. One should not forget that the melting points of optical enantiomers are the same.

The most common of these diagrams (Fig.2.16) are those with the maximum corresponding to the racemic variety, which is an ordered crystal. It forms only two-phase mixtures with enantiomers.

There are about 200 known phase diagrams of the other type (Fig.2.5). Their eutectic point corresponds to the 1:1 composition. Here only two-phase mixtures of enantiomers are possible, and there is no racemic crystal.

Since the melting temperatures of the components are the same, it can be shown that solidus line in the phase diagram in Fig.2.8 degenerates for enantiomorphs into a straight line parallel to the concentration axis. The mixtures of enantiomers have the same melting point. Camphor can serve as an example of such a behavior.

There are also phase diagrams of continuous solid solutions of enantiomorphs with a maximum and a minimum. At an extremum point the solidus and liquidus curves merge. These diagrams resemble those depicted in Figs.2.9,10, but for enantiomorphs they are symmetrical and have an extremum at the 1:1 composition. Only a few examples of such diagrams are known: carvoxim has a curve with a maximum. There is only one known example of the diagram with a minimum: in such a rare case the racemic compound is a solid solution of the simplest stoichiometric 1:1 composition, and (in contrast to a racemic compound, which has a maximum on the phase diagram) it is called *pseudoracemic*.

2.4 Determination of Phase Diagrams

To obtain a phase diagram for a mixed system, a multi-dimensional space with the concentration of the components and the pressure and temperature plotted on its axes must be divided into regions. In the case of a binary system and normal (zero) pressure, the problem reduces to the division of a flat limited graph (temperature versus concentration) into single- and two-phase regions.

Of course, measurements of physical properties of the mixture can suggest a solution of the problem. KURNAKOV [2.4] introduced the term *physicochemical analysis* by which he meant the plotting of some property of the mixture versus its composition.

In this section we shall briefly consider the main methods of plotting phase diagrams. They include *differential thermal analysis* (DTA), X-ray analysis of polycrystalline substances, growth of single crystals, and the method of *contact specimens*.

The methods of physicochemical analysis are described in detail in many monographs. We draw the reader's attention to the monograph by WENDLANDT [2.5]. X-ray phase analysis is considered in detail in [2.6].

2.4.1 Differential Thermal Analysis

Differential thermal analysis (DTA) is widely used in various investigations. It was developed by Le Chatelier who seems to have been the first to use a thermocouple to study thermal properties of minerals.

For a very long time researchers obtained phase diagrams of mixtures only with the aid of one thermocouple. FENNER [2.7] was the first to use two thermocouples, a fact which is essential for DTA. In the last decades rather elegant instruments have been built.

DTA is based on the comparison of thermal properties of the mixture under consideration with a thermally inert standard. The recorded parameter is the difference in temperature ΔT which is measured during specimen heating and cooling at a constant heating rate. This dependence is sensitive to phase transitions and to the reconstruction of the crystal lattice. Thus, having obtained such curves for a series of specimens of different compositions, we can readily and almost automatically construct the phase diagram.

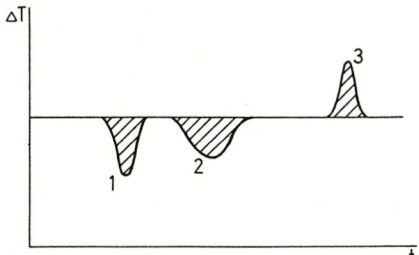

Fig.2.23. Typical DTA curves. t is the temperature of the sample and ΔT is the (small) temperature difference between the sample and the reference substance. Peaks 1,2 are endothermic phase transitions, and peak 3 is an exothermic one

When plotting the phase diagram, the researcher is mainly interested in precise measurements of the transition temperatures. Consequently, DTA instrumental design has emphasized high precision in measurements of t (abscissa in Fig.2.23) at the expense of careful determination of the heats of transition (proportional to the areas cross-hatched in Fig.2.23). Thus one must take into account the possibility of errors in the classification of the recorded transitions. In this respect, DTA cannot compete with the methods described in the following sections. Recently, differential scanning calorimetry (DSC) (a high precision DTA method) has attempted to measure both t and ΔT to comparable precision, but is still inferior to traditional calorimeters in determining heats of transition.

2.4.2 X-Ray Phase Analysis of Polycrystalline Specimens

A homogeneous specimen of definite composition should be prepared for each mixture. The (powder) specimen is given the shape of a plate or a cylinder depending on the X-ray camera used for the phase analysis.

The method of specimen preparation is very important. If a two-phase mixture corresponds to a given composition, the conditions of specimen preparation affect the grain size only and thereby the width of X-ray diffraction lines (Scherrer line-width broadening). Specimens with grains of different size give the same system of lines in the X-ray pattern. But it is useful to obtain sharp solid lines: for this the grains must be at least $10^{-4} - 10^{-5}$ cm in size.

In investigating solid solutions, one must be very careful to avoid errors. The crystallization of a melt during its cooling starts at temperatures corresponding to the liquidus curve. With a temperature decrease, the composition of the melt changes, as well as the composition of the solid phase. The specimen obtained as a result of solidification of the melt is not homogeneous. Usually fast quenching is preferred; for example, the melt is trickled off dropwise into liquid nitrogen. Since the two-phase region between the liquidus and solidus is traversed instantaneously, we can expect that the solid specimen will consist of crystallites which have the same composition as the melt. There are other methods of quenching from the melt. In some cases it is expedient to sputter liquid metal in the gas flow, while in other cases it is better to shoot a metal drop at a massive copper substrate. Another method consists in the injection of a thin jet of the substance into the space between rotating rollers. The rate of cooling can reach 10^{10} K/s.

Specimens of low-melting mixtures can be prepared from solution. Naturally, we must have a common solvent for all the components. If after the dissolution of the components, the liquid solution is slowly evaporated, a homogeneous solid of the same composition as the initial mixture can be obtained.

After obtaining a specimen of the homogeneous solid mixture, the researcher performs the X-ray analysis at different temperatures, including the melting point. Of course, before recording the diffraction pattern, the specimen should be aged at the temperature of analysis, so that the crystal lattice reaches the equilibrium state for that temperature.

Today, researchers possess sensitive X-ray cameras of large radii and automated X-ray diffractometers to record with high precision the diffraction lines of one or several phases existing at the given temperature.

If the point corresponding to the mixtures moves to the interface between phases as the temperature increases, this will be seen as a change in line intensity for a given phase, or as a line shift in the case of lattice parameter changes.

Beginning in the early twenties, the X-ray method received wide application and practically replaced all the other methods used to investigate phase diagrams of metal alloys.

X-ray powder methods are difficult to use in the case of crystals with low symmetry. Here, as a rule, we cannot readily index the X-ray pattern and identify the lines of the different phases.

For the organic compounds which are emphasized in this book, it is practically impossible to obtain phase diagrams by the above methods. However, there are two methods which provide excellent results: these are discussed in the next two sections.

2.4.3 Growth of Mixed Single Crystals

MYASNIKOVA [2.8] and KOLOSOV [2.9] described a method of obtaining phase diagrams by growing single crystals from the melt of a specified concentration under the conditions approximating as much as possible the thermodynamic equilibrium conditions of crystallization. A crystallizer (a test tube with a stopper pierced by a thin metal needle with a sharp tip) is placed into a thermostated oven. A 3-4 g mixture with a specified concentration of compounds is placed into the crystallizer. Since many organic compounds sublime with ease, the tube is closed. Upon heating, the mixture in the crystallizer melts and the melt is stirred manually with the metal needle.

The next and most important step is the choice of temperature of crystallization. If the needle is in the upper part of the crystallizer, the plug being in, it cools down due to heat exchange and serves as the center of crystallization while plunged into the melt: several nuclei form on it which either grow (if the temperature in the oven is lower than the temperature of the liquidus point for the given concentration of components) or melt (if the temperature is higher than the liquidus point). It is possible to choose a temperature in the oven such that the difference between the melting point of the crystal nuclei in the melt and the temperature of its growth does not exceed $0.1 - 0.2°C$. Under these conditions it is possible for the needle to hold only one nucleus. The temperature at which the nucleus starts melting determines the point of the liquidus curve for the given concentration. At a temperature of $0.1 - 0.2°C$ below that of equilibrium, the growth of a nucleus large enough for study (up to 0.2 - 2.0 mm) takes several hours (sometimes days).

Thus, after the determination of the point T_1 of the liquidus curve for the given concentration of the components, two of five crystals can be grown in the same melt with a minimum of supercooling. The supercooling is kept constant during the whole experiment. The total weight of the crystals grown from 3-4 g of the melt does not exceed 0.010 - 0.015 g, i.e., about 0.3% of the total weight of the mixture.

Other experiments are used to verify that the crystal-melt system really approximates, as close as possible, the conditions of thermodynamic equilibrium corresponding to the impurity distribution in the crystal. These conditions are achieved thanks to the slow rate of crystallization and the small size of the growing crystal.

The crystals extracted from the melt are used first to construct the solidus line. For that purpose it is possible to use microanalytical methods which permit determining the percentage of some chemical element in the grown crystals contained in only one of the components. More universal is a method which is based on the exact knowledge of the liquidus line.

Since the single crystals of a solid solution have a different composition than the melt from which they have been extracted, they melt in a different temperature range. The crystals are ground and placed into a sealed capillary tube inside the oven. At the temperature corresponding to the point on the liquidus curve for the melt from which the crystals have been extracted, the first stage of "liquifaction" can be seen in the tube. Raising the temperature step by step and slowly enough for the equilibrium between liquid and solid phases to be kept, the temperature at which the last solid grain in the capillary melts can be found with sufficient accuracy. Such a method permits the recording of a reliable liquidus curve.

If at a given concentration the crystals of two phases grow simultaneously, they can be distinguished by their different facets.

The method of single-crystal growth is very laborious. It takes several months to obtain a genuine equilibrium phase diagram. However, the method is justified as it gives high accuracy and reproducibility.

MYASNIKOVA et al. [2.10] have carried out a special investigation of the capability of this method for the system acenaphthene-alpha-nitronaphthalene. They compared the liquidus and solidus curves obtained by this method with others. It turned out that the liquidus curve can be determined by other methods, whereas the points smoothly filling the solidus curve can only be obtained by the method of single crystal growth. Only this method provides the melting and crystallization curves which can be confidently related to the thermodynamic equilibrium. To confirm this fact, a crystal of a solid solution with a given composition was grown at different degrees of supercooling. It was shown that the supercooling is strongly influenced by the percentage of impurities in the crystal of the solid solution. Hence, it is very important to carry out the growth as slowly as possible, trying not to supercool by more than $0.1 - 0.2°C$ (Fig.2.24).

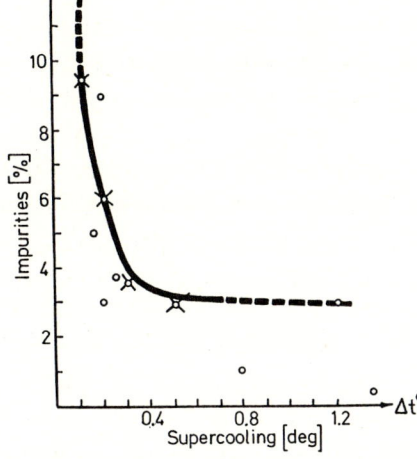

Fig.2.24. Percentage of impurities $i(\%)$ versus degree of supercooling ($\Delta t°$) for acenaphthene-alpha-nitronaphthalene

Numerous experiments show that a smooth reproducible curve requires very accurate measurements. A new portion of the mixture of the same concentration is prepared each time to obtain every point of the curve. The initial conditions are chosen so that the crystal nucleus on the seed needle is the first and the only one in the melt during the whole growth process. Crystals of acenaphthene, for instance, grow in the form of an elongated rhombic prism. When the prism reaches $1.0 - 1.2$ mm in length (approximately $0.1 - 0.3$ mm^2 in cross section) the needle with the crystal on it is taken from the melt, the oven is switched off, and the whole system is cooled down slowly. The surface of the grown crystal is washed with some solvent to remove the remainder of the solidified melt from its surface. The impurity content in the crystal is established through determination of the endpoint temperature of melting.

The curve shown in Fig.2.24 suggests that if the equilibrium conditions could be approached more closely, the impurity level in the crystal would be slightly more than 10%. Thus, the attainment of the equilibrium conditions of crystallization is far from simple.

The opinion that at relatively high supercooling (i.e., at a high rate of crystallization), the crystal of a solid solution contains more impurities than that formed during slow crystallization is thus *not correct*. Furthermore, the converse opinion, i.e., that fast crystallization promotes the formation of purer compounds, is however also wrong. Figure 2.24 shows that if the composition of the mixture corresponds to the concentration necessary for the formation of substitutional solid solutions, some impurities (in our case about 3%) will be caught in the matrix crystal. The practical conclusion is that in such systems it is impossible to purify the substance by recrystallization without significant losses of starting material.

This is a purely practical problem which is encountered, for example, in industrial purification of "raw anthracene" obtained from the by-products of the coke industry, containing anthracene, carbazole, and phenanthrene in approximately equal quantities, i.e., about 20-30% of each component. When the carbazole impurity in anthracene does not exceed 3-4%, the conventional methods of crystallization from typical organic solvents are no longer effective, because up to 4% of carbazole can be dissolved in the anthracene matrix (the molecules are similar in shapes and dimensions) [2.11]. Hence, it is impossible to remove the remaining carbazole from the solution by any method of joint crystallization. As the methods of melt refining (zone melting, Bridgman technique) used in an industrial scale are very expensive and give large losses of raw material, one should use separate crystallization of the components.

The question now arises as to whether the impurity distribution is the same in all the crystals grown by the method discussed above. This has been checked on the basis of precise X-ray measurements [2.11]. The differences appear to be negligibly small, which proves the overwhelming advantages of the method.

2.4.4 Method of Contacting Specimens

Very often a series of problems must be solved which do not require such a labor-intensive process as the single crystal growth. It is often quite sufficient to establish only the type of phase diagram of the system under investigation and to decide whether it can give solid solutions with noticeable concentrations of components.

To solve the problems of this kind we can recommend the method developed by L. and A. KOFLER [2.12] for the investigation of transparent organic compounds with a sufficiently low melting temperature. They suggest using a special heating stage as an attachment to the conventional polarizing microscope to study specimens in the transmitted light during their heating or cooling.

The investigation is carried out in two stages. The first one consists of the determination of the phase diagram. For this, it is sufficient to prepare one contact specimen. The procedure is as follows: The microscope slide and the cover glass are placed in the heating stage which is then heated to a temperature 5-10°C higher than the melting point of the higher melting component. Several grains of the substance are placed near the edge of the cover glass, where they melt and are sucked in between the slide and the cover glass by capillary forces. When about half the area under cover glass is filled with the melt, the specimen (the slide and the melt on it) is put on a cooling block (a small reservoir whose horizontal surface is cooled by circulating water). Being cooled, the melt crystallizes. Then, at a still lower temperature (but higher than the melting point of the low-melting component) the remaining part of the slide under the cover glass is filled with the melt of the second component, so that it is in *contact* with the solidified edge of the first component. Then there occurs a partial dissolution of the boundary layer of the first component in the melt of the second component, i.e., there is a partial redistribution of the concentrations of the compounds in the *contact zone*. The specimen is then quickly cooled.

The microscopic investigation of the contact zone during heating allows the determination of the phase diagram. The contact zone has indeed the complete concentration range of the components: from the pure component A at one end of the cover glass to the pure component B at the opposite end.

If the system has a eutectic phase diagram, the eutectic is the first to melt in the process of heating: a narrow black band (under crossed Nicols) appears along the whole contact zone at a certain temperature of the stage.

In the case of a peritectic diagram, melting proceeds from one end (that of the pure low-melting component). The melt front moves towards the mixed zone (contact zone), and here it is very important to detect the temperature of the "peritectic reaction", i.e., the temperature at which an almost instantaneous reconstruction occurs and the second component starts melting. This is a very difficult problem, especially if the crystals are almost similar in shape.

L. and A. KOFLER [2.12] gave a detailed description of the method, with examples. Without going into details, we mention here that their method determines only the temperatures of the special points (melting point, temperature of phase transition), but not the concentrations.

If one wishes, one can then proceed to the construction of phase diagrams point by point. Special mixtures with different concentrations of the components should be prepared; each mixture should be melted separately in the heating stage, covered with its own cover glass, and cooled quickly. Then during the second heating each mixture should be studied under the microscope, and the temperatures at which melting of the crystalline film starts and ends should be recorded. The scatter of the recorded temperatures is about 1.5 - 2.0°C (compare it with the accuracy of 0.1°C obtained in the method of single crystal growth!).

The method of contacting specimens makes it easy to choose quickly the appropriate system. On the whole, in the case for which the growth of single crystals is very difficult (say, for high molecular-weight compounds), the Koflers' method makes it possible to investigate, though not with a very high accuracy, various organic compounds and their mixtures, provided they are crystalline and do not decompose in the process of melting.

3. Particle Packing in a Crystal

The branch of science which deals with regularities in the arrangement of particles (atoms, ions, and molecules) in a crystal is called crystal chemistry. This term is obviously a misnomer; it seems more appropriate for a science dealing with chemical transformations in the crystalline state. But, strange as it may seem, scientists working in solid-state chemistry do not call themselves crystal chemists. As a matter of fact, crystal chemistry can be considered as a part of solid-state physics. But its descriptive character and the necessity of dealing with structural data on thousands of compounds are the obvious reasons why solid-state physicists refuse to have anything to do with the packing of particles. They believe the methods used in crystal chemistry to be closer to chemistry than to physics.

Regularities of particle packing in a crystal refer to the character of the interaction between the constituent atoms and molecules. Unfortunately, there are no simple rules that prove, even a posteriori, that precisely the given arrangement realized in the crystal is responsible for the local minimum of the free energy of the crystal, let alone predict the crystal structure. The regularities of metal structures are the most difficult ones to understand. The situation with ionic crystals is somewhat better, and, strange as it may seem at first glance, the problem has been solved satisfactorily for neutral organic crystals.

In most cases we can only enumerate the three main factors influencing the atomic arrangement. First, special attention is usually paid to valence bonds which provide directional interaction of neighboring atoms and link them into islands, chains, or nets. Next, attempts are made to take into account the electrical charges of atoms and ions. Third, in the case of metals or semiconductors, it is important to know how the energy levels are occupied by the electrons.

The main problem for crystal chemistry is the geometric arrangement of atomic centers. Some scientists restrict themselves to detailed classification, and search for empirical rules which would permit relating a substance, by analogy, to a certain structural type. Others try to bridge the gap between geometry and the above three factors. Usually these attempts are not satisfactory, and the "rules" suggested by these scientists have so many exceptions that it is hardly possible to speak of their heuristic value.

The only reason why crystal chemistry can be considered as somewhat more than a descriptive science is the possibility of using a very rough, but undoubtedly rather useful model, making it possible, with the aid of certain rules, to assign to the particles constituting the crystal a certain shape and certain dimensions and to consider the crystal structure as the packing of these particles, i.e., as a system which prevents particles from either "hanging in the air" or from overlapping.

Such an approach is called *geometric*; we shall try to prove that the geometric model of particle packing in a crystal has all the properties of the models used by physicists.

Before considering the well-known methods of describing crystal structures, we should like to recall some basic data from the *theory of crystal lattices* (which, incidentally, is also a stepchild of chemistry and physics). The second part of this chapter deals with the fundamentals of the geometry of ionic crystals, metals, and molecular crystals. There are many papers devoted to these fields of science. We bring to the reader's attention the monographs [3.1-4] on the structure of metals and the books [3.5,6] on ionic crystals. Possible general references to organic crystal chemistry are [3.7-9].

3.1 Elements of the Theory of Space Lattices

The following fundamental statement can be taken as the definition of a *crystal*: the distribution of matter in a crystal can be represented by a three-dimensional periodic function. In other words, a crystal is a space lattice. A parallelepiped built on three translations (in the general case an oblique one) can be chosen in an infinite number of ways. An ideal crystal can be constructed by means of three translation vectors. Such a crystal is called *ideal*, since the definition does not take into account various *defects*, such as *dislocations* and *vacancies*, which are always present in some quantity in any real crystal.

The vertex of the parallelepiped, called its *origin*, can be chosen at any lattice point. Its choice is only a matter of convenience. The parallelepiped under consideration is known as a *unit cell*. The smallest possible unit cell is the *primitive cell*.

In some cases, it is expedient to choose a unit cell of a larger volume than that of a primitive cell. Usually, three types of nonprimitive unit cells are considered: *body-*, *base-*, and *face-centered* unit cells.

To characterize a crystal it is sufficient to indicate the coordinates of all the atoms in the unit cell. If one chooses molecules with a given atomic structure as the building blocks from which a crystal is built, the crystal can be described by specifying the "arrangement" of the centers of the molecules and indicating their orientation relative to the lattice axes.

Any lattice possesses a certain *symmetry*. This means that under various linear transformations the crystal can be brought into coincidence with itself. The set of all symmetry transformations inherent in a lattice is called its *space group*. The major property of a space lattice, its periodicity, limits the number and character of symmetry transformations. The rigorous derivation of the space groups presented by FEDOROV [3.10] and independently by SCHOEN-FLIES [3.11] at the end of last century proved that there are 230 space groups.

At present, two space group notations are used; in the first, suggested by Schoenflies, the symbol for a space group is a symbol of a point group (see below) with a superscript which enumerates, in a special way, the space groups belonging to the given point group.

The second, so-called *international notation* is more rational. Its symbol first indicates whether the unit cell is primitive or centered and then describes the symmetry elements, whose knowledge is sufficient to obtain all the characteristics of the space group. Here, we will not dwell on all the symbols and refer the reader to the textbooks, for example [3.12-14]. In the course of our presentation we shall recall the meaning of some symbols and shall consider some examples in detail.

A transformation is called *closed* (or macroscopic) if all the lattice points coincide with themselves after a finite number of identical operations, i.e., symmetry transformations. The remaining symmetry transformations (that do not form a finite group) are called *open*.

Rotation and *inversion axes* are closed elements of symmetry. It can be proved that only the following nontrivial axes are possible in a crystal: 2, 3, 4, 6, $\bar{1}, \bar{2}$, and $\bar{4}$. Here a bar denotes the inversion axes. They represent a rotation and subsequent inversion with respect to a center. The one-fold inversion axis is called a center of symmetry or a *center of inversion*. The letter m always stands for axis $\bar{2}$: a two-fold inversion axis is equivalent to a mirror reflection plane.

The elements of symmetry can intersect at a point, thus forming a so-called *point group*. Each group can be characterized by its *multiplicity*, i.e., the number of equivalent points which arise as a result of the action of the group of elements of symmetry on an arbitrary point. The number of possible point groups is 32. If a point lies on a symmetry element, i.e., it is only affected by other symmetry elements, its position is called *special*. Of course, the multiplicity of a special position is several times less than that of a general position.

Such a special position occurs, for example, for the point group $2/m = C_{2h}$ when a point lies on the axis or the plane of symmetry. In the first case (y axis is along 2), the transformation is

(1) $0 y 0 \rightarrow$ (2) $0 \bar{y} 0$,

in the second (xz being the reflection plane)

(1) $x 0 z \rightarrow$ (2) $\bar{x} 0 \bar{z}$.

In both cases the multiplicity of the special position is 2, whereas a general position in that point group has multiplicity 4.

In groups with one special point there is only one special position which does not undergo any transformation by any symmetry element (in group 2/m such a point lies on the intersection of 2 and m). The multiplicity of such a special position is *one*.

The maximum multiplicity of a general point group position is equal to 48. This is the multiplicity of the cubic point group O_h (m3m).

The existence of 32 possible point groups in a lattice leads to the classification of all the crystals into 32 *crystal classes*.

We have mentioned above that the choice of the lattice point and the form of a unit cell are determined mainly by the crystal class. If there are only centers of inversion in the lattice, the unit cell has the form of an oblique parallelepiped. In this case it is convenient to choose a primitive cell, and the crystal is said to be *triclinic* (or to belong to the triclinic system). The presence of 2 and $\bar{2}$ axes makes it possible to choose one of the unit-cell axes perpendicular to two others. In this case, the crystal is *monoclinic*. The *orthorhombic* system is characterized by the presence of at least two mutually perpendicular two-fold axes. The lattice with a four-fold axis has two periods of equal length in the plane perpendicular to it; thus, a *tetragonal* unit cell is chosen. In the presence of three- and six-fold axes it is expedient to choose *hexagonal* and *rhombohedral* unit cells. The *cubic* system possesses the highest symmetry; it contains the point groups 23, m3, 342, $\bar{4}3m$, and m3m.

Choosing the directions of a unit-cell axis in accordance with the lattice symmetry, we readily arrive at 14 possible translations or *Bravais lattices*.

In a monoclinic system, there are two types of unit cells: primitive and base-centered (two opposite faces are centered). An orthorhombic system possesses all four possible types of unit cells. A tetragonal system has only two, while the cubic system, in addition to the primitive unit cell, also has body- and face-centered cells.

Now, if open symmetry elements are taken into account, the whole realm of crystals is divided into 230 space groups. The open symmetry elements are *glide-reflection planes* which reflect a point and then shift it for half a translation or a quarter of a translation. *Screw axes* rotate a representative point by a certain angle, corresponding to the axis symbol, and shift it along the axis for a fraction of a period.

Below, we describe in some detail several space groups necessary for our further presentation. The figures given below are taken from the fundamental reference, the *International Tables for X-Ray Crystallography* [3.15].

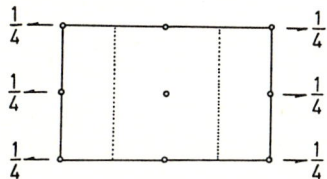

Fig.3.1. Graphic representation of the $P2_1/c$ space group

Figure 3.1 is a graphic representation of the space group $P2_1/c$. Its Schoenflies symbol is C_{2h}^5. The letter P stands for a primitive cell; 2_1 means that a two-fold screw axis passes through the lattice. The symbol c denotes the glide-reflection planes with the translation for half the period along the axis c. The full international symbol $P1\,2_1/c\,1$ indicates that the 2_1 axis is parallel to the b axis, and that the c-glide is perpendicular to b, and that only identity symmetry operators are parallel to a and c. How are the glide planes positioned? This can also be seen from the symbol: a slanted stroke indicates that the glide-reflection plane is perpendicular to the screw axis. The same is seen from the figure. If the glide-reflection plane is represented by a dashed line, the gliding occurs in the direction perpendicular to the plane of the page. The group in Fig.3.1 belongs to the symmetry class 2/m.

The possible positions in the space group are denoted by lower case Roman letters. Then, the symmetry of the position is given, followed by the coordinates of the points related by symmetry. The space group $P2_1/c$, for instance,

53

can be described as follows (Table 3.1):

a) The origin is chosen at a point with the symmetry $\bar{1}$.

b) A representative point in the general position e has a trivial symmetry 1 and corresponds to four equivalent points. In other words, this position has a multiplicity equal to 4 (as shown in column 1 of Table 3.1).

c) In addition to the general four-fold position, the space group has four two-fold positions: a, b, c, and d. The coordinates of the positions show that the representative point possesses a center of inversion, i.e., has no degrees of freedom.

Table 3.1. Coordinates of the points for the space group $P2_1/c$

4	e	1	x, y, z;	$\bar{x}, \bar{y}, \bar{z}$;	$\bar{x}, \frac{1}{2}+y, \frac{1}{2}-z$;	$x, \frac{1}{2}-y, \frac{1}{2}+z$;
2	d	$\bar{1}$	$\frac{1}{2}, 0, \frac{1}{2}$;	$\frac{1}{2}, \frac{1}{2}, 0$;		
2	c	$\bar{1}$	$0, 0, \frac{1}{2}$;	$0, \frac{1}{2}, 0$;		
2	b	$\bar{1}$	$\frac{1}{2}, 0, 0$;	$\frac{1}{2}, \frac{1}{2}, \frac{1}{2}$;		
2	a	$\bar{1}$	$0, 0, 0$;	$0, \frac{1}{2}, \frac{1}{2}$.		

If the space group and the number of particles per unit cell have been determined experimentally, it is possible to arrive at certain conclusions about the mutual arrangement of atoms and molecules permitted by the space group symmetry. What can be said about the space group under consideration?

First, it is clear that the number of particles in a crystal with the space group $P2_1/c$ can only be even. Suppose it is 4; then two cases are possible: the particles are either in the four-fold position or in any two two-fold positions. If there are 6 particles, again two cases are possible: the particles occupy either the four-fold position and one of the two-fold positions or else three two-fold positions.

In the case of an atomic crystal, the particle can be placed either in position 1 or in position $\bar{1}$. In the case of a molecular crystal the restrictions are more strict because the symmetry of the molecule itself should also be taken into account. It is clear that in the case of an ordered crystal nonsymmetrical molecules cannot be placed in two-fold positions.

In the space group $P2_1/c$ the special positions have no degrees of freedom (their coordinates are fixed), and therefore their number is limited. No restrictions are imposed on the number of particles in the general position or in special positions with one or two degrees of freedom, except for an obvious restriction due to the ratio between the size of the unit cell and the molecular dimensions.

We shall give a short explanation of the illustrations in the International Tables [3.15]. Figure 3.1 shows the lattice projection which passes through the screw axes. Small circles denote the positions of the centers of inversion; the fraction 1/4 placed next to the symbol of a screw axis indicates that the axis passes between the centers of inversion. The projection is always the xy projection with the origin in the upper left-hand corner, the axis horizontal (across the page) and the axis vertical (down the page).

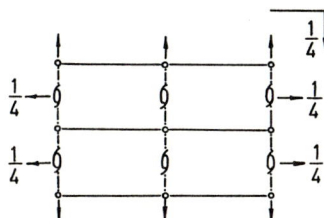

Fig.3.2. Graphic representation of the D_{2h}^7 = Pmna space group

Figure 3.2 depicts one orthorhombic space group of the class mmm. The Schoenflies symbol is D_{2h}^7. The international symbol Pmna means that the mirror plane m is perpendicular to the axis a, the glide plane along the diagonal n (with translation $\frac{a}{2}+\frac{c}{2}$) is perpendicular to the axis b, and the glide plane along the axis a (i.e., with translation a/2) is perpendicular to the axis c. It is seen from the drawing that the glide plane parallel to the projection passes at the height of z = 1/4; the direction of glide is also indicated. Glide mirror planes are designated by solid lines; a dashed line shows a plane of diagonal glide.

The group has 9 types crystallographic positions (one general, eight special): the coordinates of the respective points are given in Table 3.2.

The origin is at the point with the symmetry 2/m. The multiplicity of the general position (i) is 8; there is also a four-fold position h with two degrees of freedom. The representative point for this position has mirror symmetry. There are three more four-fold positions e, f, g, possessing only one degree of freedom. The representative point for the position g lies on the two-fold axis parallel to the y axis. In the four-fold position f, the point

Table 3.2. Coordinates of the point for the space group Pmna

8	i	1	x, y, z;	$x, \bar{y}, \bar{z}$;	$\frac{1}{2} - x, y, \frac{1}{2} - z$;	$\frac{1}{2} - x, y, \frac{1}{2} + z$;
			$\bar{x}, \bar{y}, \bar{z}$;	$\bar{x}, y, z$;	$\frac{1}{2} + x, \bar{y}, \frac{1}{2} + z$;	$\frac{1}{2} + x, y, \frac{1}{2} - z$;
4	h	m	$0, y, z$;	$0, \bar{y}, \bar{z}$;	$\frac{1}{2}, y, \frac{1}{2} - z$;	$\frac{1}{2}, \bar{y}, \frac{1}{2} + z$;
4	g	2	$\frac{1}{4}, y, \frac{1}{4}$;	$\frac{3}{4}, \bar{y}, \frac{3}{4}$;	$\frac{3}{4}, y, \frac{1}{4}$;	$\frac{1}{4}, \bar{y}, \frac{3}{4}$;
4	f	2	$x, \frac{1}{2}, 0$;	$\bar{x}, \frac{1}{2}, 0$;	$\frac{1}{2} + x, \frac{1}{2}, \frac{1}{2}$;	$\frac{1}{2} - x, \frac{1}{2}, \frac{1}{2}$;
4	e	2	$x, 0, 0$;	$\bar{x}, 0, 0$;	$\frac{1}{2} + x, 0, \frac{1}{2}$;	$\frac{1}{2} - x, 0, \frac{1}{2}$;
2	d	$\frac{2}{m}$	$0, \frac{1}{2}, 0$;	$\frac{1}{2}, \frac{1}{2}, \frac{1}{2}$;		
2	c	$\frac{2}{m}$	$\frac{1}{2}, \frac{1}{2}, \frac{1}{2}$;	$0, \frac{1}{2}, \frac{1}{2}$;		
2	b	$\frac{2}{m}$	$\frac{1}{2}, 0, 0$;	$0, 0, \frac{1}{2}$;		
2	a	$\frac{2}{m}$	$0, 0, 0$;	$\frac{1}{2}, 0, \frac{1}{2}$.		

lies on the two-fold axis parallel to the x axis. The third two-fold position e is similar to the previous one: it is on another system of two-fold axes, parallel to the x axis. Also, there are four two-fold positions a, b, c, and d. The representative point has sufficiently high symmetry: through this point pass a two-fold axis and a mirror plane perpendicular to it.

In this case, as in the previous one, it is possible to draw some valuable conclusions if the number of particles per unit cell is known, and the space group is determined correctly. It is also possible to reject some doubtful experimental data, because the number of particles per unit cell and the lattice symmetry are related.

In the case of a molecular crystal with two molecules in a unit cell crystallizing in the space group Pmna, the symmetry 2/m should be ascribed to the molecules. In case of four molecules per unit cell the symmetry cannot be chosen unambiguously: two variants are possible, 2 and m.

Now we shall describe one of the space groups that belongs to the point group 4/mmm of the tetragonal system. Its symbols are D_{2h}^{14} and P4/mnm. The symmetry elements are shown in Fig.3.3. Perpendicular to the paper, we find the screw axes 4_2 which are the combinations of a four-fold screw axis [rotation by 90° followed by a translation for a quarter of a period (c/4) along the c axis] and simple two-fold rotation axes. All the points related by the 4_2 screw axis can be obtained by rotating two points connected by a simple two-fold axis ("dumb-bell") through a right angle and then translating them by a quarter of a period along c. One should pay attention to the fact that the two-fold axes parallel to the tetragonal axis pass through the center of inversion. Mirror planes are perpendicular to the four-fold axis which is indicated by the symbol of two lines crossing at right angles depicted on the right-hand upper part of the figure. Besides these, there are also vertical mirror planes passing through two-fold axes. Two mutually perpendicular systems of diagonal glide planes (dotted-dashed line) make an angle of 45° with the mirror planes. They alternate with glide planes passing through the four-fold axes. These planes reflect a point and shift it in the plane of the paper (dashed line). The positions of the two-fold rotation and the screw axes are seen in the picture and need no comment.

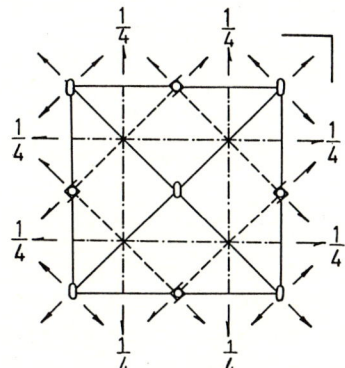

Fig.3.3. Graphic representation of the D_{2h}^{14} = P4/mnm space group

This space group has eleven types of crystallographic positions. The multiplicity of the general position is 16. Either of the two most symmetrical positions (a and b at the bottom of Table 3.3) - with no degrees of freedom - leads to a tetragonal body-centered unit cell. All crystallographic positions possible in this group and the coordinates of the equivalent points are listed in Table 3.3.

If a group with such a high symmetry is unambiguously established, it is possible to obtain important information concerning possible and impossible locations of a certain number and kinds of atoms per unit cell.

Is it possible to locate in the unit cell of this space group a compound of the composition A_2B_3? It depends. If the number of atomic groups is odd, it is impossible. If it is even, several possibilities arise. Two groups can occupy only one two-fold position. If there are four groups, the B atoms should

Table 3.3. Coordinates of the points for the space group P4/mnm

16	k	1	x, y, z;	$\bar{x}, \bar{y}, z$;	$\frac{1}{2}+x, \frac{1}{2}-y, \frac{1}{2}+z$;	$\frac{1}{2}-x, \frac{1}{2}+y, \frac{1}{2}+z$;
			$x, y, \bar{z}$;	$\bar{x}, \bar{y}, \bar{z}$;	$\frac{1}{2}+x, \frac{1}{2}-y, \frac{1}{2}-z$;	$\frac{1}{2}-x, \frac{1}{2}+y, \frac{1}{2}-z$;
			y, x, z;	$\bar{y}, \bar{x}, z$;	$\frac{1}{2}+y, \frac{1}{2}-x, \frac{1}{2}+z$;	$\frac{1}{2}-y, \frac{1}{2}+x, \frac{1}{2}+z$;
			$y, x, \bar{z}$;	$\bar{y}, \bar{x}, \bar{z}$;	$\frac{1}{2}+y, \frac{1}{2}-x, \frac{1}{2}-z$;	$\frac{1}{2}-y, \frac{1}{2}+x, \frac{1}{2}-z$;
8	j	m	x, x, z;	$\bar{x}, \bar{x}, z$;	$\frac{1}{2}+x, \frac{1}{2}-x, \frac{1}{2}+z$;	$\frac{1}{2}-x, \frac{1}{2}+x, \frac{1}{2}+z$;
			$x, x, \bar{z}$;	$\bar{x}, \bar{x}, \bar{z}$;	$\frac{1}{2}+x, \frac{1}{2}-x, \frac{1}{2}-z$;	$\frac{1}{2}-x, \frac{1}{2}+x, \frac{1}{2}-z$;
8	i	m	$x, y, 0$;	$\bar{x}, \bar{y}, 0$;	$\frac{1}{2}+x, \frac{1}{2}-y, \frac{1}{2}$;	$\frac{1}{2}-x, \frac{1}{2}+y, \frac{1}{2}$;
			$y, x, 0$;	$\bar{y}, \bar{x}, 0$;	$\frac{1}{2}+y, \frac{1}{2}-x, \frac{1}{2}$;	$\frac{1}{2}-y, \frac{1}{2}+x, \frac{1}{2}$;
8	h	2	$0, \frac{1}{2}, z$;	$0, \frac{1}{2}, \bar{z}$;	$0, \frac{1}{2}, \frac{1}{2}+z$;	$0, \frac{1}{2}, \frac{1}{2}-z$;
			$\frac{1}{2}, 0, z$;	$\frac{1}{2}, 0, \bar{z}$;	$\frac{1}{2}, 0, \frac{1}{2}+z$;	$\frac{1}{2}, 0, \frac{1}{2}-z$;
4	g	mm	$x, \bar{x}, 0$;	$\bar{x}, x, 0$;	$\frac{1}{2}+x, \frac{1}{2}+x, \frac{1}{2}$;	$\frac{1}{2}-x, \frac{1}{2}-x, \frac{1}{2}$;
4	f	mm	$x, x, 0$;	$\bar{x}, \bar{x}, 0$;	$\frac{1}{2}+x, \frac{1}{2}-x, \frac{1}{2}$;	$\frac{1}{2}-x, \frac{1}{2}+x, \frac{1}{2}$;
4	e	mm	$0, 0, z$;	$0, 0, \bar{z}$;	$\frac{1}{2}, \frac{1}{2}, \frac{1}{2}+z$;	$\frac{1}{2}, \frac{1}{2}, \frac{1}{2}-z$;
4	d	$\bar{4}$	$0, \frac{1}{2}, \frac{1}{4}$;	$\frac{1}{2}, 0, \frac{1}{4}$;	$0, \frac{1}{2}, \frac{3}{4}$;	$\frac{1}{2}, 0, \frac{3}{4}$;
4	c	$\frac{2}{m}$	$0, \frac{1}{2}, 0$;	$\frac{1}{2}, 0, 0$;	$0, \frac{1}{2}, \frac{1}{2}$;	$\frac{1}{2}, 0, \frac{1}{2}$;
2	b	mmm	$0, 0, \frac{1}{2}$;	$\frac{1}{2}, \frac{1}{2}, 0$;		
2	a	mmm	$0, 0, 0$;	$\frac{1}{2}, \frac{1}{2}, \frac{1}{2}$.		

be placed at least over three crystallographically nonequivalent positions, thus forming three kinds of B atoms. If a particle is a rigid non-planar molecule, we need 16 molecules to satisfy the symmetry of the space group.

Unit cells with 30 atoms occur very frequently among the intermetallic compounds. This number can be obtained in space group P4/mnm only by a limited number of combinations, for example, 16, 8, 4, and 2 or 8, 8, 8, 4, and 2. An eight-fold position can be replaced by two four-fold positions. It is impossible, however, to manage without one two-fold position.

We should like to underline once more the importance of the unambiguous choice of a space group in structure investigations. The necessary data can be obtained only if the X-ray structure investigation is carried out for a single crystal of the substance. The results obtained for more or less complicated polycrystals (powder data) are not very reliable.

Finally, consider the highly symmetrical space group O_h^5 - Fm3m. This cubic face-centered unit cell is so rich in symmetry elements that it is practically impossible to represent it schematically. We limit ourselves to giving the list of crystallographic positions. The whole set of coordinates can be obtained by adding to each group of point coordinates the following values:

$$0,0,0; \quad 0,\tfrac{1}{2},\tfrac{1}{2}; \quad \tfrac{1}{2},0,\tfrac{1}{2}; \quad \tfrac{1}{2},\tfrac{1}{2},0.$$

The left column indicates the multiplicity of the position. The description of the space group is found in Table 3.4.

<u>Table 3.4.</u> Coordinates of the points for the space group Fm3m

192	ℓ	1	x,y,z;	z,x,y;	y,z,x;	$x,z,\bar{y}$;	$y,x,\bar{z}$;	$z,y,\bar{x}$;
			$x,\bar{y},\bar{z}$;	$z,\bar{x},\bar{y}$;	$y,\bar{z},\bar{x}$;	$x,\bar{z},\bar{y}$;	$y,\bar{x},\bar{z}$;	$z,\bar{y},\bar{x}$;
			$\bar{x},y,\bar{z}$;	$\bar{z},x,\bar{y}$;	$\bar{y},z,\bar{x}$;	$\bar{x},z,\bar{y}$;	$\bar{y},x,\bar{z}$;	$\bar{z},y,\bar{x}$;
			$\bar{x},\bar{y},z$;	$\bar{z},\bar{x},y$;	$\bar{y},\bar{z},x$;	$\bar{x},\bar{z},y$;	$\bar{y},\bar{x},z$;	$\bar{z},\bar{y},x$;
			$\bar{x},\bar{y},\bar{z}$;	$\bar{z},\bar{x},\bar{y}$;	$\bar{y},\bar{z},\bar{x}$;	$\bar{x},\bar{z},\bar{y}$;	$\bar{y},\bar{x},\bar{z}$;	$\bar{z},\bar{y},\bar{x}$;
			$\bar{x},y,z$;	$\bar{z},x,y$;	$\bar{y},z,x$;	$\bar{x},z,y$;	$\bar{y},x,z$;	$\bar{z},y,x$;
			$x,\bar{y},z$;	$z,\bar{x},y$;	$y,\bar{z},x$;	$x,\bar{z},y$;	$y,\bar{x},z$;	$z,\bar{y},x$;
			$x,y,\bar{z}$;	$z,x,\bar{y}$;	$y,z,\bar{x}$;	$x,z,\bar{y}$;	$y,x,\bar{z}$;	$z,y,\bar{x}$;
96	k	m	x,x,z;	z,x,x;	x,z,x;	$\bar{x},\bar{x},\bar{z}$;	$\bar{z},\bar{x},\bar{x}$;	$\bar{x},\bar{z},\bar{x}$;
			$x,\bar{x},\bar{z}$;	$z,\bar{x},\bar{x}$;	$x,\bar{z},\bar{x}$;	$\bar{x},x,z$;	$\bar{z},x,x$;	$\bar{x},z,x$;
			$\bar{x},x,\bar{z}$;	$\bar{z},x,\bar{x}$;	$\bar{x},z,\bar{x}$;	$x,\bar{x},z$;	$z,\bar{x},x$;	$x,\bar{z},x$;
			$\bar{x},\bar{x},z$;	$\bar{z},\bar{x},x$;	$\bar{x},\bar{z},x$;	$x,x,\bar{z}$;	$z,x,\bar{x}$;	$x,z,\bar{x}$;
96	j	m	$0,y,z$;	$z,0,y$;	$y,z,0$;	$0,z,y$;	$y,0,z$;	$z,y,0$;
			$0,\bar{y},\bar{z}$;	$\bar{z},0,\bar{y}$;	$\bar{y},\bar{z},0$;	$0,\bar{z},\bar{y}$;	$\bar{y},0,\bar{z}$;	$\bar{z},\bar{y},0$;
			$0,y,\bar{z}$;	$\bar{z},0,y$;	$y,\bar{z},0$;	$0,\bar{z},y$;	$y,0,\bar{z}$;	$\bar{z},y,0$;
			$0,\bar{y},z$;	$z,0,\bar{y}$;	$\bar{y},z,0$;	$0,z,\bar{y}$;	$\bar{y},0,z$;	$z,\bar{y},0$;
48	i	mm	$\tfrac{1}{2},x,x$;	$x,\tfrac{1}{2},x$;	$x,x,\tfrac{1}{2}$;	$\tfrac{1}{2},x,\bar{x}$;	$\bar{x},\tfrac{1}{2},x$;	$x,\bar{x},\tfrac{1}{2}$;
			$\tfrac{1}{2},\bar{x},\bar{x}$;	$\bar{x},\tfrac{1}{2},\bar{x}$;	$\bar{x},\bar{x},\tfrac{1}{2}$;	$\tfrac{1}{2},\bar{x},x$;	$x,\tfrac{1}{2},\bar{x}$;	$\bar{x},x,\tfrac{1}{2}$;
48	h	mm	$0,x,x$;	$x,0,x$;	$x,x,0$;	$0,x,\bar{x}$;	$\bar{x},0,x$;	$x,\bar{x},0$;
			$0,\bar{x},\bar{x}$;	$\bar{x},0,\bar{x}$;	$\bar{x},\bar{x},0$;	$0,\bar{x},x$;	$x,0,\bar{x}$;	$\bar{x},x,0$;

Table 3.4. (continued)

48	g	mm	$x, \frac{1}{4}, \frac{1}{4}$;	$\frac{1}{4}, x, \frac{1}{4}$;	$\frac{1}{4}, \frac{1}{4}, x$;	$x, \frac{1}{4}, \frac{3}{4}$;	$\frac{3}{4}, x, \frac{1}{4}$;	$\frac{1}{4}, \frac{3}{4}, x$;
			$\bar{x}, \frac{1}{4}, \frac{1}{4}$;	$\frac{1}{4}, \bar{x}, \frac{1}{4}$;	$\frac{1}{4}, \frac{1}{4}, \bar{x}$;	$\bar{x}, \frac{1}{4}, \frac{3}{4}$;	$\frac{3}{4}, \bar{x}, \frac{1}{4}$;	$\frac{1}{4}, \frac{3}{4}, \bar{x}$;
32	f	3m	x, x, x;	$x, \bar{x}, \bar{x}$;	$\bar{x}, x, \bar{x}$;	$\bar{x}, \bar{x}, x$;		
			$\bar{x}, \bar{x}, \bar{x}$;	$\bar{x}, x, x$;	$x, \bar{x}, x$;	$x, x, \bar{x}$;		
24	e	4mm	$x, 0, 0$;	$0, x, 0$;	$0, 0, x$;	$\bar{x}, 0, 0$;	$0, \bar{x}, 0$;	$0, 0, \bar{x}$;
24	d	mmm	$0, \frac{1}{4}, \frac{1}{4}$;	$\frac{1}{4}, 0, \frac{1}{4}$;	$\frac{1}{4}, \frac{1}{4}, 0$;	$0, \frac{1}{4}, \frac{3}{4}$;	$\frac{3}{4}, 0, \frac{1}{4}$;	$\frac{1}{4}, \frac{3}{4}, 0$;
8	c	$\bar{4}3m$	$\frac{1}{4}, \frac{1}{4}, \frac{1}{4}$;	$\frac{3}{4}, \frac{3}{4}, \frac{3}{4}$;				
4	b	m3m	$\frac{1}{2}, \frac{1}{2}, \frac{1}{2}$;					
4	a	m3m	$0, 0, 0$.					

Placing the atoms into the last four-fold position in our list, we obtain a lattice typical of many metals. If the atoms occupy two four-fold positions, the structure is of the NaCl type. Very often atoms are situated in 8- and 24-fold positions. The atomic arrangements with maximum multiplicities (96- and 192-fold positions) are rare, being less favorable in terms of crystal free energy.

3.2 Geometric Model

3.2.1 Close-Packed Spheres

In a layer A of close-packed spheres each of them has six neighbors. There is only one way by which the second layer B can be placed over the first one: the centers of the spheres of the second layer must be placed over the voids of the first one. The third layer can be stacked in two ways, differing in symmetry. The spheres can be placed so that their centers are over the centers of the A spheres (ABA packing); it is also possible to use the other system of voids of the layer B. The third layer will be shifted relative to the previous ones. Therefore it should be denoted by a new letter, say C (ABC packing). If the procedure is continued, all the following layers will repeat one of the layers already considered. The simplest packing can be represented by the symbol ...ABABAB.., and three-layered packing by the symbol ...ABCABCABC.... It is possible to imagine structures of any periodicity. Multilayered packings of the same periodicity can be realized by different methods; their number can readily be calculated.

It can be proved that, irrespective of period, all the variations of close-packed spheres can be classified in eight possible space groups [3.16], of which seven belong to the hexagonal system and one (with a face-centered unit cell (*fcc* or *ccp*)) to the cubic system. Close-packed layers are always perpendicular to the three- or six-fold axes. In a face-centered lattice, this corresponds to a spatial diagonal of the cube.

Each of the close-packed spheres is surrounded by 12 nearest neighbors. The fraction of "empty" space is 26%. It is said that the *packing coefficient* of the system is 0.74.

Two types of *voids* are possible in close packings. If the center of the triangle formed by the spheres of a layer is covered by a sphere of another layer, a void surrounded by four spheres is formed. Another type of void can be obtained if three spheres of a layer are covered by three spheres of another layer turned by 60° relative to the first one. Such a void is surrounded by six spheres. The voids of the first type are called *tetrahedral*, those of the second type *octahedral*.

Each sphere in a layer is surrounded by six voids. If a layer is imposed over another layer, they form three tetrahedral and three octahedral voids. The same occurs if one more layer is placed under the first layer. Also, the sphere under consideration will cover two more voids, one in each neighboring layer, forming two more tetrahedral voids. Thus, each sphere is surrounded by 6 octahedral and 8 tetrahedral voids.

Each octahedral void is surrounded by 6 spheres, and each sphere is surrounded by 6 octahedral voids. Therefore in an infinite structure, each sphere is associated with one octahedral void. Each tetrahedral void is surrounded by 4 spheres, and each sphere, in turn, is surrounded by 8 tetrahedral voids. Only a quarter of a void belongs to a sphere; consequently there are two tetrahedral voids per sphere.

3.2.2 Representation of Atoms by Spheres

There is a tendency to represent atoms of all the chemical elements by spheres. Numerous tables exist for *atomic* and *ionic radii*, some of them allegedly universal, others valid only for some particular cases or for certain groups of structures. A system of *tetrahedral radii* was proposed by PAULING and HUGGINS

[3.17], and a system of radii for compounds with the β-W structure was compiled by GELLER [3.18]. WASASTJERNA [3.19], GOLDSCHMIDT [3.20], PAULING [3.21], and ZACHARIASEN [3.22] suggested a system of radii for ionic crystals. There is also a system of radii for *metals* calculated by TEATUM et al. [3.23]. The list could be continued.

The derivation of such a system of radii begins with the analysis of a vast amount of experimental material. Then some corrections are introduced for coordination numbers, atomic charge, and the fraction of covalent bonds.

In some cases, the atom cannot be represented by a sphere. This becomes obvious in the examples of *graphite* and *diamond*: it is impossible to represent a carbon atom by a sphere 1.4, 1.5, or 1.8 Å in diameter. To use packings of spherical atoms as a geometric model for a crystal, no doubt, will encounter certain difficulties, or rather the consequences of the use of such a model will often lead to erroneous conclusions.

If a structure is a three-dimensional network of atoms, such as the *diamond structure*, it can be depicted as a system of touching atoms. But since the distances between valence-bonded atoms are much less than intermolecular distances, packing into relatively open networks takes place. Let us calculate the sum of volumes of the spherical atoms distributed over the unit cell. Assuming the atomic radius to be equal to half the valence interatomic distance, we divide the sum by the unit-cell volume. The coefficient of such a packing for diamond is about 0.25!

Some elements form *chain structures* (selenium), others form *net structures*, still others have *island structures* (sulphur). If an atom in these structures is represented by a sphere, the radius of which equals half a valence distance, the packing coefficients will again be very small. Moreover, chains, nets, and islands would "hang loose" in the crystal, i.e., would not be in contact with each other.

If, on the contrary, such structures are depicted by spheres whose radii are equal to half the intermolecular distance, the atoms will overlap.

Probably the most reasonable procedure consists of the following. The *atomic volume* is calculated on the basis of density and molecular weight data, and then the obtained value is multiplied by a factor of 0.74, the packing coefficient of closest-packed spheres. The value obtained is taken to be the volume of the sphere V_{eff} representing the atom. But in this case, too, the geometric model is meaningless: the term "packing" loses its meaning because spheres are partially overlapping and partially "hanging loose" in the crystal. As to the quantity V_{eff}, its usefulness is rather doubtful, since by employing it we try to characterize by one number the atom's participation in valence and nonvalence bonded interactions, and, besides, neglect the role of the electron distribution in the crystal of a given chemical element.

Thus, we come to the conclusion that if we want to preserve the geometric model, we must give up the representation of atoms by spheres in all the cases, except those in which the atomic centers coincide with the centers of the spheres building the close packing.

Thus, the representation of the atoms of elements as spheres with the subsequent calculation of their atomic radii is fully justified only in cases in which at least one polymorphic modification can be represented by *close-packed spheres*, i.e., when the substance either crystallizes in a hexagonal unit cell with the c/a ratio of the axes being equal to 1.63 or forms crystals with a cubic face-centered unit cell. If polymorphic modifications of such structures exist at high temperatures, one can carry out the extrapolation of the unit-cell volume to absolute zero. This can readily be done even when the volume expansion coefficients have not been determined experimentally: for the elements with simple structures the differences in these coefficients are insignificant.

If one adheres consistently to the geometric model of elements whose atoms form chains or nets, the atoms should be represented by truncated spheres. This representation seems to be better applicable in the case of hexagonal unit cells with a c/a ratio distinctly different from 1.63.

3.2.3 Representation of Atoms by Truncated Spheres

The *radius of a spherical atom* participating simultaneously in valence and nonvalence bonded interactions should be taken to be equal to half the distance to one of the most distant of the 12 neighboring atoms. The choice of a radius of the sphere requires a consideration of the entire crystal architecture. The arrangement of a group of atoms bonded by valence forces into an island, a chain, or a net can be very intricate, and intermolecular distances between several pairs of atoms from different groups may slightly differ. Of course, the geometric model is somewhat conventional. It seems to be most logical to take as a radius of the sphere, i.e., the *intermolecular radius*, half of the shortest intermolecular distance. The repulsive forces increase rapidly with the decrease of the distance while the attraction forces change much slower. Thus, it is natural that some contacting atoms prevent other atoms from approaching and forming contacts.

Having chosen an intermolecular radius R, we must establish the valency of the atom. It determines the number of *truncations* to be performed on spheres.

The normals to the planes of the truncations are directed along the lines of valence bonds.

Valence-bonded atoms of radius R are at a distance d from one another. By intersecting, each sphere loses a spherical segment of height $h = R - d/2$. The volume of the cut segment can be expressed through its height h and the radius of the sphere R:

$$\frac{1}{3} \pi h^2 (3R - h) \quad . \tag{3.1}$$

The volume of the truncated sphere, i.e., the volume per atom is

$$v = \frac{4}{3} \pi R^3 - n \frac{1}{3} \pi h^2 (3R - h) \quad , \tag{3.2}$$

where n is the number of truncations.

The same considerations are applicable to chemical compounds, the atoms of which are bonded by valence forces. The volume per atom of radius R bonded by valence forces with several atoms whose intermolecular radii are R_i is given by the expression

$$v = \frac{4}{3} \pi R^3 - \sum_i \frac{1}{3} \pi h_i (3R - h_i) \quad . \tag{3.3}$$

Now, the segments have different heights, namely

$$h_i = R - \frac{R^2 + d_i^2 - R_i^2}{2d_i} \quad . \tag{3.4}$$

In the Se crystal, atoms form helical chains with the interatomic distance within the chain equal to 2.32 Å. The shortest interatomic distances between the atoms from different chains are 3.46 Å. Thus, a Se atom should be represented by a sphere of radius 1.73 Å with two truncations of height 0.59 Å. The chains of Se atoms are preserved in the melt, although with increasing temperature the chain decomposes into atoms. Yet, it is not expedient to characterize a Se atom only by an atomic radius. The situation is similar in the layered As crystals. Within the layer, atoms are positioned at a distance of 2.51 Å, the nearest atoms in the adjacent layers are at a distance of 4.0 Å. The atom is a sphere of radius 2.0 Å with three truncations at 0.75 Å in height, the normals of which form a pyramid.

In a crystal of graphite formed by flat layers spaced by 3.35 Å, the carbon atom is modelled by a sphere of radius 1.80 Å with three truncations whose normals lie in a plane and form angles of 120°. The intralayer interatomic distance is 1.42 Å. The truncated segments are 1.1 Å in height.

The above-considered model always leads to high values of *packing coefficients*. No absurd values of packing coefficients (such as 25%) are obtained if carbon atoms in diamond are considered as spheres of 0.77 Å in radius. The reader can verify that the packing coefficients in sulphur, selenium, and arsenic, and in general, in all the valence compounds, differ only slightly from the value 0.74 for the close packing of spheres in the cut-sphere approximation.

3.2.4 Hexagonal Packing

Close packing according to the scheme ...ABABAB... leads to hexagonal symmetry and to the unit cell with the c/a ratio equal to 1.63. This is called the "ideal" hexagonal close-packed (hcp) structure. About 30 chemical elements have structures similar to this ideal one. The symmetry is the same, but the c/a ratio deviates to both sides of the value of 1.63. Thus, this ratio equals 1.57 for Be, 1.86 for Zn, and 1.89 for Cd.

To be rigorous, we must admit that the geometric model fails here because the distances between the neighboring atoms within the layer and in neighboring layers are slightly different.

The geometry of the structure no doubt reflects the character of the interactive forces. Valence states are also present, to some extent, in these structures. Therefore, in accordance with Sect.3.2.3., the attempt to treat such structures as a system of spheres with small truncations is justified.

Now, let us introduce a parameter ε by the formula $c/a = \varepsilon \cdot 1.63$. The distance between the nearest neighbors from the adjacent layers is

$$D = a\sqrt{\frac{1}{3} + \frac{2}{3}\varepsilon^2} \quad . \tag{3.5}$$

If we insist on the usefulness of the representation of atoms by spheres, the following situation arises. If $\varepsilon > 1$ and the diameter of the sphere is taken equal to a, the adjacent layers are not in contact. If one assumes that the spheres spaced by D are in contact, the spherical particles of hexagonal layers overlap.

The usual procedure is to choose the average of these radii for a particle. It seems, however, more logical to represent the structure by a packing of

65

truncated spheres of radius D/2 (otherwise the layers would "hang loose"). The height of the segments which must be cut off from the sphere is $2h = D - a$. If $\varepsilon > 1$, there are three cuts. Their normals lie in one plane and form angles of 120°.

Using these formulas for a Zn crystal ($\varepsilon = 1.14$), we readily obtain the radius of a sphere equaling 1.45 Å. The height of the segments cut off from the sphere is 0.12 Å, their volumes being 0.075 Å. The volume of the sphere of radius 1.45 Å is equal to 12.8 Å^3. Cutting out segments, we reduce the volume insignificantly, down to 12.6 Å^3. Note that the packing coefficient of Zn particles (equal to the ratio of the volume of the truncated sphere to the atomic volume) has essentially increased to 0.83 (more than for cubic or hexagonal closest packing). Thus, for the model under consideration the deviation from the ideal c/a ratio results in an increase of packing density.

Next consider the case $\varepsilon < 1$. The diameter of a sphere representing the particle is taken to be equal to the lattice parameter a, also in order to prevent particles from "hanging loose". The distance between the nearest spheres of the neighboring layers is less then a but is expressed by the same formula (3.5). Now there are 6 cuts, with their normals forming angle α. The normals to the cuts form two regular triangles-pyramids supported by three spheres from two adjacent hexagonal planes. The angles α between the pyramid edges are given by

$$\sin \frac{\alpha}{2} = 2 \left(\frac{1}{3} + \frac{2}{3} \varepsilon^2 \right)^{-1/2} . \tag{3.6}$$

The largest deviation from the ideal value of the c/a ratio is observed in Be ($\varepsilon = 0.96$, $a = 2.28$ Å, $c/a = 1.57$). In this limiting case the angle is slightly different from 60°, $\alpha = 58°$. The height of a segment cut off from the sphere of radius a/2 equals 0.04 Å. The volume of the segment is small. The truncations reduce the volume of the sphere by only 0.7%. Naturally, in this case as well the deviation from the ideal c/a value results in a higher packing coefficient, namely 0.79.

The loss of simplicity of the close packing of spheres is compensated by the increased density of packing of nonspherical particles.

It seems to be obvious that the representation of atoms by truncated spheres reflects the deviation from the central interaction in Li, Be, Zn, Cd crystal atoms. It is an interesting question but, as we know, it has not yet attracted the attention of researchers.

Finally, the major differences between the interatomic distance within the layer and those between the two layers do not exceed several per cent. Thus, for rough estimates, these structures can be considered as the close packing of atoms of "average radius". The model yields an interesting result: the more the deviation from the ideal c/a value, the higher the packing density.

3.2.5 Body-Centered Cubic Structure

The body-centered cube (bcc) is a frequently encountered simple structure of elements and binary compounds. The coordination number is 8. If the atoms are represented by identical spheres (in the case of chemical elements), the packing coefficient is 0.68. At the same time, it is difficult to find any correlation between the physical properties and the values of packing coefficients for metals with packing coefficients of 0.68 and 0.74.

There are no obvious indications that crystals with a bcc lattice must be "worse" from the standpoint of their tendency to form the most dense structure, as compared to the lattices where spherical atoms can readily be packed. It is natural to assume that the coordination number 8 corresponds rather to the presence of *directional interactions* than to a worsening of the packing. Let us discuss the following, rather naive, consideration. Suppose that the same atoms can form a cubic close-packed (fcc) structure, the atomic volume being preserved. This condition can be written in the following way:

$$a_{fcc}^3/4 = a_{bcc}^3/2 \quad, \text{i.e.,} \quad a_{fcc} = 1.26 \, a_{bcc} \quad,$$

where fcc means face-centered cubic.

In the assumed fcc lattice the radius of the spheres forming a close-packed structure with the coordination number 12 is

$$\sqrt{2} \cdot a_{fcc}/4 = \sqrt{2} \cdot 1.26 \, a_{bcc}/4 = 0.445 \, a_{bcc} \quad.$$

The distance between the centers of the nearest spheres in the body-centered lattice is

$$\sqrt{3} \cdot a_{bcc}/4 = 0.435 \, a_{bcc} \quad.$$

Then, the height of the segment which should be cut off from the bcc sphere to form the fcc structure is $0.01 \, a_{bcc}$. The volume of a sphere taken for our model is $(4\pi/3) \cdot 0.445^3 \cdot a_{bcc}^3 = 0.368 \cdot a_{bcc}^3$. The volume of eight segments which should be cut off from the spheres constitutes only 0.3%. An insignificant value indeed! But again, as in the case of hexagonal crystals, the model under consideration strongly affects the calculated packing coefficient for the bcc structure: it equals 2 · 0.368, i.e., the same 0.74 as for the close packing of spheres in the fcc (ccp) structure!

3.2.6 Ionic Crystals

Geometric concepts are often useful for the understanding and, sometimes, for the prediction of the structure and properties of ionic crystals. As it has been mentioned, the rules of building ionic structures are essentially simpler than the general rules used for the explanation of structures of metallic systems. Yet, the slow decrease of the electrostatic potential for $r \rightarrow \infty$ makes it difficult to understand the structure of ionic crystals. Therefore, the situation in organic chemical crystallography is even simpler.

The situation with ionic crystals is relatively simple only in the case when one can safely accept that atoms have transferred all the valence electrons to their neighbors, and there is no electron gas. Relatively simple are also the rules governing ionic crystals acting as insulators, while the exchange of electrons and the proximity of electrical properties to those of semiconductors complicate the picture for other ionic crystals.

It is natural to approximate by a sphere an ion which has lost or acquired an electron so that a filled electron shell has formed. But if we want to make the ionic radius a universal constant, it seems also expedient, in some cases, to use truncated spheres. Indeed, the equilibrium state in ionic crystals depends on the balance between the attractive and repulsive forces. Therefore, the coordination number affects interatomic distances. It is, of course, possible to use spheres of different radii for different cases. But a model of cut spheres can also be used.

As shown by SHANNON [3.24], a significant number of ionic crystals built from monoatomic ions can be described as close-packed anionic spheres, the voids of which are filled by cations with smaller ionic radii.

Each anion can be surrounded by 8 small ("tetrahedral") and 6 large ("octahedral") cations. It is easy to calculate possible dimensions of the cations, i.e., of the spheres which can be inscribed into the voids.

A small sphere in the void between the main spheres of the packing does not "fit loosely" if it is in contact with all the main spheres forming the void. The center of the small sphere is equally spaced from the centers of all the neighboring main spheres, i.e., it is situated in the center of a tetrahedron or an octahedron (with an edge a) formed by the lines connecting the centers of the large spheres surrounding the void. The condition of contact by the spheres has the form $r + R = d_{r-R}$, r being the radius of a sphere placed into a void, R the radius of the main sphere of the packing, and d_{r-R} the distance between the center of a tetrahedron (or octahedron) and its vertex.

For the sphere placed into an octahedral void,

$$r = \sqrt{2}\,\frac{a}{2} - \frac{a}{2} = 0.414\,\frac{a}{2} \quad,$$

for a sphere in a tetrahedral void,

$$r = \frac{1}{\sin(109°28'/2)}\,\frac{a}{2} - \frac{a}{2} = 0.225\,\frac{a}{2} \quad,$$

where a is the distance between the centers of the main spheres ($a/2 = R$ is the radius of the main sphere) and $109°28'$ is the angle between the altitudes of the regular tetrahedron.

Naturally, we have calculated the maximum possible radii of the spheres in voids. If the radius of a sphere in a void exceeds the value obtained, the packing becomes *distorted*; if it is smaller than this value, the sphere can move loosely in the void. It is worth noting that the total volume of the spheres filling voids is rather small. In fact, a hexagonal prism with the volume of $3a^3\sqrt{2}$ is associated with 6 main spheres, and consequently, 6 x 1 "octahedral" and 6 x 2 "tetrahedral" voids. The volumes of the corresponding spheres are $6(4\pi/3)(0.414 \cdot a/2)^3$ and $12(4\pi/3)(0.225 \cdot a/2)^3$. Thus, the spheres filling tightly all the octahedral voids occupy only $\pi\sqrt{2}(0.414)^3 \cdot 100/6 = 5.255\%$ of the space, and the spheres filling all the tetrahedral voids occupy $\pi\sqrt{2}(0.255)^3 \cdot 100/3 = 1.687\%$ of the packing space.

Most minerals can be described fairly well as *close-packed anionic spheres* with the voids occupied by cations. Each mineral is classified according to the type of close packing, the kind and the number of the voids occupied by cations, and also by the design of the occupied voids.

The simplest case is realized when all the voids of one kind are occupied. Most of the compounds with the formula Mex (NaCl, KBr, etc.) are close packings built by halogen ions, all the octahedral voids being occupied by the cations of alkali metals. In the structure Li_2O the lithium ions (Li^+), whose number is twice that of O^{-2} ions, occupy all the tetrahedral voids in the close packing of the oxygen ions.

If not all the voids of the packing are occupied, then the description of the structure, and in particular its spherical representation, become very difficult to visualize. Large anions filling 74% of the crystal space screen the cations, i.e., just the part of the picture which is responsible for the diversity of the world of minerals.

The best way out of the situation is the representation of structures by the so-called coordination polyhedra. This was done by Linus Pauling many years ago. His method is not used in this book.

The principles of the close packing are widely used, and not only in the investigation of ionic structures. They can also be applied to interstitial phases. Here the close packing is built by metal atoms, various nonmetallic atoms occupying the voids, for example, B, Si, C, H, N, and O, which play the role of a "packing basis" in ionic crystals. In some cases the close packing is built by larger metal atoms, and the voids are also occupied by metal atoms, but of smaller size (alloys).

There are cases of ionic crystals in which a large cation replaces a sphere of the main packing. It can occur when the cation "requires" the coordination number 12 or 8. The example of the former case is $CaTiO_3$ where Ca replaces O.

If the rules of close packing are violated, it is still possible to use the coordination polyhedra for the space representation, i.e., to consider trigonal prisms, cubes, etc. But in this case the method is no longer predictive and must be considered only as a convenient method for representing structures.

However, even for the simplest cases when ions can be represented by spheres of the same radii, the predictability of the above-described method is rather limited. There may be cases when cation and anion radii, the derivation of which is rather conventional and is carried out by different methods, are equal, for example, the radii of K^+ and F^-. In such cases, it is natural to expect that cations and anions occupy lattice points corresponding to the close packing of spheres. Indeed, such cases occur. Of course, some doubts arise in the case where the radius of a cation is smaller than that of an anion, but exceeds the value of $0.414 \cdot a/2$. There is not enough space for the cation even in an octahedral void; when replacing the "anion" of the main packing it either "fits" loosely or, on the contrary, pushes its way through the spherical anions. Such is the classical case of NaCl (the radius of the Cl anion is 1.81, that of the cation 0.98).

Now compare the structures of three Li halogenides. The cation radius is taken equal to 0.78 Å; that of an octahedral void in an iodide sublattice is 0.91, in bromide 0.81, and in chloride 0.75 Å. Calculating packing coefficients for these structures, i.e., the fraction of space occupied by the spherical ions divided into a quarter of the unit-cell volume, one has 0.82 for iodide, 0.79 for bromide, and 0.75 for chloride. There is no obvious correlation between these values. When there is not enough space for the cation, the packing coefficient has the minimum value. Of course, the situation can be improved by the use of truncated spheres. But the ionic interaction is of central character and the use of truncated spheres is not suitable. To verify the increase in density which occurs during molecule formation is impossible because the system has only one component, and there are no crystals built by anions alone or cations alone. The corrections for "compressibility" of ions can hardly be supported experimentally. Thus, the only possibility left is to recognize the roughness of the geometric model and not to expect from it more than it can really give.

Consider four oxides with the NaCl structure. Tables give the value of 1.32 Å for the radius of a oxygen dianion; Mg, Fe, Ca, and Ba cations have radii 0.78, 0.83, 1.06, and 1.43 Å, respectively. In the last case, the size of the cation is larger than that of the anion. Nevertheless, the structural type is the same. If one assumes that the anions form close packing, then the radius of octahedral voids is equal to 0.55 Å. Thus, all the cations (if they are considered to be rigid spheres) move the anions apart. Calculating packing coefficients, we obtain 0.64, 0.60, 0.55, and 0.51, respectively. The dependence of the packing coefficient on radius is exactly what we expected. The larger the cation radius, the less the density of the packing. But the degree of space filling is too low. Thus, in this case, we again come to the conclusion that the geometric model fails. Simple structures, where anions form a close packing of spheres with cations filling in some way or another its voids, are found rather often not because ions are packed as spheres, but rather because such a highly symmetrical arrangement results in the minimum value of electrostatic energy of interaction.

The energy calculations for a crystal lattice built by monatomic ions are fairly simple if we assume that there are no valence bonds. Such calculations will be considered later.

3.2.7 Molecular Crystals

The boundaries of all organic molecules can be described by the so-called intermolecular radii. In molecules, atoms bound by valence forces are closer to each other than to the atoms in neighboring molecules. Intramolecular forces are much stronger than intermolecular ones, and therefore molecules can be characterized by interatomic distances. It is natural that the sum of intermolecular radii at points of contact between neighboring molecules significantly exceeds the distance between atoms bonded by valency. As is seen from Fig.3.4, due to the numerous valence bonds between spherical atoms with some intermolecular radius, each atom in the molecule is associated with only a certain fraction of a sphere left after cutting from it several (1-4) segments. The number of segments equals the valency of a given atom. The principle for the construction of a model of a molecule is clear from Fig.3.4 which presents a molecule of 2,6-dichloronaphthalene. This is a rather simple case since the molecule is flat. But there are many cases where a special investigation should be done prior to the consideration of molecule packing in the crystal with the aim of finding the optimal shape of the molecule.

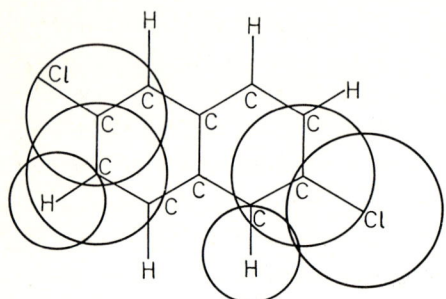

Fig.3.4. The principle of constructing molecular models

Organic chemists were successful in solving the problem of the atomic arrangement in molecules even for very complicated multiatomic molecules back in the years when physical methods could not provide these data. Chemical experience proved that some molecules behave as rigid systems while others are flexible. A molecule can acquire an intricate shape due to the fact that its separate parts are linked by single bonds and can rotate about these bonds. Each molecule can have many such bonds, and therefore, the molecule has a great number of degrees of freedom. Such changes in molecular shape are called *conformational changes*. The problem of finding an optimum molecular conformation can be solved in different approximations. The simplest approximation is the representation of atoms as *hard rigid spheres*. Then rotations which bring the spheres into contact are permitted.

This problem can be solved quantitatively with the help of the *atom-atom potential method*. Some examples are given in [3.8].

In some cases the problems of finding an optimum macromolecular conformation and the best packing of molecules are interrelated. In other words, crystallization can change the conformation of a molecule. But we are not going to consider these complications now. Assume that the shape of a molecule is known and consider the question of how molecules are packed in crystals. Experience shows that close packing is realized in such a way that *the bulk of one molecule fits into the hollow of the neighboring one*. This principle, formulated by the author [3.7,8,25], is a natural law which has no exceptions. In all the investigated organic structures, neighboring molecules have never been related by a mirror plane, as this would contradict the above-formulated principle. In molecular crystals the principle of close packing permits only a small number of space groups, and therefore, the world of organic substances is characterized only by a limited number of space groups. The principle of close packing allows us to understand why nearly half of all the organic substances crystallize in the space group $P2_1/c$ considered in Sect.3.1. Space groups which are most appropriate for the close packing of molecules can be derived, though not very strictly, from general geometric considerations [3.7,8,25].

To what extent the theoretical predictions are fulfilled can be judged from the statistical treatment of structural data for about 2000 molecular crystals listed in Table 3.5 borrowed from [3.26]. It is seen that half the structures belong to only two space groups, and the ten most frequently encountered space groups describe about 80% of all the crystalline structures.

Let us consider some other important conclusions about the mutual distribution of molecules in a crystal, which follow from the principle of close packing. First, a free molecule containing among other symmetry elements a center of inversion should occupy a site in the crystal such that the center of symmetry is preserved. Only such an arrangement optimizes the packing density.

Table 3.5. Distribution of about 1850 known crystal structures among the ten most frequent space groups

Group	Number of structures	%
$P2_1/c$	717	38.8
$P2_12_12_1$	183	10.0
$P\bar{1}$	176	9.5
$C2/c$	143	7.8
$P2_1$	81	4.4
Pbca	69	3.7
Pnma	58	3.1
$Pna2_1$	45	2.4
$P2_1/c$	23	1.2
Cc	19	1.0

Only some very rare cases are known where the molecule in a crystal preserves the point group with two or three mirror planes. At the same time, it is possible to combine, to some extent, the requirements of close packing and the preservation by a molecule in a crystal of one symmetry axis or one mirror plane.

In the vast majority of cases, molecules occupy one position in a crystal, either a general or a special position. But the trend to form close packing can lead to an increase in the number of molecules in a unit cell so that the molecules occupy two or more crystallographically independent positions. This fact is of special interest to us because in substitutional solid solutions the molecule can "choose" one of the two or more variants of entering the lattice.

The structure of a molecular crystal can be characterized, similar to the packing of spheres, by the coordination number. It is obvious that the concept of a coordination number in this case is more vague because molecules can realize contacts through different numbers of atoms. Therefore, the coordination numbers 14, 12, and 10 are encountered most often. There is no precise meaning in these figures; they are not related to the directional character of molecular interactions. The explanation lies in a specific molecular shape.

The calculations of the packing coefficient serve as an excellent confirmation of the close packing of organic molecules in solids. The calculation of the volume of a molecule is carried out by (3.3). Volume increments for the atoms which are typical of organic molecules can be easily calculated. The most typical increments are listed in [3.7,8]. Thus, the volume of an aromatic carbon atom is 8.4 $Å^3$, and that of the aromatic CH group (the sum of two increments) is 14.7 $Å^3$. Using these two values one can calculate the volumes of the molecules of aromatic hydrocarbons. Thus, the volume of a naphthalene molecule is $(2 \cdot 8.4 + 8 \cdot 14.7)$ $Å^3$. Dividing this value into half the unit-cell volume of a naphthalene crystal (there are two molecules in the unit cell), one can readily obtain the packing coefficient.

For most organic compounds the packing coefficients are close to 0.74. Sometimes they are larger, while in the case of molecules whose shape is not convenient for close packing, the packing coefficient can drop to 0.65.

3.3 Structure of Single-Phase Mixed Crystals

3.3.1 Ordered Crystals

If a solid consists of a single-phase mixture of molecules or atoms, it can be obtained as a single crystal which usually cannot be distinguished by its appearance from crystals built by only one kind of atoms or molecules. Crystals which are the mixtures of molecules or atoms can be divided into ordered and disordered systems.

We speak of *ordered crystals* when the structures of all the unit cells throughout the crystal are strictly identical. In other words, when translated by any lattice translation, all the atoms of a unit cell are brought into coincidence with the identical atoms of another unit cell. This means, if one applies elements of symmetry of the space group of a mixed crystal to atoms or molecules as a whole, these atoms or molecules will coincide with the identical atoms or molecules situated in the same or any other unit cell of the crystal.

From the standpoint of structural crystallography, an *ordered binary crystal* does not differ in any way from a single-component crystal. There is no essential difference in the amounts of intrinsic defects in these crystals. The number of *dislocations* per unit volume is the same; the number of *vacancies* increases quite insignificantly, if at all. Both theory and experience show that in an ordered atomic crystal, the fraction of vacancies at temperatures close to the melting point is about $10^{-3} - 10^{-4}$. The number of vacancies quickly decreases with a decrease in temperature. In molecular crystals, where the processes of self-diffusion are slower, the number of vacancies is insignificant. Then the defects are restricted to dislocations, i.e., to disturbances of the regular structure at the mosaic block boundaries.

There are molecular crystals formed from two components in a simple stoichiometric ratio. In most cases these are the simplest compositions AB and A_2B. Binary metal alloys can have ordered crystals with rather unusual component ratios, such as A_5B_{21} or AB_{16}.

One interesting type of ordered crystal should be singled out: the fully ordered solid solutions with some degree of long-range order (defined in Sect.3.3.3). If, in a crystal with a *superstructure*, particles (molecules or atoms) of one component are replaced by particles of the other component, the resulting crystal will have a structure only slightly different from that of a single-component crystal. The mutual arrangement of atoms will be practically the same, and the lattice parameters will differ insignificantly. This fact

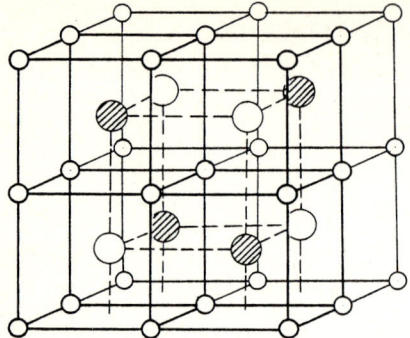

Fig.3.5. A primitive unit cell of Heusler alloy built of three kinds of atoms

justifies the term "superstructure". Thus, Fig.3.5 shows a primitive unit cell of a Heusler's alloy built from three kinds of atoms. If the atoms are distributed in a *random* manner and each site is occupied by the same "*average*" atom, the unit cell is reduced by a factor of 16. Then the large cube shown in the picture can be replaced by 8 small body-centered cubes. A primitive unit cell obtained from a superstructure as a result of mixing atoms over various positions is called a *subcell*.

We should emphasize once again that a crystal of an ordered solid solution possessing superstructure is *not* distorted. All the unit cells are equivalent, and can be converted into each other by the same translation vectors. Figure 3.5 does not prove that the term "superstructure" is justified: in Fig.3.5 we can only see an ordered crystal built by three particles. The term "superstructure" is justified only in the case when, along with the structure shown in Fig.3.5, there exists a disordered structure.

3.3.2 Disordered Crystals

In a phase diagram special points correspond to single-phase ordered crystals. There is a significant difference between *molecular compounds* (intermediate phases, complexes, etc.) and *superstructures*. On the one hand, molecular compounds have a unique melting point, a temperature at which a molecular compound is in equilibrium with the melt or liquid solution. On the other hand, crystals with superstructure are within the existence domain of solid solutions. This superstructure is formed by a temperature decrease in a completely or partially disordered solid solution. (Disordered single-phase crystals are also present only within the existence domain of solid solutions).

In various crystals (metal, molecular, etc.), *disorder* can occur in three ways. First, multiple crystallographic positions within the same cell may be only partially occupied. For example, a fourfold position in the unit cell may be occupied only by two or three particles. Secondly, positions may be fully occupied only in some unit cells. Thirdly, chemically different particles may occupy the same crystallographic position.

If a particle represents a molecule and not an atom, disorder can be due to the fact that a particle can have different conformations or orientations in the crystal. Of course, it is possible to obtain formal expressions to characterize the *degree of disorder* (or the *degree of order*) relative to the parameters mentioned above. However, we will not burden this text with such formulas.

If the probabilities of finding the particles in a certain state at all points forming a translation lattice are not equal, but are different from the corresponding probabilities for a random distribution of particles at these points, then the system is said to possess *partial long-range order*.

The structures of all the unit cells of a solid solution without long-range order are different. Experience (X-ray data) leads us to an *average unit cell*, which is the Nth part of the overall cell, obtained by superposition of all the N unit cells of the domain of the crystal under consideration. In this average unit cell, for a crystal with partial long-range order, the positions occupied by particles are filled by "average" molecules. Such particles are formed by A and B molecules with weights which correspond to their concentrations, the multiplicity of the crystallographic position, the possible orientations and conformations of the molecules. Differences in the *degree of long-range order* affect the structure of an "average" unit cell and "average" molecule.

3.3.3 Degree of Long-Range Order

For partially ordered solid solutions one may define a *parameter* σ, which represents the *degree of long-range order*. This can be made in several ways, but σ must be equal to unity for superstructures ($\sigma = 1$) and to zero for the completely disordered structure ($\sigma = 0$). One may introduce a parameter of long-range order using any criterion: orientational, conformational, positional, etc.

Consider the simplest case. The crystal lattice is divided into two *sublattices* whose *nodes* are occupied by the particles A and B in fractional concentrations c_A and c_B, such that $c_A + c_B = 1$. The particles can be of different chemical species, or of different conformation or orientations.

In the case of full order, the particles with concentration greater than 0.5 are defined to occupy one of the sublattices. In the case of full disorder, they are distributed over the sites of the sublattice in proportion to their concentrations.

If a fraction of particles A in one sublattice, or the probability p_A of finding the particles A in the sublattice, differs from c_A, i.e., $p_A - c_A \neq 0$, then the system possesses a partial order. Then the *long-range order parameter* σ can be defined by the expression

$$\sigma = (p_A - c_A)/(p_A' - c_A') \quad , \tag{3.7}$$

where primed quantities relate to the stoichiometric solid solution in a fully ordered state.

For the alloy AB the degree of long-range order is

$$\sigma = (p - c)/0.5 \quad ; \tag{3.8}$$

for the alloy of the composition AB_3 it is

$$\sigma = (p - c)/0.25 \quad . \tag{3.9}$$

(The above formulas are written for the particles A.)

It is obvious that the *crystal free energy* must be a function of the long-range order parameter. As temperature increases the superstructure reveals elements of disorder, the degree of long-range order decreases, and at a certain *critical temperature* the long-range order completely disappears. In Chap.4 a simple theory will be considered which permits obtaining the temperature dependence of the long-range order parameter and determining the critical temperature for a simple model of a solid solution.

3.3.4 Static Concentration Waves

A very interesting formalism has been developed by KHACHATURYAN. It permits a very convenient and elegant calculation of thermodynamic functions, phase transitions, X-ray scattering of solid solutions with long-range order, etc. A recent review of his work can be found in [3.27]. This new approach is called the *method of static concentration waves* and consists of the following. Suppose there is a completely ordered crystal whose particles A and B are distributed over t crystallographic positions. If partial disorder occurs, the particles B (or, for that matter, A) can be found with probabilities n_1, n_2, ..., n_t, at any of t positions.

Now, define the function $n(\underline{r})$, the probability of the presence of the particle (A or B) at any crystallographic position of any unit cell of the crystal. In fact, we deal with a series of discrete probabilities n_k, but for the purpose of calculation it is more convenient to deal with a continuous function. Such a function of the probability of the presence of the atom at the point $\underline{r}$ can be expanded into a Fourier series as a superposition of static concentration waves

$$n(\underline{r}) = c + \frac{1}{2} \sum_j \left[Q(\underline{k}_j) \exp(i\underline{k}_j \underline{r}) + Q^*(\underline{k}_j) \exp(-i\underline{k}_j \underline{r}) \right] \quad . \tag{3.10}$$

The number of waves is equal to the number of positions which can be occupied by the particle. Here c is the fraction of the atoms whose partial disorder is under examination. The vector $\underline{k}/2\pi$ can be drawn from the origin to any point of the reciprocal space. If the tip of the vector reaches a superlattice node, then $\underline{k}_j = 2\pi \underline{H}_j$, i.e., the vector is equal to a vector of the discrete reciprocal lattice (Sect. 6.1).

Later we shall need a somewhat different form of the sum of concentration waves. Equation (3.10) can be rewritten in the form

$$n(\underline{r}) = c + \sum_s \sigma_s E_s(\underline{r}) \quad , \tag{3.11}$$

where

$$E_s(\underline{r}) = \frac{1}{2} \sum_s \left[\gamma_s(j_s) \exp(i\underline{k}_{j_s} \underline{r}) + \gamma_s^*(j_s) \exp(-i\underline{k}_{j_s} \underline{r}) \right] \quad . \tag{3.12}$$

Here, the sum over s is singled out. It combines vectors $\underline{k}_{j_s}$ related by symmetry operations inherent in a crystal of a fully disordered solid solution, σ_s are long-range order parameter, and the numbers γ_s determine the symmetry of the superstructure.

For example, consider a binary alloy AB whose atoms in a disordered state are distributed over the vertices and centers of cubes; in the ordered state let the particles A occupy the vertices whereas particles B occupy the centers of cubes; then the scalar product $\underline{H} \cdot \underline{r}$ is equal to $x+y+z$ (the factor 2π was introduced for the sake of simplicity). Then the probability function acquires the simple form

$$n(\underline{r}) = c + \frac{1}{2} \sigma \exp\left[i2\pi(x+y+z) \right] \quad . \tag{3.13}$$

Thus the probability of the presence of the atoms at the vertices of a cube equals $c + \sigma/2$ and at the centers of the cube $c - \sigma/2$. This expression does not require that the numbers of A and B atoms be equal.

The concentration waves' amplitudes are determined unambiguously by an X-ray pattern of the partly ordered solution. The scattered X-ray amplitude equals the amplitude of the concentration wave of the corresponding *superstructure lattice point*.

3.3.5 Short-Range Order

The degree of long-range order is not the only characteristic of the order of the crystal in solid solution. Long-range order can be absent, and yet the crystal, which is built up of different particles, can essentially be ordered. It is another kind of order, a *short-range order*, which is due to the tendency of particles A to surround themselves by particles of the same or of different species.

Consider the crystallographic points occupied by the particles A in a fixed orientation and consider also a thin spherical layer within which are located the centers of the particles B, being in contact with the given particles A. Let us call Z the coordination number of these nearest neighbors B. If c_B is the concentration of particles B, then the number of the nearest neighbors of a particle B is, on the average, $c_B Z$ in the case of complete disorder. If it is different from $c_B Z$ and is equal to n_B, then a short-range order exists which can be described by the parameter α:

$$\alpha \equiv 1 - n_B/c_B Z \quad . \tag{3.14}$$

This parameter can have different signs. A positive α value testifies to the tendency to segregation, i.e., the particles of one species form groups.

Only in rare cases have researchers succeeded in describing short-range order in more detail. Therefore, the strict formulation of the problem of short-range order is of comparatively small interest. Such a formulation should contain the derivation of formulas for calculating a correlation function $\varepsilon_{ik}(\underline{R})$ on the basis of experimental data. This function determines the probability of finding two particles A, or two particles B, or a particle A and a particle B at the ends of vector $\underline{R}$, which joins two certain points of two given crystallographic positions i and k. If necessary the correlation function can be defined to indicate the states of the particles.

The analytical expression for the correlation function can be written in terms of the variables $\gamma(0)$ and $\gamma(\underline{R})$, which are both equal to unity if particles A (in a certain state) are at the ends of the vector $\underline{R}$ and equal to zero if particles B are at the ends of the vector.

The calculation of the correlation function $\varepsilon_{ik}(\underline{R})$ in the general form is rather difficult. If the particles are translationally identical, e.g., particles occupying the vertices and centers of cube faces, then

$$\varepsilon(\underline{R}) = <\gamma(\underline{R})\ \gamma(0)> - c_A^2$$

and

$$\varepsilon(0) = c_A(1 - c_A) \quad . \tag{3.15}$$

Note that the average values $<\gamma(\underline{R})>$ and $<\gamma(0)>$ correspond to the percentage of the component A.

The parameters of correlation and short-range order are related by the expression

$$\alpha(R) = \frac{\varepsilon(\underline{R})}{c_A(1 - c_A)} \quad . \tag{3.16}$$

With $\underline{R} = 0$ the short-range order parameter is unity. For this simple case (i.e., for translationally identical particles) short-range order parameters are directly related to quantities determined from the measurements of diffuse scattering of X-rays and neutrons. They can be calculated if the scattering intensities are known for the points of a reciprocal unit cell. Some examples of such calculations will be given in Chaps.8,12. For readers who are not interested in the theory of X-ray scattering we only note that such calculations are time-consuming, since they require the measurement of hundreds of scattered intensity data. The theory of order-disorder in alloys is discussed in the monographs [3.28-30].

3.3.6 Size and Shape Factors

Researchers in crystal chemistry look for rules which could help them predict structures for new compounds. Unfortunately, such predictions have been possible only in some rare cases where the substances under investigation were unambiguously related to the substances with known structure (yet, failure is possible even in these cases).

In studies of binary systems one always wonders about the role of a *geometric factor* in the formation of a solid solution or a molecular compound. Molecular compounds and ordered solid solutions can be regarded as *regular crystals*. Lattice distortions are absent. Then the geometry of the atomic arrangement is usually analyzed as follows. Interatomic distances AA and BB in crystals of pure chemical components are compared with the distances AA, BA, and BB in the compound built by the atoms of both species A and B. The researcher is interested in whether the formation of the compound is accompanied by a *decrease or an increase in the density* of the structure. There are long tables and many papers concerning this problem, as summarized in [3.1-6]. LAVES [3.31] carried out a very interesting and detailed analysis of this kind for simplest compounds of the compositions AB, AB_2, and AB_3.

Let us define the parameter Δ which describes the *increase in density*:

$$\Delta \equiv \frac{mV_A + nV_B}{V_{A_mB_n}} - 1 \quad . \tag{3.17}$$

The atomic volumes of the components V_A, V_B and compounds ($V_{A_mB_n}$) should be calculated on the basis of precise measurement of unit-cell parameters (although density and atomic weight data can also be used). One wants to know whether the increase in density occurs from a better use of space (for example, in the case of the incorporation of small interstitial atoms) or from the formation of stronger bonds. This analysis requires that the packing coefficients are known. This, in turn, requires the consideration of particle shape.

The formation of the compound A_mB_n implies that its free energy is lower than that of both the disordered solid solution of the same composition and the corresponding two-phase system. Of course, it would be interesting to study how the parameter Δ is involved in the decrease of free energy, but this seems to be a difficult problem.

The role of *particle size* in the formation of solid solutions has been investigated very intensively. First of all, what is the *maximum difference* in particle dimensions that allows any noticeable solubility of one component in another? If substitutional solid solutions are meant where particle B is replaced by one particle A, the difference in sizes may not exceed 15 - 20%. For metals, this statement is called the Hume-Rothery *rule* [3.32] and is formulated in terms of atomic radii. Chapter 8 is devoted to substitutional solid solutions for metals and there we shall analyze this rule in detail. Remembering Sect.3.2.3, it seems hardly worth comparing the atomic radii: it is more correct to compare atomic volumes. The Hume-Rothery rule allows the differences in *specific volumes* up to 50%; later, we shall see whether it is really so.

For all crystals built by the particles which exchange their electrons readily, the comparison of the particle sizes can give only negative data, i.e., such a comparison can only confirm that the formation of solid solutions is impossible. Positive information is absent, because numerous examples are known when there is no solubility of components even with equal atomic volumes.

The situation is much better for mixed crystals of organic compounds. Here the *geometrical factor* plays the decisive role. If the particles are not bonded into the framework by hydrogen bonds, and if the dipole moments of the molecules are compensated, the close *similarity of the shapes of molecules* is not only the necessary but also the sufficient condition for component solubility. The similarity of the shape of molecules can be evaluated by the method suggested in [3.7]. Superimpose the molecules of the components so as to maximize intermolecular overlap. Let the volume of the nonoverlapping parts be Γ. If the volume of overlapping parts is r, the parameter

$$\varepsilon = 1 - \Gamma/r \qquad (3.18)$$

describes the *degree of isomorphism* and can be defined as the *coefficient of geometrical similarity*. Experience shows that the minimum values of ε for solubility are close to 0.85.

The formation of solid solutions can be accompanied by the appearance of short-range order. Therefore, it is not clear beforehand which of the two following factors is more important, the degree of isomorphism ε or the parameter $\eta = (V_A - V_B)/V_A$ (V_A and V_B are unit-cell volumes such that $V_A > V_B$). The latter parameter (η) takes into account the difference in the close packing density of the molecules of the components, which can be rather significant. This important topic will be repeatedly discussed in the chapters devoted to organic substances.

The geometrical analysis of the structure of solid solutions brings out another interesting question: the extent of nonlinear dependence of the unit cell volume $V(x)$ on concentration x. The value of the parameter Δ describing the increase in density,

$$\Delta = \frac{(1-x)V_A + xV_B}{V(x)} - 1, \qquad (3.19)$$

provides the answer: if the $V(x)$ versus x curve is linear then Δ is equal to zero. If the formation of the solid solution is accompanied by a decrease in density, Δ is negative.

In physical metallography one can find references to Vegard's *law* [3.33], which postulates the *linear dependence of unit-cell parameters in solid solutions*. For low symmetry crystals, where the unit-cell dimensions are characterized by three of six parameters, one must consider the concentration dependence of all the lattice parameters. These dependences can supply some interesting data on the process of substitution. But, if we seek general rules, it is simpler to consider the dependence of the specific unit-cell volume of the solid and the parameter Δ on concentration.

In conclusion, it should be underlined that the investigation of the packing of particles in disordered solid solutions is, strictly speaking, impossible without studying the short-range order, which has been done only in a few cases. Therefore, the geometric model of solid solution, built only on the basis of the data on size and shape of molecules and unit cells, should be used with extreme caution.

4. Free Energy of a Solid Solution

It is practically impossible to predict the structures and properties of solids on the basis of even the most detailed data derived from the atomic and electronic structures of the constituent atoms or molecules. This is true not only for binary systems, but also for the simplest monoatomic crystals.

Such a prediction requires the calculation of the free energy of all possible structures and the selection of the structure with the minimum free energy. Since, in principle, any space group and any number of particles in a unit cell are possible, it is impossible to solve the problem even resorting to the best methods of energy calculations and powerful computers.

Experience proves, however, that the number of the most common structure types is rather *limited*, especially for organic crystals. Also rare are cases where the number of formula units per unit cell is very large. In organic crystals *only one general position* in a unit cell is usually occupied. Therefore, a theory that can predict the correct structure (if only within the framework of some arbitrary assumptions) would be of great importance for solid-state physics and especially for the problems discussed in this book. Unfortunately, there is no such universal theory.

Nevertheless, crude theories or theories important only for a limited class of substances, or even concepts which bridge the structure and properties of solids can be quite useful even in technological applications. Of great importance is also any theory which can calculate the free energy of a solid of known structure.

The reader can find the fundamentals of solid-state theory in numerous monographs [4.1-6] and the theory of mixed crystals in [4.7-9].

In this chapter we shall consider only some simple models, useful for pragmatic investigations. We shall see, first, how our theory can predict the *mutual solubility* of substances, second, how it can predict the *formation of chemical compounds*, third, how it can help us understand the *degree and the character of ordering* in solid solutions, and finally (in the next chapter), how it explains the formation and *microstructure* of *heterophase systems*.

4.1 General Formula for the Free Energy

The difficulties encountered in the rigorous solution of the above-discussed problems become evident if we recall that the expression for the Helmholtz free energy of perfect crystals contains four terms:

$$F = NV_0 + F_{vibr} + F_e + F_{conf} , \qquad (4.1)$$

where F_{vibr} is the vibrational free energy, F_e is the electronic free energy, and F_{conf} is the conformational contribution to the free energy. These terms are discussed further below.

The first term on the right-hand side of (4.1), NV_0, is the energy of the interaction of particles at equilibrium. The structure of a crystal is determined primarily by this term.

Some 60 or 70 years ago it was shown by many investigators, first of all by BORN [4.4], that one can obtain remarkable results by using the concept of the *spherically symmetric pairwise interaction* of atoms and by assuming the potential energy to be the sum of such interactions. The general shape of the interaction curve is quite obvious: it has a minimum at the equilibrium configuration, it rises sharply at smaller distances, and asymptotically approaches the horizontal (zero interaction energy) as the distances go to infinity.

The analytical expression for such curves can be selected when striving to obtain a proper coincidence with experiment. However, there is sufficient theoretical background which favors the so-called 6 - 12 and 6-exp potentials. The two-parameter potential $Ar^{-6} + Br^{-12}$ [6 - 12] was first used by Lennard-Jones, the three-parameter potential $Ar^{-6} + B \exp(-\alpha r)$ [6 - exp] by Buckingham.

Single-atom noble gas crystals such as argon, neon, and krypton appear to be the most suitable objects for testing the method of pairwise interactions. In fact, extremely sophisticated studies of these substances are being carried out at present with this method.

Furthermore it has been shown that for ionic crystals of the sodium chloride or cesium chloride type, the paired potentials also yield excellent results. For the ionic crystals, the attractive potential term is (instead of Ar^{-6}) the electrostatic (Coulomb or Madelung) energy, which is inversely proportional to the first power of the interionic distance.

The long-standing investigations started by this author (summarized in [4.1]) and successfully pursued by a great number of scientists in many countries have demonstrated that the idea of pairwise interactions can be applied to organic molecular crystals. The interaction of the molecules should be regarded as the sum of the interactions of their constituent atoms.

At first glance it may seem that the problem becomes complicated by the necessity of employing not one, but several interaction energy curves. Even for hydrocarbon we need three atom-atom potential curves for C-C, C-H, and H-H interactions respectively. However, these complications are not only for hydrocarbons compensated, but impart to the atom-atom potential method an entirely new dimension not found in the application of the theory to a crystal such as argon. Indeed, we have at our disposal an enormous number of hydrocarbons, and the three selected potentials can be checked against an almost unlimited amount of experimental material.

Let us discuss the connection between the atom-atom potential and the packing approach. In the packing approach we consider the molecules to be rigid, i.e., interacting via a *hard-sphere atom-atom potential*. The close packing, i.e., the fitting of one molecule to another according to the "*the projection-to-the-hollow*", "*the key-to-the-lock*" principle is an immediate result of assuming the additivity of interactions between the atoms forming the molecules.

The success in applying the packing approach to the structure of crystals shows that rough quantitative results can be obtained if *one-parameter potentials* are used for organic atoms. The author has suggested such a potential, the only parameter being the *equilibrium distance* between the valently non-bonded atoms [4.10]. The potential assumes the form:

$$V = 3.5 \left[- 0.04/Z^6 + 8.6 \cdot 10^3 \exp(-13Z) \right] \quad , \tag{4.2}$$

where $Z = r/r_0$, and r_0 is the equilibrium distance.

Let us return to consider the other terms of (4.1). The second term is the free energy of the lattice vibrations, which is known to have the following formula

$$F_{vibr} = kT \sum \ln \left[1 - \exp(-h\nu/kT) \right] \quad .$$

Here the summation must be carried out over the frequencies ν of all the normal lattice vibrations. It is possible to calculate the vibration (phonon) spectrum from structural data. Such calculations have been done for single-component crystals. We bring to the readers' attention the papers on atomic crystals [4.11-13] and to studies by PAWLEY [4.14,15] and MUKHTAROV [4.16,17] on the calculations of vibration frequencies in molecular crystals. The chapter on lattice dynamics in the author's monograph [4.1] can be considered as an introduction to that part of the theory which deals with molecular crystals. As to the calculations of vibrational spectra of solid solutions, one encounters great difficulties. In principle, however, the numerical value of F_{vibr} can be obtained by the measurement of the phonon spectrum.

The most difficult problem is to calculate the electronic component F_e of the free energy (4.1), because it can be determined (not at all trivially) only if the Brillouin zone structure, i.e., the distribution of electrons over the allowed energy levels, is known.

Crystal defects considerably complicate the calculation of F_e. In the general case, the presence of impurities in the *host lattice* affects all the three terms of the free energy F_{vibr}, F_e, and F_{conf}. If the concentration and the values of the free energy of the components are known, it is sufficient to calculate the free energy of mixing:

$$f_M = \Delta U_M - T\Delta S_M \quad . \tag{4.3}$$

The presence of impurities in the crystals of the host not only changes F_{vibr}, F_e, and F_{conf}, but also leads to the appearance of a new term, namely, the configurational free entropy equal to $T\Delta S_M$. The configurational entropy, in the simplest case, has the form

$$-k \left[(1 - x) \ln (1 - x) + x \ln x \right] \quad . \tag{4.4}$$

Let us recall the derivation of this formula. It follows immediately from Boltzmann's formula $S = k \ln W$, where W, the thermodynamic probability function, is here the number of realizable possibilities of a mixed crystal by permutation of the particles. Consider a certain site in the lattice which can be occupied either by the particle A or by the particle B. Then $N_A/N = 1 - x$, $N_B/N = x$, and in Boltzmann's formula, W has the form $N!/N_A!N_B!$. Using Stirling's formula $\ln N! \approx N \ln N$, we obtain

$$\ln W = N \ln N - N_A \ln N_A - N_B \ln N_B \quad . \tag{4.5}$$

Finally, expressing N_A and N_B through $(1 - x)$ and x, we obtain (4.4).

To study mixed crystals, we may start from an idealized system where the heat of mixing is equal to zero, and the changes in entropy correspond to a total disorder in the distribution of particles over lattice sites. The lattice of this ideal solid solution is not distorted and the vibration spectrum is not changed.

A better "reference model" is a *regular solution*, i.e., a system in which the entropy of mixing is represented by (4.4), the influence of the impurity on lattice vibrations is neglected, but the heat of mixing is assumed to be nonzero. Such a definition of a regular solution is not unambiguous, because different approximations can be used to calculate the changes in the interaction energy.

In any case, the free energy f of a solid solution per particle can be written in the form:

$$f = (1 - x)f_A + xf_B + \Delta V + kT\left[(1 - x) \ln (1 - x) + x \ln x\right] + \varphi(x, T) \quad , \tag{4.6}$$

where $\varphi(x, T)$ is given by:

$$\varphi(x, T) = \Delta f_{dist} + \Delta f_{vibr} + \Delta f_{ord} + \Delta f_{conf} + \Delta f_{el} \quad . \tag{4.7}$$

It is possible to carry out numerical calculations of the free energy, neglecting the term $\varphi(x, T)$, and to use (4.6) to predict the solubility. But sometimes one cannot neglect $\varphi(x, T)$ and its five contributions: Δf_{dist} due to lattice distortion, Δf_{vibr} caused by changes in the lattice vibrational structure, Δf_{ord} due to partial ordering, Δf_{conf} that includes conformational energy changes, and Δf_{el} that describes changes in the electronic distribution. One must consider the variations in the distortion of the lattice (Δf_{dist}) which manifest themselves in any solid solution (appearance of diffuse scattering of X-rays or neutrons, a change in the lattice parameters and in the intensity of X-ray reflections). Further, one must include in Δf_{vibr} the changes in the vibrational spectrum of the lattice. Partial ordering and the formation of clusters also result in changes in the free energy (Δf_{ord}). The next term (Δf_{conf}) appears when the particles building the crystal are not atoms or ions but molecules which can change their conformation and the conformation of the neighboring molecules on the matrix. Finally, (Δf_{el}) the last term in (4.7) reminds us of the possible changes in the electron distribution which can essentially influence the solubility.

The presence of impurity particles (atoms and molecules) in the host lattice changes the lattice energy. This change can be divided into three effects: a change in the interaction potentials, a change in the short-range order, and a change caused by the lattice distortion. The term ΔV in (4.6) denotes the change of the first kind. In other words, ΔV is the change in the energy which would occur if the distribution of impurities were statistically disordered and did not distort the lattice. It is likely that in many cases ΔV is of greater importance than $\varphi(x, T)$ in (4.6), and the estimation of ΔV may lead to reasonable semiquantitative conclusions about the possible solubility of impurities.

We can calculate ΔV numerically if we assume, first, that the energy of interaction of particles creating the crystalline lattice consists of pair-wise interactions, second, if we know the crystal structure of the lattice accommodating the admixture particle and, third, if we know the interaction potentials.

If the particle occupies a position in the lattice that would be empty before that (called an "*interstitial*" solution by metal physicists), then

$$\Delta V = x \, V_{AM} \, , \qquad (4.8)$$

where V_{AM} is the energy of interaction of the admixture molecule with the environment, i.e.,

$$V_{AM} = \sum_{i,k} V_{A_i M_k} \, , \qquad (4.9)$$

where $V_{A_i M_k}$ are the pairwise potentials. The indices run over all atoms in the molecules A and M.

In principle, the summation should be performed until convergence is achieved. In organic substances the summation can be limited to nearest neighbors, and even in the presence of long-range interactions, the substitution is usually governed by the short-range part of the potential.

In the simplest cases of the close packing of spheres the term "nearest" has rigorous meaning, because the neighborhood can be divided into coordination spheres. In more complicated atomic and molecular structures, the term "nearest" must be defined. As in Sect.3.3.5, neighboring particles can mean the nearest particles whose geometric centers are situated inside a spherical layer of a certain thickness and a certain diameter. Other definitions are also possible.

If an impurity particle replaces a particle in the host lattice, the energy of a solid solution with a concentration x is equal to

$$x \, V_{AM} + (1 - x) \, V_{MM} \, ,$$

and the energy of substitution is equal to

$$\Delta V = x(V_{AM} - V_{MM}) \equiv x\gamma_{AM} \quad . \tag{4.10}$$

The values of γ_{AM} are expressed by (4.9) in terms of pairwise potentials.

The calculation of ΔV by the above-described procedure can be carried out in a more complicated way: the quantity ΔV can also be made to account for a change in the lattice energy due to a short-range ordering. Later on we shall discuss this solution in an analytical form. It should be noticed, however, that the use of modern computers makes it possible to carry out computations by working out simple programs. A computer program sorts out all possible types of neighbors around the impurity molecule, estimates the respective energies of the interaction between the particle and its neighbors, and determines the relative probability of every possible neighbor. Then we shall know the character of the short-range order for a given concentration and can calculate the value of ΔV to account for the short-range order. As far as we know such calculations have not yet been done. There are two reasons for this. First, until recently only metallographers have shown interest in crystals of solid solutions. They are interested mainly in alloys produced on the basis of simple face-centered and body-centered cubic lattices. Formulas from Sect.4.4 are valid in these cases. The cases when these formulas cannot be easily used (complicated atomic and molecular structures) were insufficiently studied: in these a straightforward computation of the numerical effect of short-range order on the potential energy would be essential. And, second, we should bear in mind that the influence of the short-range order on the thermal motion may require an additional term Δf_{ord} in (4.6) to account for the character and degree of order.

4.2 Solid Solution Characterized by One Interaction Parameter

In this section we present the principle of the first *theory of solid solutions* developed by BRAGG and WILLIAMS [4.18] and, independently, by GORSKY [4.19]. The theory was developed by GUGGENHEIM [4.8] whose book is the best introduction to the theory of mixtures.

The consideration of this simplest version of the theory is essential because it provides an explanation for all phenomena observed during the formation and decomposition of solid solutions, and for the wide variety of structures which can be observed.

4.2.1 Free Energy of Mixing

It is possible to obtain the analytical expression for ΔV in the case of simplest-packing spheres, containing only one *parameter* W which can be numerically estimated if the potentials of interaction between the atoms are known.

Consider a binary alloy consisting of spherical atoms A and B, in the *nearest neighbor approximation*. Then only the interaction energy of the contacting spheres V_{AA}, V_{BB}, and V_{AB} are taken into account. It is significant that the properties of the solid solution are determined only by the combination

$$W = V_{AA} + V_{BB} - 2V_{AB} \quad . \tag{4.11}$$

A change in the number of AA contacts by the value x results in a change in the number of BB contacts by the same value and in the number of AB contacts by -2x.

For an *ideal solution* W = 0. Nonideal solutions can be divided into two types: the first with W < 0 (a strong interaction between atoms of different kinds), and the second with W > 0 (a strong interaction between atoms of the same kind). There is an essential difference between these two cases. If W < 0, the solution becomes ordered at low temperatures; if W > 0, decomposition may occur as the temperature is lowered.

It is not difficult to obtain the analytical expression for ΔV. The numbers of contacting spheres AA, BB, and AB are equal to $0.5 \, NZ(1-x)^2$, $0.5 \, NZx^2$, and $NZx(1-x)$ respectively, where Z is a coordination number. The potential energy of a crystal consisting of N atoms has the form

$$-NZ \left[(1-x) V_{AA} + x V_{BB} - x(1-x) W \right] \quad .$$

The last term within the square brackets is precisely the heat or energy of mixing. For one atom it is equal to $Zx(1-x)W$.

Hence, if (x, T) is neglected [see (4.6)] the free energy of mixing has the form

$$Zx(1-x) W + kT \left[(1-x) \ln(1-x) + x \ln x \right] \quad . \tag{4.12}$$

The dependence on x is represented by a curve symmetrical at about $x = 1/2$. If W > 0, the curve has a maximum at $x = 1/2$ and two symmetrical minima at, say, x_m and $1 - x_m$. When temperature increases, x_m tends to 1/2 and the central maximum becomes less pronounced. At a certain critical temperature, T_c, there appears a symmetrical "sagging" curve: above this critical temperature a continuous series of solid solutions can be formed.

We can also use (4.12) for the case of limited solubility. Since (4.12) is symmetrical relative to the concentration $x = 1/2$ and there is a horizontal tangent to the two minima of the saddle-shaped curve, we can obtain the following equation

$$\frac{1}{2x - 1} \ln \left(\frac{x}{1 - x}\right) = \frac{WZ}{kT} ,$$

which has a solution at, say, x_ℓ and $1 - x_\ell$.

For dilute solutions the limit of solubility is equal to

$$x_\ell = \exp(-WZ/kT) . \qquad (4.13)$$

This simple equation can sometimes be useful, especially if there are some experimental data or theoretical indications concerning the dependence of the parameter W on temperature.

Thus, if the *interaction parameter* W is positive, the particles of the same kind tend to group together. At temperatures below the critical temperature T_c given by

$$T_c = \frac{ZW}{2k} , \qquad (4.14)$$

the solid solution decomposes into small crystals of different compositions. However, the structural type will be the same (because above T_c we have a region of continuous solid solutions), and hence, the small crystals of the newly formed phases can differ only in lattice parameters. The decomposition of a solid solution is preceded by formation of more or less large groups of particles of the same kind (*clusters*). If the crystal particles are large and it is difficult for them to change places, then the size and position of the clusters will not change. But we cannot be sure whether it is really so, because the presence of *microcracks* and *dislocations* permit *lattice reconstructions* even if the *mobility* of the particles is small.

The processes preceding the decomposition, and the decomposition itself of crystals consisting of particles with a low mobility should probably be considered as a *growth of nuclei* of two phases at the defects. In such cases there are no oriented relationships between new crystals and the matrix of a solid solution.

We do not touch upon these phenomena in the present section: they will be considered in the next chapter. As to the case when the interaction parameter W is negative, i.e., when the particles A tend to be surrounded by particles B, the above-described crude theory predicts the main phenomena which can occur.

4.2.2 The Ordering Model

In the case $W < 0$ in (4.11,12) the ordering of the solid solution consists in a gradual increase of the long-range order parameter. Of course, this change is possible only if the size of the particles does not prevent diffusion. The process of ordering is possible because as temperature decreases, the entropy term in the expression for the free energy becomes less and less important, and the interaction energy becomes more important. Consequently, particles tend to leave "foreign" sites and go to their own sites.

BRAGG and WILLIAMS [4.18] and GORSKY [4.19] suggested expressing the free energy of the crystal through the long-range interaction parameter σ determined in Sect.3.3.3 and then establishing the dependence of σ on temperature from the condition of minimum of the free energy.

To simplify the calculations, let us consider a solid solution with an equal number of particles A and B. Then (3.7) becomes $\sigma = 2p - 1$. If we assume a complete disordering in the distribution of particles A and B over two crystallographic sites, the expression for the free energy of the crystal becomes

$$F = -NZ\left[W + p(1 - p)\,W\right] + NkT\left[p \ln p + (1 - p) \ln (1 - p)\right] \quad . \quad (4.15)$$

Making some transformations, differentiating with respect to σ, and equating the derivative to zero, we obtain

$$\ln \frac{1 + \sigma}{1 - \sigma} = \frac{W\sigma}{kT} \quad . \quad (4.16)$$

We rewrite this expression in the form

$$(1 + \sigma)/(1 - \sigma) = \exp(\alpha\sigma/2) \quad , \quad \text{where} \quad \alpha \equiv 2W/kT \quad .$$

Then we carry out the following transformations:

$$(1 + \sigma)/(1 - \sigma) - 1 = 2\sigma/(1 - \sigma) = \exp(\alpha\sigma/2) - 1 \quad ;$$

$$(1 + \sigma)/(1 - \sigma) + 1 = 2/(1 - \sigma) = \exp(\alpha\sigma/2) + 1 \quad .$$

Dividing the two equations term by term, we obtain

$$\sigma = [\exp(\alpha\sigma/2) - 1]/[\exp(\alpha\sigma/2) + 1] \quad .$$

Then we multiply the numerator and the denominator by $\exp(-\alpha\sigma/4)$;
$\sigma = [\exp(\alpha\sigma/4) - \exp(-\alpha\sigma/4)]/[\exp(\alpha\sigma/4) + \exp(-\alpha\sigma/4)] = \tanh(\alpha\sigma/4)$, where tanh is a hyperbolic tangent. Defining $\alpha\sigma/4 \equiv \xi$, we obtain finally

$$\sigma = \tanh\xi \quad \text{and} \tag{4.17}$$

$$\sigma \equiv 4\xi/\alpha = (2kT/W)\xi \quad . \tag{4.18}$$

To establish the temperature dependence of $\sigma = \sigma(T)$, we must find the existence domain of a simultaneous solution of the two equations (4.17) and (4.18). Let us do it graphically. Figure 4.1 shows the dependences $\sigma = \tanh\xi$ and $\sigma = (2kT/W)\xi$ (curves 1 and 2, respectively).

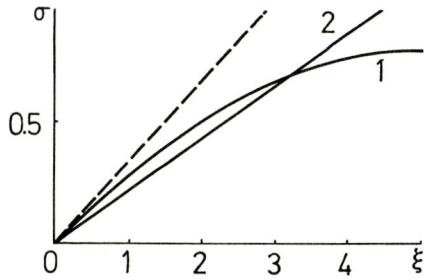

Fig.4.1. The dependences $\sigma = \tanh\xi$ and $\sigma = (2kT/W)\xi$ (curves 1 and 2, respectively)

The slope of curve 2 is equal to $2kT/W$ for all ξ. The slope of curve 1 varies with ξ. At $\sigma = 0$ the first derivative of curve 1 is equal to 1 (the broken line in Fig.4.1).

Our simultaneous solutions of (4.17,18) are nonzero if the slope of the curve 2 in Fig.4.1 at point $\xi = 0$ is less than the slope of the curve 1, i.e., if $2kT/W < 1$. If $2kT/W = 1$ the value of σ becomes zero and does not change if $2kT/W > 1$.

Thus, there exists a *minimum temperature* T_{cr} at which $\sigma = 0$. This temperature can be determined from the condition

$$T_{cr} = W/2k \quad . \tag{4.19}$$

According to this theory, a temperature increase in the alloys of the β-brass type causes a continuous decrease of the equilibrium value of the

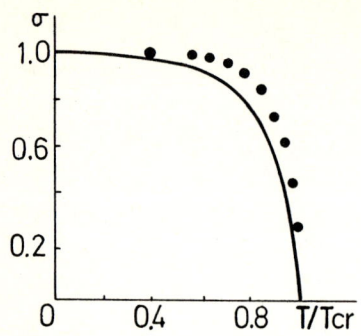

Fig.4.2. A continuous decrease in the equilibrium value of the degree of long-range order for alloys of the γ-brass type

degree of long-range order from 1 to 0 at $T = T_{cr}$. This dependence is shown graphically in Fig.4.2. At the point $T/T_{cr} = 1$ the long-range order disappears. The solid line has been calculated. The experimental points have been obtained from neutron diffraction data.

In other alloys, for instance in Cu_3Au, the disordering occurs with a phase transition of the first kind, and the degree of order (as well as the main thermodynamic properties such as entropy, volume, etc.) changes abruptly at the transition point.

4.3 Limited Solubility

If we know the pairwise potentials of the particle interactions, we can solve several problems and obtain more detailed results than those given in Sect.4.2. For dilute solutions, the calculations become more simple, and it is not difficult to take short-range order into account. It is more difficult to account for the lattice distortion and the changes in the vibrational spectrum occurring as a result of the formation of a solid solution. A simple phase diagram is also very helpful.

In this section we shall consider the eutectic phase diagram of components with a limited solubility (see Fig.2.12). At the eutectic temperature, the crystals of α and β phases are in mutual equilibrium; we assume that α has the type of crystal structure of pure A, and that β has the type of structure of pure B.

Two possibilities for theoretical calculations should be considered. First, we can determine the relationship between the energies of mixing, limiting concentrations, and free energies of the components. The formulas we shall derive will, unfortunately, require the knowledge of the free energies of the components; to obtain these values we must know the heat capacities as a function of temperature, beginning with absolute zero. Such measurements are not complicated, but require modern equipment and much time.

In Sect.4.3.2 we shall demonstrate the possibilities of considering theoretically the solubility without requiring the knowledge of the components' free energies. However, such a consideration also has a shortcoming. It becomes necessary to derive formulas anew, for every structure formed by the solid solution.

4.3.1 Relationship Between the Energy of Mixing and Solubility

The equilibrium between the phases α and β of a solid solution with the limiting concentrations x_α and x_β at eutectic temperature is expressed by (2.11,12). Let the pure component A correspond to the concentration $x = 0$. Then the expression for the free energies can be represented as

$$f^\alpha = (1 - x)f_A + xf_B + \Delta^\alpha(x) \quad ,$$

$$f^\beta = (1 - x)f_A + xf_B + \Delta^\beta(x) \quad .$$
(4.20)

The *excess energies of mixing* $\Delta^\alpha(x)$, $\Delta^\beta(x)$ for dilute solutions are expressed by (4.10). They are equal to $x\gamma_{BM}$ and $(1 - x)\gamma_{AM}$. Thus,

$$\Delta^\alpha(x) = x\gamma_{BM} + kTx \ln x + kT(1 - x) \ln (1 - x) \quad ,$$

$$\Delta^\beta(x) = (1 - x)\gamma_{AM} + kTx \ln x + kT(1 - x) \ln (1 - x) \quad .$$
(4.21)

By differentiating f^α and f^β with respect to x we can express (2.11) in the form

$$\gamma_{BM} + \gamma_{AM} = kT \ln \frac{x_\beta(1 - x_\alpha)}{x_\alpha(1 - x_\beta)} \quad .$$
(4.22)

Thus, there is a direct relationship between the limiting concentrations and the energies of the substitution. To obtain the numerical values of x_α and x_β by (2.12), we must know the values of f_A and f_B.

4.3.2 The Calculation of Solubility by a Given Substitution Scheme

Let the particles occupy two crystallographically inequivalent sites B_1 and B_2. We must calculate the free energy of the system which consists of N_A basic particles and N_B impurity particles:

$$N = N_A + N_B \quad \text{and} \quad N_B = N_{B_1} + N_{B_2} \quad .$$

The value of the interaction energy of the system depending on the degree of ordering of impurity particles can be written as

$$E(B_1 B_1 B_1 \ldots B_1) = N \cdot z \cdot 1/2 [m_A V_{AA} - m_B(V_{AA} - 2V_{AB_1})] = E_0 \quad ;$$

$$E(B_2 B_1 B_1 \ldots B_1) = E_0 + 1 \cdot z(V_{AB_2} - V_{AB_1}) \quad ;$$

$$E(B_2 B_2 B_1 \ldots B_1) = E_0 + 2 \cdot z(V_{AB_2} - V_{AB_1}) \quad ; \qquad (4.23)$$

$$E(B_2 B_2 B_2 \ldots B_2) = E_0 + N_B \cdot z(V_{AB_2} - V_{AB_1}) \quad ,$$

where $m_A = N_A/N$; $m_B = N_B/N$; z is the number of the nearest neighbors; V_{AA}, V_{AB_1}, V_{AB_2} are potentials of the molecular interaction; $E(B_1 B_1 B_1 \ldots B_1)$ is the interaction energy of the system when all impurity particles are ordered and each of them occupies the position B_1; $E(B_2 B_1 B_1 \ldots B_1)$ is the interaction energy of the system when one impurity particle is in position B_2 and the others in positions B_1, etc. Then the calculation of the statistical sum (*partition function*) Z reduces to the calculation of the sum of the series

$$Z = \frac{N!}{N_A! N_B!} \left\{ \exp\left[-\frac{E(B_1 B_1 B_1 \ldots B_1)}{\Theta}\right] + \frac{N_B!}{1!(N_B - 1)!} \exp\left[-\frac{E(B_2 B_1 B_1 \ldots B_1)}{\Theta}\right] \right.$$

$$+ \frac{N_B!}{2!(N_B - 2)!} \exp\left[-\frac{E(B_2 B_2 B_1 \ldots B_1)}{\Theta}\right] + \ldots +$$

$$\left. + \exp\left[-\frac{E(B_2 B_2 B_2 \ldots B_2)}{\Theta}\right]\right\} \quad . \qquad (4.24)$$

Equation (4.24), after the substitution of the interaction energy spectrum (4.23), can easily be calculated and is equal to

$$Z = \frac{N!}{N_A! N_B!} \exp\left(-\frac{E_0}{\Theta}\right) \left\{ 1 + \exp\left[-\frac{z(V_{AB_2} - V_{AB_1})}{\Theta}\right] \right\}^{N_B} \quad . \qquad (4.25)$$

The logarithm of (4.25) yields the free energy of the system:

$$F = -\Theta \ln Z = N\left(\frac{z}{2}\left[m_A V_{AA} - m_B(V_{AA} - 2V_{AB_1})\right]\right.$$

$$\left. - \Theta m_B \ln\left\{1 + \exp\left[\frac{-z(V_{AB_2} - V_{AB_1})}{\Theta}\right]\right\} + \Theta m_A \ln m_A + \Theta m_B \ln m_B\right) \quad . \tag{4.26}$$

Formula (4.26) can be easily generalized for a random number of possible nonequivalent positions of the impurity particle:

$$F = N\left(\frac{z}{2}\left[m_A V_{AA} - m_B(V_{AA} - 2V_{AB_1})\right] - \Theta m_B \ln\left\{\sum_{p=1}^{k} \exp\left[-\frac{z(V_{AB_p} - V_{AB_1})}{\Theta}\right]\right\}\right.$$

$$\left. + \Theta m_A \ln m_A + \Theta m_B \ln m_B\right) \quad , \tag{4.27}$$

where k is the number of inequivalent positions of impurity molecules.

The free energy of α-phase, calculated by (4.27) with k = 2, has the following form:

$$F_\alpha = N_\alpha f_\alpha = N_\alpha \left(Ax_A^\alpha + B + \Theta x_A^\alpha \ln x_A^\alpha + \Theta x_B^\alpha \ln x_B^\alpha\right) \quad , \tag{4.28}$$

where x_B^α is the concentration of A-component in α-phase; x_B^α is the concentration of the impurity in α-phase;

$$A = z\left(\frac{V_{A_1A_1} + V_{A_1A_2}}{2} - \frac{V_{A_1B_1} + V_{A_2B_1}}{2}\right)$$

$$+ \Theta \ln\left\{1 + \exp\left[-\frac{z}{\Theta}\left(\frac{V_{A_1B_2} + V_{A_2B_2}}{2} - \frac{V_{A_1B_1} + V_{A_2B_1}}{2}\right)\right]\right\} \quad ;$$

$$B = -\frac{z}{2}\left(\frac{V_{A_1B_2} + V_{A_1A_1}}{2} - 2\frac{V_{A_1A_1} + V_{A_2B_1}}{2}\right)$$

$$- \Theta \ln\left\{1 + \exp\left[-\frac{z}{\Theta}\left(\frac{V_{A_1B_2} + V_{A_2B_2}}{2} - \frac{V_{A_1B_1} + V_{A_2B_1}}{2}\right)\right]\right\} \quad .$$

The free energy of the β-phase, calculated by (4.27) with k = 1, is equal to

$$F_\beta = N_\beta f_\beta = N_\beta\left(Lx_B^\beta + M + \Theta x_B^\beta \ln x_B^\beta + \Theta x_A^\beta \ln x_A^\beta\right) \quad , \tag{4.29}$$

where x_B^β is the concentration of B-component in β-phase; x_A^β is the concentration of the impurity in β-phase;

$$L = z\left(\frac{V_{B_1B_1} + V_{B_1B_2}}{2} - \frac{V_{AB_1} + V_{AB_2}}{2}\right) ;$$

$$M = \frac{z}{2}\left(\frac{V_{B_1B_1} + V_{B_1B_2}}{2} - 2\frac{V_{AB_1} + V_{AB_2}}{2}\right) .$$

(4.30)

Now we can use thermodynamic equilibrium conditions (2.11) and (2.12) for a two-phase system. The first condition determines the relationship between the concentration of the component in the first and the second phases:

$$x_A^\alpha = \frac{x_A^\beta}{K + x_A^\beta(1 - K)} ,$$

(4.31)

where $K = \exp[(A+L)/\theta]$. The second condition determines the equilibrium values of the impurity concentrations in the α- and β-phases in an explicit form:

$$x_A^\beta = \frac{\exp\left(\frac{B + A - M}{\theta}\right) - \exp\left(\frac{A + L}{\theta}\right)}{1 - \exp\left(\frac{A + L}{\theta}\right)}$$

and

(4.32)

$$x_B^\alpha = \frac{\exp\left(\frac{L + M - B}{\theta}\right) - \exp\left(\frac{A + L}{\theta}\right)}{1 - \exp\left(\frac{A + L}{\theta}\right)} .$$

It is unwise to expect that this crude model can give quantitative results. However, at low temperatures and neglecting electron exchange between the particles, these simple considerations may be of some help for estimating the solubility. An application of such calculations will be given in Sect.12.1.

4.4 Permutation of Particles in a Rigid Lattice

We now formulate the problem of calculating the Hamiltonian and the free energy of a solid-solution crystal, taking into account partial long-range order but neglecting the changes in vibrations and the distortions of the lattice caused by the impurity. A rigorous calculation can be done if we use the pairwise interaction approximation and assume that the permutations of particles A and B occur within the rigid frame of crystallographic positions related by the symmetry elements of the space group.

The free energy expression allows all possible permutations of the particles. Thus, the problem of ordering a solid solution can be solved rigorously. In the general case, however, calculations for complicated structures are very tedious. Therefore, they were done only for the Bravais lattices.

The procedure described below works if metallurgists are interested in "interstitial" alloys, alloys where one of the components occupies part of the tetrahedral or octahedral voids. This was pointed out by KHACHATURYAN [4.2]. The voids here play the role of the second component. But we shall not dwell on this specific case and in further considerations we shall deal only with substitutional solid solutions.

The Hamiltonian of a solid-solution crystal can be written in the form

$$H = \frac{1}{2} \sum_{\underline{r}\,\underline{r}'} V_{AA}(\underline{r}\,\underline{r}')\, c_A(\underline{r}) c_A(\underline{r}') + V_{BB}(\underline{r}\,\underline{r}')\, c_B(\underline{r}) c_B(\underline{r}') + 2 V_{AB}(\underline{r}\,\underline{r}')\, c_A(\underline{r}) c_B(\underline{r}') \quad . \tag{4.33}$$

The quantities $c_A(\underline{r})$ and $c_B(\underline{r})$ describe the ordering of particles in the lattice. These random quantities have the following meaning:

$$\begin{aligned}
c_A(\underline{r}) &= 1 \quad \text{if the particle A is at the point } \underline{r} \\
c_A(\underline{r}) &= 0 \quad \text{if the particle B is at the point } \underline{r} \\
c_B(\underline{r}) &= 1 \quad \text{if the particle B is at the point } \underline{r} \\
c_B(\underline{r}) &= 0 \quad \text{if the particle A is at the point } \underline{r} \quad .
\end{aligned} \tag{4.34}$$

The following equality is obvious:

$$c_A(\underline{r}) + c_B(\underline{r}) = 1 \quad .$$

It permits the exclusion of c_B from (4.33).

The expression (4.33) reduces to

$$H = H_0 + \frac{1}{2} \sum_{r\,r'} V(r\,r')\, c(r)\, c(r') \quad , \tag{4.35}$$

where the Index A of vector r is omitted.

The pairwise interaction approximation used in (4.35) should not be considered as a very stringent limitation. For metal systems, this approximation was used by HARRISON in the *pseudopotential* method [4.20]. This approximation is also valid for crystals of monoatomic ions. Finally, the author of this book has demonstrated that it is possible to use the additive scheme of the atom-atom interaction for organic molecules [4.1].

Nevertheless, the calculation of the statistical sum by (4.35) is not so simple and can be done only by approximation methods. The simplest of them were suggested by GORSKY [4.19] and BRAGG and WILLIAMS [4.18] fifty years ago. The method of *quasichemical equilibrium* developed by GUGGENHEIM [4.8], and the methods of BETHE [4.21], PEIERLS [4.22], and KIRKWOOD [4.23] are widely used. All these well-known theories have several shortcomings which have been overcome in the past ten years by the method of static concentration waves worked out by KHACHATURYAN [4.2,24], who developed a general approach valid in the framework of a *self-consistent field* theory. Using this method, we can account for correlations in any superstructures where the interaction of an arbitrary number of coordination spheres takes place. We briefly summarize the results of Khachaturyan's calculations.

The problem of the structure of a binary solid solution will be solved if one can find the expression for the probability of finding the particle A (or B) in the crystallographic position r. This probability is equal to the *mean value* of the random function $c(r)$ introduced by (4.34):

$$n(r) = \langle c(r) \rangle \quad . \tag{4.36}$$

Let us assume that the rigid frame where the particles are distributed is a Bravais lattice. The self-consistent field equation for the function $n(R)$ has the form

$$n(R) = \left\{ \exp\left[-\frac{\mu}{\Theta} + \sum_{R'} \frac{V(R - R')}{\Theta} n(R') \right] + 1 \right\}^{-1} \quad . \tag{4.37}$$

Here, μ is a Lagrangian multiplier which serves as a chemical potential; R is a radius vector of the sites in the Bravais lattice. Formula (4.37) is driven to self-consistency.

The internal energy is equal to

$$U = \frac{1}{2} \sum_{\underline{R}\,\underline{R}'} V(\underline{R} - \underline{R}') \, n(\underline{R}) \, n(\underline{R}') \quad .$$

The entropy can be written as

$$S = -k \sum_{\underline{R}} \{n(\underline{R}) \ln n(R) + [1 - n(\underline{R})] \ln [1 - n(\underline{R})]\} \quad . \tag{4.38}$$

The free energy has the form

$$F = U - TS - \mu \sum_{\underline{R}} n(\underline{R}) \quad . \tag{4.39}$$

Equation (4.37) gives several extrema for the free energy to which correspond disordered solutions as well as ordered systems. If the composition and temperature are fixed, we obtain a particular solution corresponding to the stable state.

How can (4.37) be solved? It is difficult to obtain the direct solution. For this purpose we can use Khachaturyan's method of *concentration waves* described in Sect.3.3.3. Using (3.10) we obtain a system of transcendental equations

$$c + \sum_{s} \sigma_s E_s = \left(\exp\left\{ \left[-\mu + \Phi(0)c + \sum_{s} \Phi(\underline{k}_s) \sigma_s E_s \right] / \Theta \right\} + 1 \right)^{-1} \quad , \tag{4.40}$$

which can be solved more easily than the initial equation.

Here,

$$\Phi(\underline{k}) = \sum V(\underline{r}) \exp(i\underline{k}\underline{r})$$

is the Fourier transform of the potential of the particle interaction. The *internal energy* of the system can be expressed through the same energy parameters $\Phi(\underline{k})$:

$$U = \frac{1}{2} \left[\Phi(0)c^2 + \sum_s \sigma_s^2 \, \Phi(\underline{k}_s) \right] \quad . \tag{4.41}$$

In [3.28] the solutions of this equation for different cases are considered. Some conclusions from the theory will be considered in the chapter devoted to metal alloys.

The essential shortcoming of the self-consistent field approximation is that it does not take into account the *correlations* in the mutual arrangement of particles. BADALYAN and KHACHATURYAN [4.25] demonstrated how this correlation can be accounted for without the limitation of nearest neighbor interactions. The method of concentration waves permits the problem to be solved in this case, too.

The main advantage of Khachaturyan's method is the possibility to predict superstructures arising in an alloy at different concentrations and temperatures. For cubic packing detailed calculations are performed.

4.5 Lattice Distortions

It is evident that placing a particle A into a basic crystal B causes a lattice distortion. The particles A and B have different volumes and shapes. Under such conditions, rigorous periodicity (inherent in the ideal crystal) is impossible.

It was already mentioned more than once that the unit cells in the crystals of a disordered solid solution are not identical. *Nodal lines* and *planes* of such lattices are distorted. Deviations from straight lines and planes occur with equal probability in all directions. X-ray measurements provide us with the dimensions of an "average" unit cell, whose parameters essentially differ from the parameters of the unit cells of the components. For a *substitutional* solution an increase and a decrease in the unit-cell volume or any unit-cell parameter (sides, angles) are equally possible.

If the particle enters a crystallographic site which was previously vacant (*interstitial* solution), the size of the cell must increase. In a rigid particle model it seems natural to expect that the lattice volume depends linearly on the concentration, thus the forming of a continuous series of solid solutions. In the case of a cubic lattice a linear concentration dependence of unit-cell edge is quite possible.

As we shall see later, there are cases when this simple rule works well enough. But more often we deal with non-linear (convex or concave) curves, and in some cases the dependence of lattice parameters on concentration is expressed by curves with maxima or minima. This is not surprising.

Indeed, the problem of the lattice distortion is not at all simple. To begin with, the crystal is not ideal. The impurity can enter the boundary between mosaic blocks without changing the dimensions of a unit cell. Examples of such behavior will be considered in Chap.11. Then changes in the form and size of the cell with concentration depend considerably on concentration-dependent differences in the interaction energies V_{AA}, V_{AB}, and V_{BB}. In simple models particles A and B are said to have different compressibilities. And finally, the character of a short-range order is important, i.e., whether or not there is a tendency to segregate.

There have been attempts to determine the distortion of the lattice of a solid solution by studying the changes in the intensity of X-ray reflections. As it is shown in Chap.6, the formula for diffracted intensity includes a Debye-Waller *factor*:

$$\exp(-M) \equiv \exp\left[-B\left(\frac{\sin\Theta}{\lambda}\right)^2\right] \equiv \exp\left[-8\pi^2 u^2\left(\frac{\sin\Theta}{\lambda}\right)^2\right] \quad , \tag{4.42}$$

where Θ is the Bragg angle and λ is the X-ray wavelength. Here u^2 is the *mean-square displacement* of an atom from its position in the corresponding nodal plane. Unfortunately, this displacement appears not only due to the lattice distortion, but also due to thermal vibrations. Therefore, the mean-square displacement is the sum of two terms: one depending on the lattice dynamics and the other on static distortions. To distinguish between them, u^2 must be measured at different temperatures. It seems that static distortions should not depend on temperature. If this is the case, the quantity characterizing the lattice distortion can be singled out. The calculation of u^2 is possible only for simple lattices, because the Debye-Waller factor enters the formula for the scattering of radiation by the atoms of each crystallographic species. In complicated crystals, the scattering formula contains a sum over each atomic species. One measures the total intensity of the atomic scattering. In such cases the calculation of mean-square deviations is difficult (although not impossible, if many X-ray reflections from a single crystal are measured). Nevertheless, satisfactory experimental studies of this type are few.

In this chapter we do not consider the results obtained for individual systems. We only indicate that dynamic and static displacements of atoms of the same order of magnitude lie within the interval of 0.1 - 0.2 Å.

The lattice distortion of a pure substance results in an energy increase. If the distortion energy is large, the solid solution is not formed. Estimation of this quantity is of great interest because it enters the general formula (4.1) for the free energy.

A relatively small distortion is a necessary, but not sufficient, condition for solubility. From experimental data, we can say that size differences may not exceed 15 - 20%. For molecular crystals the maximum volume difference does not exceed 15%.

These difference data are not sufficiently precise. What matters is the *maximum energy of distortion*, which cannot be exceeded because it would lead to two-phase crystallization. This energy can be estimated by the method of atom-atom potentials. It can be obtained by putting the impurity particle in the undistorted lattice of a pure component. Then we can calculate the difference between the interaction energy of the host particle with its nearest (host) neighbors and the interaction energy of the impurity particle with the same neighbors of the host lattice. In this manner we obtain the value of ΔV discussed in Sect.4.1.

It is more difficult to obtain an exact value of ΔV when we are interested in the real picture of all the distortions which naturally become smaller as we move away from the defect site. These difficulties are not the same for dilute and for concentrated solutions.

Let us consider a crystal of a dilute solid solution whose particles are displaced from their positions $\underline{r}$ by a vector $\underline{P}$ as a result of a lattice distortion (in the case of a spherical particle $\underline{P}$ is a three-dimensional vector and in the case of molecules, six dimensional, to account for orientational displacements).

We are interested in the interaction energy of the impurity particle D and the host particles (numbered by index s). It can be expressed as

$$V = \sum_s v_{DM}(\underline{r}_s + \underline{P}_s) \quad . \tag{4.43}$$

If the displacements are small, we can expand the energy in a power series of the displacements, and restrict it to the first two terms. Then the interaction energy has the form

$$V = V_0 + \sum_s \underline{\beta}_s \underline{P}_s + \frac{1}{2} \sum_{ss'} \alpha_{ss'} \underline{P}_s \underline{P}_{s'} \quad , \quad \text{where} \tag{4.44}$$

$$\beta_s = \left[\frac{\partial v_{DM}}{\partial \underline{P}_s}\right]_0 \quad \text{and} \tag{4.45}$$

$$\alpha_{ss'} = \frac{\partial^2 V}{\partial \underline{P}_s \partial \underline{P}_{s'}} \quad . \tag{4.46}$$

The equilibrium conditions are expressed by a system of linear equations from which the particle displacements can be obtained:

$$\sum_{s'} \alpha_{ss'} \underline{P}_{s'} - \underline{\beta}_s = 0 \quad . \tag{4.47}$$

The number of equations corresponds to the number of the host particles whose interaction with the impurity molecule D must be taken into account.

The calculation of such displacements can be performed by two methods. KRIVOGLAZ [4.7] and FISCHER [4.26] have demonstrated that it is possible to relate the coefficients α and β to experimentally measured quantities. Thus, the constants α can be expressed in terms of the crystal elasticity coefficient, whereas the constants β can be obtained if the dependence of the unit-cell geometry on the impurity concentration is known.

The second method of calculation is more expedient. One has to resort to atom-atom potentials; for example, to the potential of Buckingham or Lennard-Jones, whose parameters can be obtained independently. Such calculations were performed twenty years ago. HALL [4.27], for instance, made calculations for a face-centered cubic lattice. The crystal was divided into concentric spherical layers around the defect and it was assumed that the atomic displacements in the lattice occurred along the radii. The results of this calculation are shown in Fig.4.3 which plots the displacements (ordinates) against the distances from the impurity atom (vacancy)(abscissa) in fractions of unit-cell parameters. Distortions oscillate and vanish at the sixth coordination sphere. The vertical bars show the coordination sphere number.

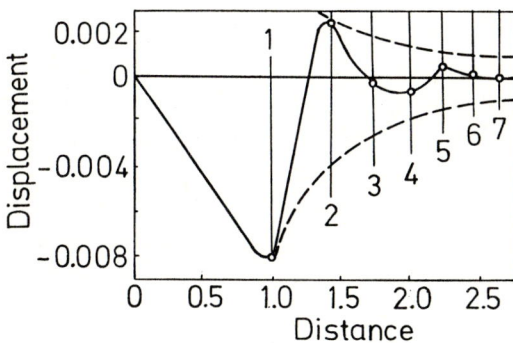

Fig.4.3. Displacements and distances from the impurity atom (vacancy) expressed as fractions of unit-cell parameters

At present, computers help us make such calculations for cases of any complexity: for particles of an arbitrary form, for six-dimensional displacement vectors, for any number of coordination spheres.

Such computations have not been done for molecular crystals yet, though the author and FEDIN [4.28] tried to use the method of nuclear quadrupole resonance for the estimation of the lattice distortion area which occurred because of the inclusion of paradibromobenzene molecules into the paradichlorobenzene lattice.

The interaction between the impurity molecules can be neglected only in dilute solutions. Even if the area of distortion does not exceed 4 - 5 periods

of the unit cell, we can disregard the interference of distortions caused by
impurity molecules only if the crystal contains less than 1% impurities. The
interaction between impurity particles can lead to the formation of various
complexes.

It should be remembered that separating the formation of a short-range order and the lattice distortion is conditional; otherwise, it can lead to wrong
conclusions. Different modes of surrounding particle A by particles B lead to
various lattice distortions. Singling out the term ΔV in the general formula
of the energy of mixing (4.6) is an artificial technique. Even more so is the
division of this term into two parts: one of which describes the character of
the host particles' surroundings in terms of the impurities and the other which
relates lattice distortions to different sizes and shapes of the particles.

At present, modern computers make it possible to simulate optimal structures
proceeding from the potentials of interaction of particles AA, AB, and BB.
Many calculations of this kind were carried out for simple cases. JOHNSON
[4.29], for instance, studied the stability of *divacancies* in the body-centered
lattice of iron. It turned out that the binding energy of the vacancies situated at the vertices of a cube was 50% greater than the energy of interaction
of two vacancies situated in the center and at the cube vertex, respectively,
even though in the second case the defects are closer to each other. Different
arrangements of three vacancies in the cubic lattice of copper were also
studied. The linear configuration along the plane diagonal turned out to be
less stable. The most stable is the tetrahedral arrangement, where vacancies
occupy a vertex and the centers of three faces of the cube. More complicated
problems were also solved which concerned the optimal arrangement of interstitial atoms and vacancies in simple lattices. The corresponding references
are given in the review by JOHNSON [4.30].

If the impurity particles tend to segregate, it is useful to consider the
lattice distortion and its contribution to the energy of the lattice using
the model of the *solid continuum*. In such cases we can omit the consideration
of particle interactions and use methods of the *theory of elasticity*. The
problems dealing with the inclusion into the continuous medium of a sphere
of a body of arbitrary shape which possesses compressibility and other mechanical properties different from those of the medium are well-known and
are described in textbooks (for instance, [4.31]).

4.6 Thermal Vibrations

The neglect of vibrations, explicitly or implicitly accepted by the authors of all solid-solution theories, can lead to serious errors, which are greater than those caused by the crude account of the degree of ordering solid solutions. The change in the vibrational energy caused by addition of an impurity into the lattice can also be very important.

GUGGENHEIM, who did much for the development of the theory of solid solutions, realized the importance of taking *thermal vibrations* into account. He noted [4.32] that although it was impossible to include the dependence of the interaction parameter on short-range order in a one-parameter theory (Sect.4.2), nevertheless its formalism accounts for the dependence of W on temperature.

We cannot quantify this dependence, and so we must use an empirical approach. FREEDMAN and NOWICK [4.33] have shown that neglecting the thermal entropy of mixing can cause serious errors. They analyzed the equilibrium solubility curves.

The most rigorous approach should consist in estimating the changes in the lattice vibration spectrum as a function of impurity concentration. Such estimates should be made at different temperatures. One can get an idea of the role played by thermal vibrations by employing the Debye model of a crystal, since it can help to determine the dependence of a *characteristic temperature* on concentration. There are many methods of measuring such a characteristic temperature. Probably, the measurements of the heat capacity and the thermal diffusion of X-rays are the most appropriate methods.

If we use the Debye *approximation* for high temperatures, then the change in the free energy, caused by the change in the characteristic temperature, has the form

$6 \, RT \ln \Theta_1/\Theta_2 \quad$ for molecular crystals [3.8] $\quad$ and $\quad$ (4.48)

$3 \, RT \ln \Theta_1/\Theta_2 \quad$ for monoatomic crystals. $\quad$ (4.49)

The variation of Θ_1/Θ_2 by 10% (which is much) changes the logarithm by only 0.04. At room temperature, the correction is of about several tenths of kcal/mol.

One may study the relation between the characteristic temperatures of the pure substance and the solid-solution by evaluating the thermal factors of the scattered intensity for several X-ray reflections.

The relation between the Debye characteristic temperature and the thermal factor of X-ray interference was shown by WALLER [4.34]. The Waller formula has the form

$$B = \frac{h^2}{mk\Theta}\left[D(\Theta/T)/\Theta/T + \frac{1}{4}\right] \quad . \quad (4.50)$$

When the temperature T approaches the characteristic temperature Θ an approximate relation holds

$$D(x) \approx x \quad . \quad (4.51)$$

However, Waller's formula is valid only for monatomic cubic lattices and isotropic thermal vibrations. Hence, (4.48,49) indicate only the probability of such correlation between the entropy changes and the changes in the lattice vibrations.

X-ray diffraction analysis permits fairly accurate measurements of the amplitudes of thermal vibrations of atoms and molecules. In the case of atomic crystals, the experiment provides directly the root mean square values of the atomic amplitude.

Now we can reason as follows. The potential energy curves for the interatomic interaction are established for a stationary lattice. Changes in the potential energy due to an increase in interatomic distances with increasing temperature are easy to determine if the curves are given. The thermal expansion of the lattice is not great and in many cases such an influence on the thermal vibrations can be neglected (though it is also easy to include it). As to the thermal entropy, we can only assume that we may replace the ratio of characteristic temperatures by the ratio of the root of the mean-square amplitudes of thermal vibration, i.e., that we may use the following expression for the changes in the free energy due to thermal vibrations:

$$1.5 \, RT \, \ln \, (u_A^2/u_M^2) \quad \text{and} \quad 3 \, RT \, \ln \, (u_A^2/u_M^2) \quad , \quad (4.52)$$

for atomic and molecular crystals, respectively.

For molecular crystals, the intermolecular vibrations are larger, at least by an order of magnitude, than the intramolecular vibrations. PAWLEY [4.35] has shown how to determine the average amplitude of the atomic vibrations due to the intermolecular vibrations.

5. Heterophase Systems

Solids that consist of a mixture of crystals of different composition and structure are, of course, more numerous than solid single-phase systems. Such a mixed body can also form in a single-component system if the substance can exist in several polymorphic modifications. But then, such two-phase systems (e.g., a mixture of yellow and red sulphur crystals, or of grey and white tin crystals) can exist in a stable state only at phase equilibrium points of a phase diagram. At a given pressure, equilibrium is attained only at a certain temperature. A two-phase region in the phase diagram can be reached only through eutectic crystallization, or when a solid solution decomposes (disproportionates). These two processes will be examined briefly in this chapter.

Two-phase polymorphic systems will not be discussed. As was mentioned in Sect.2.3.6, these interesting compounds are not low-molecular-weight solids. (Polymers will be considered in Chap.15.)

As before, we are interested only in the structure but not in the kinetics or the mechanism of formation of heterogeneous systems. Yet, what does one mean by the structure of heterogeneous systems? Some decades ago the heterophase system was considered as a disordered agglomeration of crystals. But more often than not it has been found that there is an order in the arrangement of the crystallites. The nature of this order is of considerable scientific and practical interest.

We should like to draw attention to the fact that in many cases the shape and orientation of the domains of newly formed phases are similar, for metals, ceramics, as well as for polymers. This can be clearly seen from a comparison of Figs.5.11,14 and 14.9.

5.1 Eutectic Crystallization

5.1.1 Mechanism of Uncontrolled Crystallization

At a certain point in the supercooling of a melt whose composition corresponds to the eutectic point, the liquid becomes supersaturated with the particles of both components A and B, and the growth of crystals begins. Of course, the stability of crystal nuclei of the components is different. Therefore, the growth of A and B crystals begins at different times. The microstructures of the forming solids can be different: in some cases an agglomerate of A and B crystals forms, in other cases the solid consists of two-phase, or eutectic, grains.

In the early 1900s scientists realized that the microstructure of an eutectic grain obeys certain rules. In 1909, TAMANN [5.1] suggested the model of an eutectic colony in which the eutectic grain is an agglomerate of alternating layers of crystals A and B. This model seems to be quite natural. The first crystal of A appears, grows, and the surrounding liquid becomes richer in component B, thus providing the conditions for the formation and growth of crystals of B. The process continues. Since such initial crystals arise simultaneously in many points of the melt, many eutectic grains can grow simultaneously. When they meet, the solidification ends. Typical of this process are the microphotographs illustrated in Figs.5.1-3.

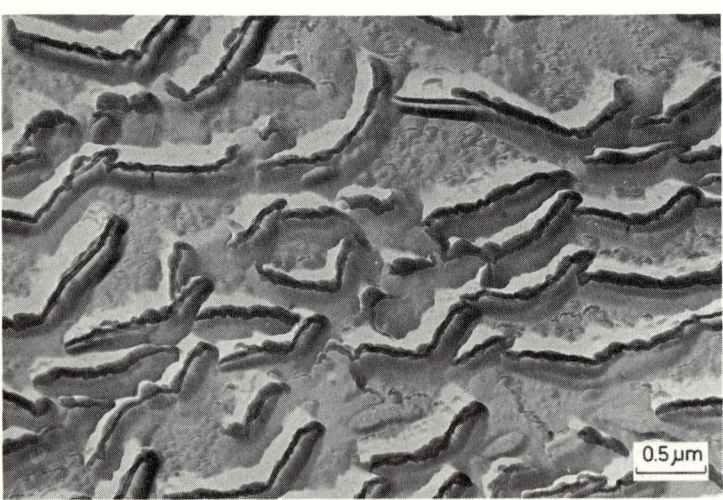

Fig.5.1. Microphotograph of a dibenzyl-diphenyl system with equimolecular composition

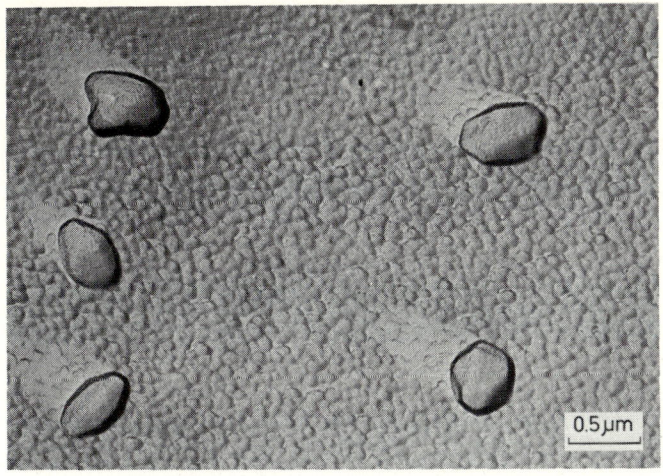

Fig.5.2. The same as in Fig.5.1 but for a 75% dibenzyl content

In the dibenzyl-diphenyl system the eutectic point corresponds to the equimolecular composition. Microphotographs (Fig.5.1) show that the solid cannot be considered as homogeneous. The diphenyl crystals are seen against the background of a small-sized crystalline alloy.

Deviation from the eutectic composition will no doubt lead to an increase in the size of the crystals. The microphotograph of an alloy containing 75% dibenzyl is seen in Fig.5.2. Figure 5.3 demonstrates an alloy containing

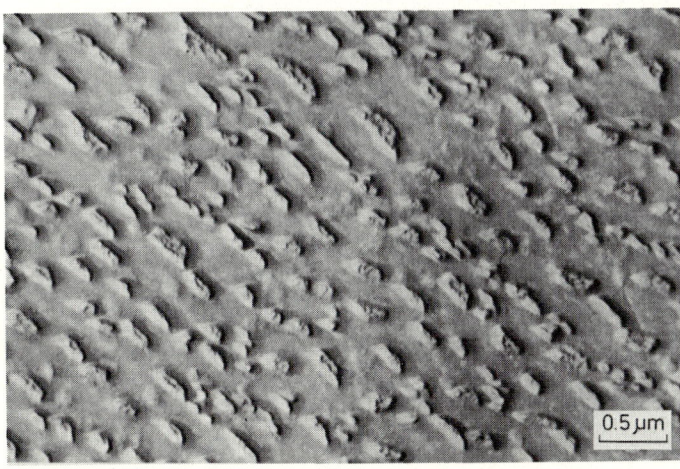

Fig.5.3. The same as in Fig.5.1 but for a 75% diphenyl content

75% diphenyl. One can see in Figs.5.1-3 that differences in the structures and properties of the components will give different visual patterns. Furthermore, such differences will also affect the selective growth rate and the number of crystallization centers per unit volume.

There is no reason to believe that the mechanisms of eutectic crystallization for metal and organic compounds are different. At the same time, it is much easier to investigate the behavior of transparent low-melting organic compounds with a microscope. BOCHVAR [5.2] studied in detail the mechanism of eutectic crystallization of the *azobenzene-piperonal* system. His conclusions, given below, establish the basis for the modern knowledge of the microstructure of solidified eutectic alloys:

(a) The rate of growth of the initial crystals of each component in the melts of various compositions decreases with the increase in the content of the second component. This is illustrated in Fig.5.4, which depicts the growth rate (mm/min) versus piperonal content. Curves 1 and 2 are for azobenzene and piperonal, respectively. The line 3 represents the growth rate of the eutectic *colony*. The supercooling is 9°C. The liquid composition affects the linear growth rate of azobenzene crystals much more strongly than the slow growth of piperonal crystals.

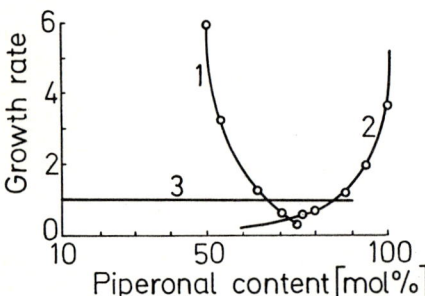

Fig.5.4. Azobenzene-piperonal system growth rate versus piperonal content

(b) Even in a supercooled liquid of exact eutectic composition the crystals of each phase nucleate spontaneously and grow separately.

(c) The layer of the liquid surrounding the growing crystal becomes strongly deficient in molecules of the given component. This can be judged by a change in color: the piperonal rich liquid surrounding the orange crystal of azobenzene becomes noticeably colorless, whereas the azobenzene-rich liquid that surrounds the colorless piperonal crystal becomes brightly colored.

(d) As the growing crystals move closer to one another, the growth rate of azobenzene crystals increases. When they come into contact, the azobenzene crystals produce needles, which quickly grow in the liquid surrounding the piperonal crystal.

(e) The eutectic crystallization proper starts only after the crystals of different phases come into contact. Eutectic crystallization proceeds with a rate significantly higher than the rates of the crystallization of separate phases in the liquid of eutectic composition.

(f) At the interface of solid eutectic with liquid, there occurs a preferential separation of one of the phases which is accompanied by an enrichment of the boundary layer with the substance making the basis of the second phase. As a result, one of the phases crystallizes, forming the framework of an eutectic crystallite, whereas the second phase forms within this framework.

Why liquid composition affects the growth rate is evident: fewer molecules of, say, component A are in contact with the surface of growing crystals of pure A, and thus fewer molecules join the crystal surface per unit time.

The crystal growth from solution is even more hindered by the depletion of the layer immediately surrounding the crystal. However, if an eutectic conglomerate forms, then the situation is quite different: the crystal of component A, during its growth, depletes the liquid of molecules of A in the layer adjacent to the growing grain, but this facilitates the growth of crystal B, which, in turn, helps the growth of crystal A. Such a diffusional separation of the liquid into components explains why the eutectic growth rate in the liquid of eutectic composition is higher than the growth rate of initial grains of either component, for any degree of supercooling.

As seen from Fig.5.4, the growth rate of eutectic grains does not depend on the liquid composition; the eutectic solid grows faster than the *initial* crystals of either component.

Azobenzene often forms "*rims*" around the piperonal crystals, whereas the piperonal never forms "rims" around azobenzene crystals. This can be explained by the stronger influence of the liquid composition on the linear crystallization rate of azobenzene. Therefore, the azobenzene crystal grows extremely fast into the piperonal-deficient liquid layer around the piperonal crystal when the two crystals come into contact. In the layer surrounding the crystal of one of the components there is no nucleation of the second component (though concentration is favorable for it) until when the separately formed initial crystals come into contact. This lack of nucleation can be explained by the heating of the liquid in the layer due to the release of the *latent heat of crystallization*. It is known that when supercooling is not very high, the *probability of nucleus formation* is very low even in a single-component melt. The lowering of supercooling affects the crystal growth rate insignificantly, which makes the "rim" formation possible. Such phenomena have also been observed in metal eutectic alloys.

Often the main phase has a more complicated structure and a higher fraction of covalent bonds. For example, in a solid solution of metals with chemical compounds (such as Fe-Fe$_3$C and Al-CuAl$_2$), the compound is the main phase. In the alloys Fe-C, Al-Si, and Sn-Bi, eutectic *colonies* appear on the crystals of graphite, silicon, and bismuth, respectively. The main phase in eutectic crystallization can be studied directly by considering the *surface energy changes* during the nucleation of one phase on the other. Such an analysis was carried out by TARAN and others [5.3] for the eutectic crystallization in the system Ag-Cu: here Ag and Cu are chemically similar, and the phases (solid solutions of Ag and Cu) have the same fcc lattice. The main phase here is the Cu phase, as the surface energy of copper at the interface with the melt is higher than the surface energy of silver at the same interface (17.7 and 12.6 J/m^2, respectively). Therefore, the formation of a silver crystal on the surface of a copper crystal is more advantageous than the formation of copper on silver.

Very interesting heterophase systems can produce *amorphous glass*. Figure 5.5 illustrates a biphase system obtained by heat treatment of a glass consisting of six oxides. As is seen in the photograph, liquation of the matrix takes place, and the separated particles have a different composition. The particles and the host matrix are amorphous. These materials, called

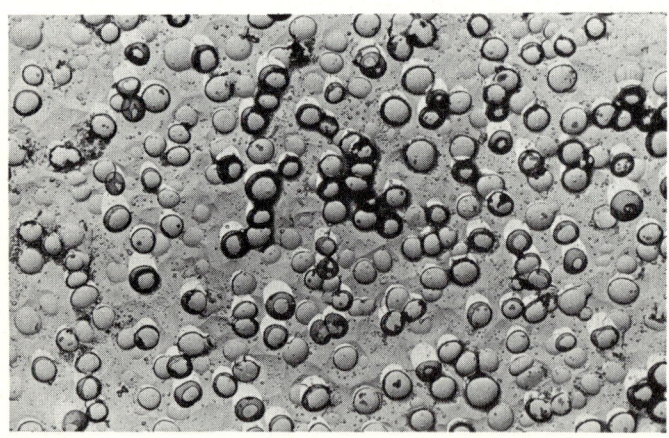

Fig.5.5. Two-phase glass system, both phases being amorphous

sitals, have become widespread in industry. Sitals represent mechanical mixtures of crystals and glass: such systems possess remarkable mechanical properties. The presence of crystals surrounded by an amorphous substance eliminates the brittleness of the materials, and the strength increases 5- to 10-fold.

There are many different ways of preparing these interesting heterophase systems. If we start from multicomponent glasses containing some silver chloride, it is possible to induce their crystallization through light irradiation by maintaining such a system at temperatures of 600° to 700°C. Typical microphotographs are exhibited in Figs.5.6,7.

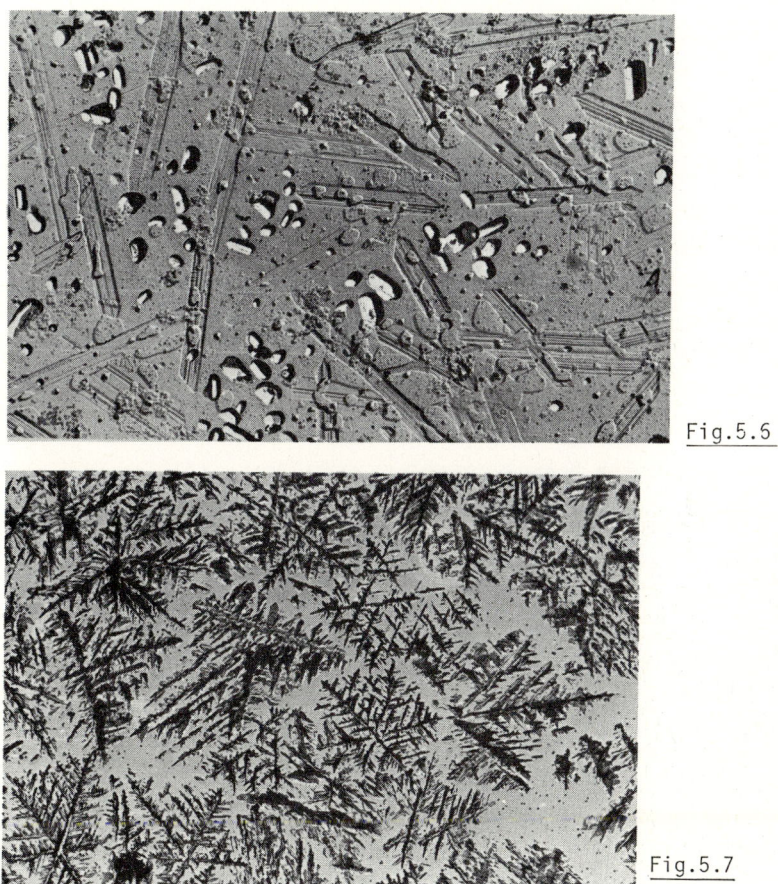

Fig.5.6

Fig.5.7

Fig.5.6. Microphotograph of two-phase crystal-glass system
Fig.5.7. Sitals based on AgCl crystals

Technologists are able to vary the percentage content of the crystalline and amorphous materials. It is possible to produce sitals by alloying the crystals and glass. Figures 5.8,9 illustrate sitals prepared on a titanium oxide base. The technologies of production of sitals differ substantially, therefore producing varied microstructures. In Fig.5.8 we see small titanium oxide crystals coated with a thin glass film. The photograph in Fig.5.9 shows small crystals intruding into the glass matrix which occupies a large volume of the material. It is a pleasure for the author to exhibit these photographs and say a few words about sitals for the reason that the interesting materials were discovered by his father, Professor I.I. Kitaigorodsky.

Fig.5.8

Fig.5.9

Fig.5.8. Sitals based on TiO_2 crystals (a thin glass film)
Fig.5.9. Sitals based on TiO_2 crystals but with a large content of glass

5.1.2 Unidirectional Crystallization

TAMANN [5.1] knew that controlling the solidification of an eutectic alloy by *directional heat removal* can produce solids with highly ordered microstructures. *Equally spaced rods* or *plates* of the second phase are placed into the matrix of the first one. The eutectic colony that results is very extended and is oriented along the direction of heat removal.

For a long time this fact attracted no attention, but recently, scientists and technologists have concentrated on the preparation of mixed systems of metals, semiconductors, and oxides on an industrial scale by means of the unidirectional crystallization of eutectics.

Here we have great possibilities for obtaining new materials with excellent magnetic, electrical, mechanical, and thermal properties. The technology of preparation of these "natural" composites is rather simple.

We have already mentioned that of highly *ordered eutectics* can form in any substances, irrespective of bonding metals, ceramics, polymers, etc. We shall now illustrate this with the help of several photographs.

Figure 5.10 shows the "platelike" eutectic of an $Al-CuAl_2$ alloy. Figure 5.11 shows a "rodlike" eutectic of an $Al-NiAl_3$ alloy. Interesting microphotographs (Fig.5.12) have been made of $TiC-TiB_2$ alloys: the eutectic represents a matrix made of titanium carbide with plates of titanium diboride rods uniformly distributed and oriented along the growth direction (the alloys were obtained by zone melting).

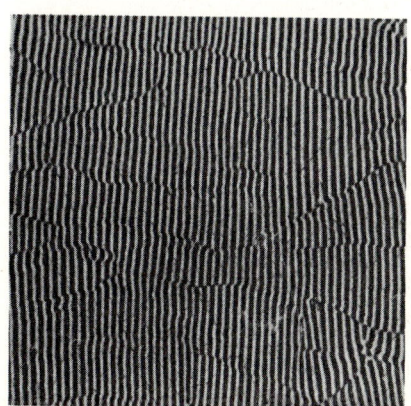

Fig.5.10. Photograph of a platelike eutectic in an $Al-CuAl_2$ alloy

Fig.5.11. A rodlike eutectic in an $Al-NiAl_3$ alloy

Fig.5.12. A matrix made of titanium carbide with plates having titanium diboride rods

The same TiC-TiB$_2$ system but with a boride content exceeding the eutectic concentration by 10% produces the structure illustrated in Fig.5.13. In this case the eutectic is the matrix. The distance between the "rods" made of primary crystals is, in this case, equal to 15 mμ. A thorough study of the directional crystallization of ceramic alloys has been carried out by STUBICAN [5.5]. The objects were produced at a cooling rate of 1 cm/h. Figure 5.14 shows the cross section of a ZrO$_2$-MgO eutectic alloy (see also [5.4]). Figure 5.15 shows the longitudinal section of a ZrO$_2$-SrZrO$_3$ eutectic specimen.

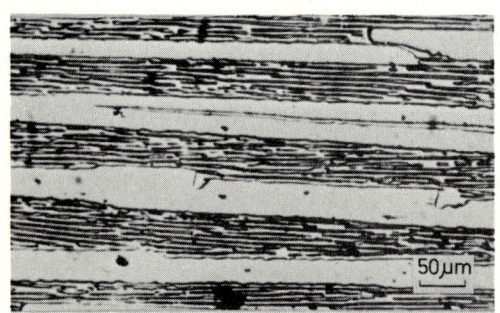

Fig.5.13. The same system as in Fig.5.12 but with boride content exceeding the eutectic concentration by 10%

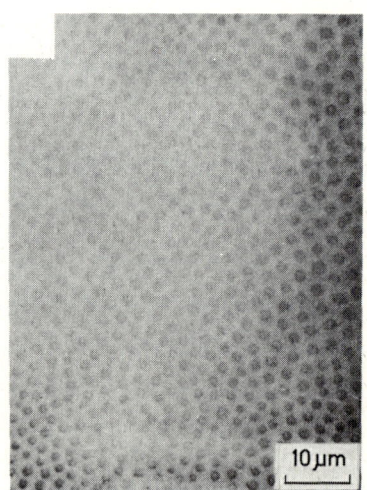

Fig.5.14. A cross section of a ZrO$_2$-MgO alloy specimen

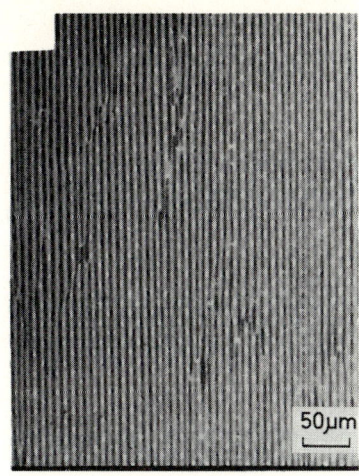

Fig.5.15. A longitudinal section of a ZrO_2-$SrZrO_3$ eutectic specimen

It is interesting to compare Figs.5.16 and 5.1. The photographs are of the same material, but Fig.5.16 was obtained by applying a small temperature gradient to the material. This proved to be sufficient for orienting the small diphenyl crystals.

Thus, under the conditions of unidirectional crystallization, oriented and regular plates or rods growing from the melt are obtained in the direction perpendicular to the crystallization front.

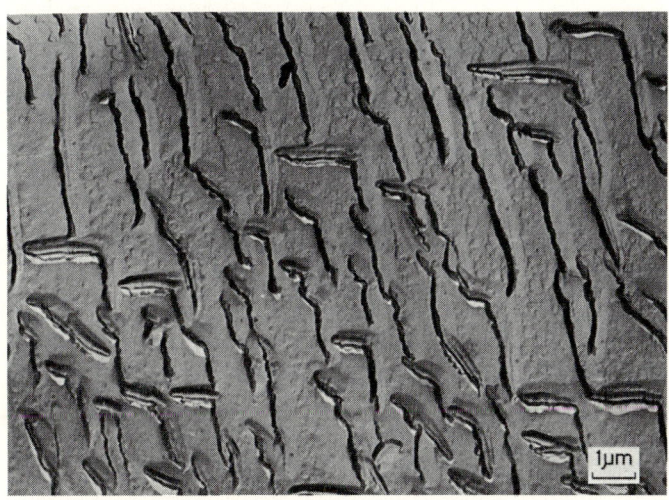

Fig.5.16. Microphotograph of diphenyl-dibenzil system obtained by presence of temperature gradient

A rough analysis of the conditions of component differentiation at the crystallization front gives a simple dependence of the distances between the particles λ on the rate of movement of the *crystallization front* c:

$$\lambda^2 c = \text{const} \tag{5.1}$$

The morphology of the eutectic structure depends on the *volume fraction* of each phase. If the *interphase surface energy* for the phases of the eutectic is isotropic and the volume fraction of the second phase is less than 0.28, then the structure of rounded rods is favored (their length can be thousands of times greater than their diameter). If the volume fraction is greater than 0.28, the lamellar structure is favored. When because of surface energy anisotropy there is only one habit plane with low surface energy, then lamellar structures will tend to form; if several planes of eutectic phases can conjugate and they have low surface energy, then the faceted rods will form.

CLINE and WALTER [5.6] have shown that coherent interfaces can form if during alloying there are only small changes in the interplanar spacings of the crystals of the conjugate phases. In the alloy 34 at .% Cr + 66 at .% NiAl the eutectic consists of two isomorphous phases: a solid solution based on Cr (bcc lattice, a = 2.88 Å) and an intermetallic compound NiAl (CsCl type, a = 2.89 Å), whose crystals have the same orientation in the eutectic. The addition of about 0.6 - 1.0 at .% Mo changes the lattice period of Cr, which becomes almost equal to that of the NiAl phase at the eutectic melting temperature. With a further increase in Mo content, the rod structure is replaced by a lamellar one; this seems to be due to a decrease coherence at the interface. At the same time, the crystallographic characteristics of the structure also change: for the rod structure the orientation of the rods and the growth direction are $[100]_{NiAl} \parallel [100]_{Cr}$, whereas for the lamellar structure the planes of the habit are $(11\bar{2})_{NiAl} \parallel (11\bar{2})_{Cr}$ and the growth directions are $[111]_{NiAl} \parallel [111]_{Cr}$.

The transition from rod growth to lamellar growth can also result from a change in the growth rate. The critical value of the volume fractions of the eutectic phases at which the morphology changes is obtained on the assumption that the distances between the particles, rods, or plates are equal. If the interplate distance is taken to be 1.5 times that of the inter-rod distance, the critical value of the volume fraction must be 0.10 and not 0.28.

Changes in growth rate can lead to especially significant structural changes in the presence of impurities or alloying additives. To prevent cellular or dendritic growth, when the interface is no longer flat, the alloys must be of high purity.

The regularity of the mutual orientation of two-phase crystals has attracted considerable attention. MINFORD et al. [5.7] proved by X-ray methods that for several *oxide eutectic* systems two criteria can be used to control interfaces. First of all, the phases are oriented so that the lattices can better "match" each other. Second, as is typical for oxides, the neutralization of ionic charges across the interfacial planes is essential. If the two-phase crystals belong to the cubic system, e.g., (MgO-MgAl$_2$O$_4$, MgO-ZrO$_2$, MgO-CaO), their axes must orient parallel to each other (the body diagonal of the cube is parallel to the heat-removal direction). The microstructures of the oxide eutectics can be predicted, from the composition and the surface energy of the crystals. For most systems investigated the distance between the plates obeys the same law (5.1).

5.2 Decomposition of Solid Solutions

The decomposition of solid solutions (as well as polymorphic transitions) is the growth of new crystals in the crystalline medium. In Fig.2.22, below the equilibrium line bka that separates the solid-solution α from the existence domain of a two-phase system $\alpha_1 + \alpha_2$, the *thermodynamic pressure*, i.e., the difference between the free energies of the stable and unstable states, promotes a change in the lattice. In various places of the crystalline medium nucleation of two new phases (α_1 and α_2) begins.

This restructuring can, in principle, occur at any point of the crystalline medium thanks to occasional fluctuations of the particles. In an ideal atomic crystal the exchange of atomic sites is possible which can result in the formation of a significantly large nucleus of the new phase. The inevitable inhomogeneity in particle distribution also promotes *lattice restructuring*. There are sites in the solid-solution crystal in which the particles are situated in such a way that only a limited number of exchanges of atomic sites are required to bring about a nucleation of a new phase.

The decisive role in the nucleation probably is played by the presence of *defect sites* in the crystal. Cracks, block and grain boundaries (for polycrystalline materials), and most importantly, the crystal surface are the regions where the least energy is needed for nuclei formation.

The *microstructures* formed in the process of decomposition of solid solution are very diverse. Practically all shapes are possible: perfectly faceted crystals, plates, needles, and spherical and shapeless formations.

If during lattice restructuring the *elastic stresses* between the crystals of a new phase and the matrix are of no importance, the growth of new phases is similar to the growth from liquid or vapor. The crystals of this new phase have an equilibrium shape which is determined by surface tension, and there are no orientational relationships between the matrix and the decomposed solid solution. There is no *coherent conjugation* of the matrix and the precipitates.

The other limiting case is a continuous transformation of the crystal planes of a new phase and the matrix into one another. Such a *coherent conjugation* requires minimum movements of the particles building the crystal lattice, but does lead to maximum internal stresses.

There are possible intermediate cases. Obviously, in most cases there is a partially coherent conjugation of phases. The diversity of the observed shapes of inclusions is explained by the rivalry of the energy of internal stresses and surface energy.

The relation between the unit-cell parameters of the matrix and precipitates, on the one hand, and the shape of these precipitates, on the other, has been well studied (we direct the reader to the monograph by KHACHATURYAN [5.8] in which several chapters are devoted to the morphology of heterophase crystals).

It is natural that coherent conjugation of lattices is seldom observed in low-symmetry molecular crystals, but may often occur for cubic crystals. Microscopy and X-ray investigations of the decomposition of a cubic solid solution into two cubic phases (differing from each other and from the matrix only in their concentration and specific volume) proved that the inclusion distribution has one dimension, two dimensions, and sometimes even three dimensions of periodicity. Such *microstructures* are often called *modulated*. Now we present an elementary theory of the phenomenon.

Suppose that a cubic crystal decomposes into cubic crystals of two types with slightly different unit-cell parameters a. All necessary conditions for the coherence of the two new phases with each other and with the matrix from which they have grown are satisfied. The decomposition transforms the single crystal into a system of alternating plates. Consider one of these plates of thickness d and area L^2. The elastic stresses are concentrated on the end faces of the platelike crystals, the elastic energy per plate being

$$\lambda g^2 \cdot e^2 \cdot d \cdot L^2 ,$$

where e is the difference in concentrations, λ is the modulus of elasticity, and g is the derivative da/dc of the unit-cell parameter a which characterizes the coherent transition of one lattice into the other (c being concentration).

Now let us analyze a fixed volume containing a certain number of plate inclusions. How many, and how thick must these plates be for the energy to be at a minimum? If we make the layers thin, the energy of each plate will decrease, but then their number must increase. With an increase in the number of inclusions, the surface energy on the boundaries of inclusions also increases and is $\alpha L^2 \cdot L/d$, where α is the coefficient of surface tension. Thus the total energy per plate is

$$\lambda g^2 \cdot e^2 \cdot d \cdot L^2 - \alpha L^3/d \quad .$$

This energy is minimized if the plate thickness is

$$d_m \sim k \sqrt{L} \quad ,$$

k being a material constant.

The same consideration is valid for the crystal plates of the second phase. But if the minimum is attained for equal thickness d_m, $d_{m'}$ of all the plates of the two new phases, the decomposition results in a periodic distribution of inclusions. The period is equal to the sum of thicknesses $(d_m + d_{m'})$ of the plates of both kinds.

This simple theory indicates that if the lattices of the solid solution and the phases into which it has decomposed are mutually coherent, then one can expect the formation of a periodic domain microstructure.

6. X-Ray Scattering

There are two main methods for investigating the structure of solids. They are *X-ray structure analysis* (study of X-ray diffraction and scattering) and microscopic investigation. The first method is rather general, i.e., it can study the atomic and molecular structure of matter and also study the microstructures of heterophase systems. Microscopy, at present, cannot do the former. Yet the advance in *electron microscopy* in the last few decades have made it the main method in the study of heterophase systems.

Neutron and electron diffractometry are valuable methods which supplement X-ray diffractometry. We shall not give these methods special consideration since their physical fundamentals do not differ from those of X-ray analysis. The only distinction consists in the different atomic scattering power and absorbance for X-rays, neutrons, and electrons. Electron diffraction is used with thin films, or in the study of surface properties. Neutron diffraction is used when hydrogen atom positions must be found precisely in the structure or when Fe must be distinguished from Co (their atoms scatter X-rays practically equally).

Thus, X-ray analysis and electron microscopy are the two main methods which must be used to study mixed systems or mixed crystals. For detailed discussions of experimental techniques and of methods for analyzing the experimental data, we direct the reader to monographs on X-ray analysis [6.1-7] and electron microscopy [6.8,9]. The results obtained by electron and light microscopy are comparatively easy to evaluate. This is not so with X-ray structure analysis. In order to read scientific papers and judge the reliability of the obtained results one must understand thoroughly the physical foundations of X-ray crystallography. Therefore, we present in this chapter, very briefly and without proof, the fundamentals of the mathematical theory of X-ray scattering which are relevant to the structure of mixed crystals.

6.1 X-Ray Scattering by a Single Crystal

On a modern diffractometer one places a crystal on a special holder (*goniometer head*) and can set the crystal at any angle to the (stationary) *incident beam*. Also, the apparatus can detect secondary (scattered) X-rays in any direction. The intensity of the scattered beam depends on the position of the crystal lattice axes relative to the incident beam. By varying the mutual position of the crystal and the detector one can measure the relative scattering at any crystal orientation. However, there are some preferred orientations at which the diffracted rays have an intensity several orders of magnitude higher than that of the rays scattered in an arbitrarily chosen direction. X-ray scattered radiation consists of two components: coherent and diffuse scattering.

In X-ray structure studies only the first component is of interest, i.e., we consider only the intense rays scattered coherently by the crystal at discrete mutual orientations of incident and scattered rays and the crystal. However, the diffuse background around each discrete reflection is of major importance for the study of solid solutions and defects. But the coherent scattering is also sensitive to these phenomena.

The term "*reflection*" is used here intentionally. In the early stages of the development of X-ray analysis, William and Lawrence Bragg (father and son) proved that the direction of a strong diffracted ray can be represented as the selective reflection of an incident beam from a system of *nodal planes* of the crystal lattice in accordance with Braggs' *law*.

$$2 \cdot d \cdot \sin\theta = n\lambda \quad , \tag{6.1}$$

where λ is the wavelength of both incident and reflected beams, n (an integer) is the order of reflection, d the spacing, i.e., the distance between two neighboring planes passing through lattice *nodes* (defined below), and $90° - \theta$ is the angle formed by the incident beam and the normal to the system of nodal planes (2θ is the scattering angle). By definition the nodal plane (often called a Miller plane) (hkℓ) in direct space has intercepts $\underline{a}/h$, $\underline{b}/k$, and $\underline{c}/\ell$ along the direct lattice axes $\underline{a}$, $\underline{b}$, and $\underline{c}$ respectively. The points hk1 = ($\underline{a}/h$, $\underline{b}/k$, $\underline{c}/\ell$) are called lattice nodes.

To describe elegantly the mutual directions of the crystal and the incident and reflected X-ray beams we introduce the concept of the *reciprocal lattice*. The parameters $\underline{a}*$, $\underline{b}*$, and $\underline{c}*$ of a reciprocal lattice (a mathematical space with linear dimensions measured in $Å^{-1}$) are defined by

$$\underline{a}^* = \underline{b} \times \underline{c}/V \quad , \quad \underline{b}^* = \underline{c} \times \underline{a}/V \quad , \quad \underline{c}^* = \underline{a} \times \underline{b}/V \quad . \tag{6.2}$$

Here $\underline{a}$, $\underline{b}$, $\underline{c}$ are translation vectors and V is the volume of a unit cell. Any vector of the reciprocal lattice

$$\underline{H} = h\underline{a}^* + k\underline{b}^* + \ell\underline{c}^* \tag{6.3}$$

is normal to the nodal plane (hkℓ) of the ("direct") crystal lattice (as can be readily shown), and hkl are integers which enumerate the nodes along the axes.

Equation (6.1) can be generalized using the new concepts just introduced. The condition of the appearance of strong coherent diffraction rays has the form

$$\underline{s} = 2\pi \underline{H} \quad . \tag{6.4}$$

Here the vector $\underline{s}$ denotes the difference

$$\underline{s} = \underline{k} - \underline{k}_0 \quad . \tag{6.5}$$

The vectors $\underline{k}$, $\underline{k}_0$ have the directions of the scattered and incident beams respectively. The absolute value of both $\underline{k}$ and $\underline{k}_0$ is $2\pi/\lambda$. The length of the vector $\underline{s}$ is

$$s = 4\pi \sin\Theta/\lambda \quad . \tag{6.6}$$

It is convenient to represent the scattering as the function of $\underline{s}$.

The scattering of X-rays by one atom can be considered, with fairly high accuracy, to be isotropic. The value of the atomic scattering amplitude f of all the elements as a function of $\sin\Theta/\lambda$ can be found in tables [6.10]. The scattering amplitude F due to a system of atoms at different lattice positions $\underline{r}_k$ has the form

$$F(\underline{s}) = \sum f_k \exp(i \underline{s} \times \underline{r}_k) \quad . \tag{6.7}$$

If we consider a crystal and are interested only in the directions of coherent rays, the summation in (6.7) should be limited only to the N atoms in the unit-cell, and the vector $\underline{s}$ should be replaced by the reciprocal lattice vector $\underline{H}$ (6.4). Then the scattering amplitude becomes

$$F_{\underline{H}} = \sum_{k=1}^{N} f_k \exp(2\pi i \underline{H} \times \underline{r}_k) \quad , \tag{6.8}$$

which is called the structure factor of the crystal.

The intensity of the coherent diffracted beam is proportional to the square of the amplitude. The proportionality factor is I_e the intensity of scattering by one electron

$$I_{\underline{H}} = I_e F_{\underline{H}}^2 \quad . \tag{6.9}$$

For an ideal crystal containing no impurity atoms or defects and whose atoms were static, the X-ray scattering would be only coherent and would occur only in directions given by (6.4). Then, for each $\underline{H}$, the intensity of the scattered beam would be represented by (6.9).

6.2 Scattering by One- and Two-Dimensional Systems

The scattering of X-rays by three-dimensional crystals is represented by a reciprocal lattice, to whose nodes we ascribe the intensity of the corresponding diffracted (reflected) beam. Peculiar effects arise (a) in the case of a *needle crystal* whose thickness is of the order of the unit-cell parameter in the direction perpendicular to the needle axis, and (b) in the case of a *lamellar crystal* whose thickness is of the order of the unit-cell parameter in the direction perpendicular to the plate. Such systems can be considered as *one-* and *two-dimensional periodic formations*. What is the reciprocal space for such ideal models?

Let us rewrite the conditions (6.5) of the appearance of a strong diffracted beam in an expanded form:

$$\underline{s} \cdot \underline{a} = 2\pi h \quad , \qquad \underline{s} \cdot \underline{b} = 2\pi k \quad , \qquad \underline{s} \cdot \underline{c} = 2\pi \ell \quad . \tag{6.10}$$

These three simultaneous equations in $\underline{s}$ (the Laue equations) define the point $\underline{H}$ in the reciprocal lattice. Now, let us assume that the scattering object is a particle which possesses long-range order in only one dimension, i.e., is a *linear chain*. The reciprocal lattice of such a particle with the parameter a along the chain is a system of parallel planes spaced by 1/a. Indeed, in this case (6.10) reduces to one diffraction condition $\underline{s} \cdot \underline{a} = 2\pi h$. The integers k and l are replaced by real numbers with any value. This means that the reciprocal lattice vector, that reaches points with nonzero scattered intensity, can have a continuous series of values, but its projections onto the axis a must be one of the discrete values a*, 2a*, 3a*, etc. The reciprocal space of a one-dimensional needle, or linear chain, is a system of *parallel planes*.

Similarily, for *two-dimensional order* only two of the diffraction conditions (5.10) remain, say

$$\underline{s} \cdot \underline{a} = 2\pi h \quad \text{and} \quad \underline{s} \cdot \underline{b} = 2\pi k \quad , \tag{6.11}$$

which indicates that the diffraction takes place only in the case where the vector $\underline{H} = \underline{s}/2\pi$ has the components $h\underline{a}^*$ and $k\underline{b}^*$ along $\underline{a}^*$ and $\underline{b}^*$ respectively. Thus, the intensity maxima, expressed as the function of coordinates of reciprocal space, lie on lines parallel to the $\underline{c}^*$ axis and passing through the points $h\underline{a}^*$ and $k\underline{b}^*$. The reciprocal space of the two-dimensional layer is a system of *parallel rods*. At first glance, these limiting cases seen unphysical. However, such effects are observed, for example, at the early stages of decomposition of aging alloys.

6.3 Thermal Vibrations

Now, let us take into consideration that atoms vibrate around their equilibrium positions. If at a certain moment the atom is displaced from its position by $\underline{\Delta}_k^L$ (L is the unit-cell number, k is the index for atom k in the unit cell), its scattering changes by a phase factor, and the atomic factor at this moment becomes

$$g_k^L = f_k \exp\left[i\underline{s} \cdot \underline{\Delta}_k^L\right] \quad . \tag{6.12}$$

Experiment yields the mean value of this factor

$$\langle g_k^L \rangle = f_k \left\langle \exp\left(i\underline{s} \cdot \underline{\Delta}_k^L\right) \right\rangle \quad . \tag{6.13}$$

Direct calculations show that

$$\left\langle \exp\left(i\underline{s} \cdot \underline{\Delta}_k^L\right) \right\rangle = \exp\left(-\frac{1}{2} u_H^2 s^2\right) \quad , \tag{6.14}$$

where u_H^2 is the mean-square value of the atom displacement from the "reflecting" plane whose normal coincides with $\underline{H}$, and s is defined by (6.6). Thus, the influence of thermal vibrations requires the replacement of the atomic scattering factor f in the formula for structure factor by combined atomic scattering and Debye-Waller temperature factors $B = u_H^2$ in the form (6.13). Expression (6.14) shows that diffraction beams will be attenuated at finite temperature and will be attenuated more as the angle between the incident and diffracted beams.

However, this is not the only effect of thermal vibrations. Even for a crystal, whose ideal structure is disturbed only by long-wavelength heat waves, a diffusion background is produced, which requires adding a second term I_D for *thermal diffuse scattering* in the expression for intensity (6.9). This second term is due to the fact that it is not sufficient to consider the movement of one individual atom. One should calculate the instantaneous amplitude and the average intensity of the scattering by the whole crystal. This instantaneous scattering amplitude in an arbitrary direction is

$$\sum_L F_L \exp(i\underline{s} \cdot \underline{R}_L) \quad , \tag{6.15}$$

where $\underline{R}_L$ is the radius vector of the node.

Let us introduce into consideration $\bar{F}$, the average scattering amplitude of the unit cell. Rewriting (6.15) in the form

$$\bar{F} \sum_L \exp(i\underline{s} \cdot \underline{R}_L) + \sum_L (F_L - \bar{F}) \exp(i\underline{s} \cdot \underline{R}_L) \tag{6.16}$$

and raising it to the second power, we obtain two terms in the formula for the intensity of scattered radiation

$$I_{TOT} = I_e \left| \sum_{k=1}^{N} f_k \exp(2\pi i \underline{H} \cdot \underline{r}_k) \left\langle \exp\left(-\frac{1}{2} u_H^2 s^2\right) \right\rangle \right| + I_D \quad .$$

The first term is the same I_H of (6.8,9) with the atomic factors multiplied by the Debye-Waller factor (6.14) and the second term describes thermal diffuse scattering background I_D, which exists at any angle of scattering and can be written in the form

$$I_D = I_e \sum_L \sum_{L'} \overline{(F_L - \bar{F})(F_{L'} - \bar{F})^*} \exp\left[i\underline{s} \cdot (\underline{R}_L - \underline{R}_{L'})\right] \quad . \tag{6.17}$$

This term I_D is much weaker than I_H and must be considered in studies of thermal vibrations. It should be remembered that (6.17) neglects compton scattering effects (where $\lambda_{scattered} > \lambda_{incident}$).

6.4 X-Ray Diffraction by Solid Solutions

6.4.1 Lattice Distortions

Valuable information about the structure of a solid solution can be obtained by studying the symmetry and the relative intensities of the X-ray diffraction of the solid solution. The dependence of reflection intensity on the impurity concentration at different temperatures makes it possible to find, according to (6.14), the mean square value of particle displacements from their equilibrium positions in the direction of the normal to the reflecting plane. These displacements are caused by *atomic thermal vibrations* and by *static distortions of the lattice*. These effects can be differentiated if measurements are made at different temperatures.

Now let us come back to (6.3). All the considerations of Sect.6.1 extend to solid solutions. But the value of the *mean atomic scattering factor* should be substituted into the expression for g_k^L. If only a fraction of the lattice positions k are occupied by atoms, this must be taken into account in the calculations of $\bar{f}_k$.

The mean structure factor has the form

$$\bar{F}_{hk\ell} = g \sum_k \exp[i2\pi(hx_k + ky_x + \ell z_k)] \tag{6.18}$$

Here the f_k values are replaced by a single, averaged g. Then the intensity expression will consist of two terms: one is proportional to $[\bar{F}_{hk\ell}]^2$ and determines sharp interferences, and the other gives the thermal diffuse scattering I_D (6.17). The structure of the solid solution influences both terms.

Next, let us consider the influence of the specific structure of the solid solution on the value of the structure factor. The quantity g can be written in the form

$$g = \bar{f}_k \, \overline{\exp\left[i(\underline{s} \, \underline{\Delta}_k^L)_t\right]} \cdot \overline{\exp\left[i(\underline{s} \, \underline{\Delta}_k^L)_d\right]} \quad , \tag{6.19}$$

where the first factor is the mean atomic scattering factor (averaged over k positions), the second factor is the Debye-Waller temperature factor, and the third factor accounts for the lattice distortions (the general displacement $\underline{\Delta}_k^L$ is divided into two corresponding parts).

The temperature factor can be written in the form (6.14), i.e., $\exp(-M)$. If the same form is given to the distortion factor we obtain

$$g = \bar{f}_k \exp(-M_t) \exp(-M_d) \quad . \tag{6.20}$$

For qualitative evaluations of the alloy structure (e.g. to check its type and the percentage of substituted or interstitial atoms) it may be sufficient to consider only $\bar{f}_k$. The value $\exp(-M_t) \cdot \exp(-M_d)$ was calculated by BEZNOSIKOVA and IVERONG'A [6.11] for Cu-Zn alloys. They showed that the factor $\exp(-M_t) \cdot \exp(-M_d)$ for all the investigated alloys (from 0% to 50% Zn) drops off faster with increasing in $\sin\Theta/\lambda$ than $\exp(-M_t)$ for pure Cu. This drop-off, naturally, becomes greater as the Zn content increases in the alloy. For β-brass with 46.4% Zn, $\exp(-M_t) \exp(-M_d)$ is almost half the factor $\exp(-M_t)$ of pure Cu at $\sin\Theta/\lambda = 0.5 \text{Å}^{-1}$.

In the decades that followed this pioneering investigation many studies were made of lattice distortions (average static displacements of atoms). Such studies were devoted mainly to metal systems. For most metal alloys the magnitude of *static distortions* is of the same order as the thermal distortions.

6.4.2 Structure of the Average Unit Cell

As has been mentioned repeatedly, all unit cells in a solid solution are different. Then in what sense can we speak of the space lattice of such a crystal? A *crystal of a solid solution* is the space lattice of equivalent "*average*" particles. The symmetry of such a crystal transfers "average" molecules from one position to another. An "average" molecule (atom) of a solid solution is the result of the superposition of the molecules of different species (in the general case the superposition may be of different orientations and conformations). The "*centers of gravity*" of average molecules form a regular lattice.

Such a lattice and its symmetry elements can be found by the same methods that are used in X-ray structure analysis of one-component, completely ordered crystals (Sect.6.6). Disorder in a solid solution consists in the composition of "average" molecules occupying all crystallographically independent positions in the lattice.

Assume there are three independent positions in a lattice, over which particles of four species are distributed (or, equivalently, particles of two species, each of which can be in two different states, for example, have different orientations or conformations). In this case the crystallographic disorder or the degree of long-range order for different species and states of the molecules can be described as follows. The first position is occupied by x% of molecules A in the first state, y% of molecules A in the second state, z% of molecules B in the first state, and w% of molecules B in the second state. Two other independent crystallographic positions can be described in a similar way.

The *occupancy factors* x,y,z, and w can be obtained either by constructing the electron density function and analyzing its *maxima* (which correspond to the "average" atoms) or else by the method of *least squares structure refinement*.

If weak solid solutions are under consideration or the scattering powers of the particles forming the solid solution differ insignificantly, the method of least squares is preferable. Very often this method, an inalienable part of any X-ray structure determination, obtains reliable data on the positions of the particles in a lattice and makes it possible to find directly the fraction of particles A or B in any position (occupancy factors), or to obtain the parameters of long-range order.

Let us illustrate this with a simple example of a two-atom alloy. Suppose that there are two crystallographic positions with multiplicities n and m and that the position of multiplicity n is occupied by a fraction r_A of atoms A and a fraction r_B of atoms B. If the atomic concentrations of A and B are denoted by c_A and c_B respectively, the structure factor has the form

$$(f_A r_A + f_B r_B) \sum_i^n \exp(2\pi i \underline{H} \cdot \underline{r}_i) + [f_A(c_A - r_A) + f_B(c_B - r_B)]$$

$$\cdot \sum_i^m \exp(2\pi i \underline{H} \cdot \underline{r}_i) \quad . \qquad (6.21)$$

Using this expression in the method of least squares, we determine the numerical values r_A and r_B, which give the best reliability factor for agreement between the calculated and observed values of structure amplitudes.

In the case of *complete disorder*, c_A and c_B are divided into equal parts ($c_A = 2r_A$, $c_B = 2r_B$). The structure factor then becomes

$$\bar{f} \sum_i^n \exp(2\pi i \underline{H} \cdot \underline{r}_i) + \sum_i^m \exp(2\pi i \underline{H} \cdot \underline{r}_i) \quad ,$$

and the mean atomic scattering factor is

$$\bar{f} = \frac{1}{2} c_A f_A + \frac{1}{2} c_B f_B \quad . \qquad (6.22)$$

The highest degree of order is achieved in the case where atoms of one species leave any one of the two positions, e.g., when r_A becomes equal to zero.

In the case of *complete order* the atoms "separate" over two positions. Therefore, the process of ordering is described in the following way: atoms leave positions that are not "theirs" and occupy "their own" positions. In the case of complete disorder only the fraction $n/(m+n)$ of atoms A occupy their own positions. If there is a partial order an additional part Δ_A of atoms A occupy their positions: $\Delta_A = [r_A - n/(m+n)]$. This results in the appearance of an additional structure amplitude

$$(f_A - f_B)\left(r_A - \frac{n}{m+n}\right) \sum_{i=1}^{n} \exp(2\pi i \underline{H} \cdot \underline{r}_i) \quad .$$

A similar addition appears due to the fact that now $\Delta_B = [r_B - n/(m+n)]$ of atoms B occupy their places instead of atom A.

The degree of long-range order is

$$\sigma = \frac{\Delta_A}{m}(m+n) = \frac{\Delta_B}{n}(m+n) \quad . \tag{6.23}$$

The structure amplitude can be written as

$$F = \left[\bar{f} + \frac{n}{m+n}\sigma(f_A - f_B)\right] \sum_{i=1}^{n} \exp(2\pi i \underline{H} \cdot \underline{r}_i) + \left[\bar{f} + \frac{n}{m+n}\sigma(f_B - f_A)\right]$$
$$\cdot \sum_{i=1}^{m} \exp(2\pi i \underline{H} \cdot \underline{r}_i) \quad . \tag{6.24}$$

It can readily be seen that at $\sigma = 0$, we obtain the structure factor for the case of full disorder, and at $\sigma = 1$ the expression for the structure factor decomposes into two sums with the factors f_A and f_B.

The transition of a solid solution from a disordered into an ordered state is accompanied by a change in the intensities of the diffraction lines. Let us note that in the case of complete disorder in the positions of atoms A and B, (6.21) will vanish for certain $\underline{H}$ values. Hence, a disordered system will give fewer diffraction lines than an ordered system.

The phenomena considered here were first observed in simple cases similar to those just discussed (two-component alloys usually have rather simple lattices); additional lines which appear in a diffraction pattern during ordering are called *superstructure lines*, and ordered solid solutions are called *superstructures*. (One should remember, though, that the general criterion of the change in the degree of order in solid solutions is a gradual change of relative intensity of the lines, and not their appearance or complete disappearance.)

Now, let us return to the case of an alloy with an arbitrary composition. Assume there is an excess of atoms in a unit cell, i.e., there are, on the average, $x_A > n$ A atoms per unit-cell.

If there is maximum order, we have

$$\Delta_A = 1 - x_A/(m + n) \quad .$$

The value of Δ_A/m which is equal to Δ_B/n is typical of the whole alloy and not of a component. This value (or another value proportional to it) should be accepted as the degree of long-range order. Using the previous definition $\sigma = \Delta_A/m\,(m+n)$, we obtain

$$\sigma_{max} = \frac{m - (x_A - n)}{m} < 1 \quad . \tag{6.25}$$

It is customary to say that an alloy with maximum order and concentration distinct from m/n possesses a *concentration disorder*. The value σ_{max} determined by an alloy composition can be accepted as a measure of concentration order.

6.5 X-Ray Diffuse Scattering by Solid Solutions

Measurements of diffuse scattering are very laborious and require great precision and skill. But the data which can be obtained are rewarding. Nevertheless, only a few binary systems have been studied by this method. From investigation of diffuse scattering of X-rays (or neutrons) by solid solutions one can establish the short-range order in the particle arrangement and determine thermodynamic quantities necessary for constructing a phase diagram. Examples will be supplied in Chaps.9 and 15. Now we proceed to the fundamentals of the method.

6.5.1 Short-Range Order

First of all, what is the character of diffuse scattering inherent in a fully disordered solid solution? Consider an alloy of N atoms consisting of two kinds of atoms: c_A of atoms A and c_B of atoms B. Let a part c_B of the $c_A N$ positions of atoms A be occupied by atoms B, and, vice versa, let a part c_A of the $c_B N$ positions of atoms B be occupied by atoms A. Thus, exchanging sites for

$2c_A c_B N$ atoms, one can obtain an ordered alloy from a disordered one. Or, vice versa, if to an ordered alloy we add a disordered system of "atoms" of different electronic densities, we shall obtain a disordered alloy. A system of such "atoms" (or particles) scatters like a gas. Therefore, X-ray patterns will also contain a diffuse background of monatomic gas scattering with the intensity

$$2c_A c_B N (f_A - f_B)^2 \quad . \tag{6.26}$$

If some quantity of atoms is incorporated into a solvent, the diffuse scattering is proportional to the number of interstitial atoms and the square of the atomic factor.

This rather approximate consideration assumes no correlations between the particles of the solid solution. Yet, (6.17) contains all the data on short-range order.

To calculate the intensity when the structure is known is very easy. The same is true of the diffuse background. If the short-range order is known, we can calculate the intensity of diffuse scattering. But we are interested in the converse problem, i.e., we wish to find a structure starting from the scattered intensities. To "obtain" the short-range order in explicit form from (6.17) is not easy. But the problem is soluble if we deal with particles of solid solution which are distributed over the points of the translation lattice.

For this purpose the scattering expression should be given in terms of functions $c_A(\underline{r})$ and $c_B(\underline{r})$ defined by (4.34). It can be shown that

$$I_D(\underline{s}) = |F_A - F_B|^2 \left\langle \sum_R [c_A(R) - c_A] \exp(-i\underline{k} \cdot \underline{R}) \right\rangle \quad . \tag{6.27}$$

This expression can be transformed into

$$I_D(\underline{s}) = N |F_A - F_B|^2 \tilde{\varepsilon}(\underline{k}) \quad , \tag{6.28}$$

where $\tilde{\varepsilon}(\underline{k})$ is the Fourier transform of the correlation coefficient defined by (3.15). Thus, for this case it is possible to express the intensity in terms of the short-range order parameters (3.14). Equation (6.28) then becomes

$$I_D(\underline{s}) = N |F_A - F_B|^2 c_A(1 - c_A) \sum_R \alpha(R) \exp(-i\underline{k} \cdot \underline{R}) \quad . \tag{6.29}$$

If the intensities of diffuse scattering are measured within one unit cell, we can use the Fourier transform of the intensity function to obtain the value of the short-range order parameter. We will not consider the details of this calculation in the general case. Later, we shall return to this problem and consider two examples: metal alloys Cu-Au and organic solid solutions of di- -phenylmercury in tolane (Chaps.8,12).

The derivation of (6.29) can be found in the monograph by KRIVOGLAZ [6.12], who not only derived (6.29), but also gave the expression for the intensity which accounts for both short-range order and lattice distortion. It seems that simultaneous determination of these two characteristics of solid solutions can be done only by a trial-and-error method.

The intensity of diffuse scattering can be measured close to any point of a reciprocal lattice. If there is a certain symmetry, the region of reciprocal space in which measurements must be made corresponds to a fraction of a unit cell. Especially important are measurements carried out close to the direct beam direction (zero scattering angle) at small angles. Compton scattering (which contributes significantly to the common background at larger diffraction angles) is practically absent. Measurements should be done in vacuum to avoid parasitic scattering of X-ray radiation.

It should be remembered that even after the Compton and the parasitic scattering are properly subtracted or eliminated, the measured intensity depends not only on the short-range order, but also on thermal motion and static distortions. Comparing the results with scattering by pure components, we can exclude thermal scattering on the assumption that the influence of thermal motion on the scattering of X-rays in crystals with impurities and without them is practically the same. (This assumption may not always be valid.) As to static distortions, the only thing which can be done is to neglect them.

6.5.2 Energy of Mixing

As was shown by KRIVOGLAZ [6.12, see also 6.13], the measurement of the diffuse scattering component responsible for short-range order permits calculating the energy of mixing. We are not going to repeat here the rather tedious derivations and give here only the final expression used in practical calculations:

$$I(\underline{k} + 2\pi\underline{H}) = \left| F_A - F_B \right|^2 c(1-c) \left[1 + c(1-c)\phi(\underline{k})/kT \right]^{-1} . \quad (6.30)$$

The scattered intensity is measured at the reciprocal space point at a distance $\underline{k}$ from the nearest point $2\pi\underline{H}$ of the reciprocal lattice. Here, F_A and

F_B are the amplitudes of scattering by particles A and B, and c is the molar fraction of component A. Experiment gives the value for the Fourier component of the energy of mixing

$$\phi(k) = \underline{w}(\underline{r}) \exp(-i\underline{k} \cdot \underline{r}) \quad . \tag{6.31}$$

Equations (3.14 and 6.29-31) assume a self-consistent field for the translation lattice (lattice distortions are not taken into account).

The energy of mixing can be expressed by

$$\underline{w}(r) = V_{AA} + V_{BB} - 2V_{AB} \quad ,$$

and for non-spherical particles by

$$\underline{w}(r) = V_{AA} + V_{BB} - V_{AB} - V_{BA} \quad .$$

To obtain the energy of mixing one should, naturally, invert (6.31). To obtain the other thermodynamic functions expressed in terms of concentration and temperature, we must know (Sect.4.4) the value of $\phi(k)$.

6.6 X-Ray Structure Analysis

Numerous monographs cited at the beginning of this chapter are devoted to this discipline. At present, X-ray structure analysis of crystals is practically completely automated. Manually driven or computer-controlled diffractometers acquire experimental data, i.e., measure the geometry of the diffraction pattern and the intensities of the scattered radiation. Computer programs can then determine the unit-cell dimensions and the crystal symmetry.

The determination of the atomic coordinates in a unit cell can be performed in two stages. First of all, the phases of structure factors must be determined. These phases are not available from experiment since the experiment provides only the intensity of the diffracted rays and not their amplitudes. Nevertheless, the phases can be inferred from experiment. The *phase problem* can be solved by analyzing the relationships between the intensities of the beams diffracted by planes whose reciprocal lattice vectors are algebraically related (the *direct method*). Another procedure used for the determination of phases of structure factors consists in studying structures of crystals containing a small number of *heavy atoms*. The intensity of X-ray scattering by an atom is proportional to the square of the number of its electrons, hence it can be accepted, in the first approximation, that the phases of the structure factors are dominated by the heavy atoms.

Using either procedure or their combination, the computer determines the first set of phases for the structure factors. Now a Fourier series for the electron density can be constructed

$$\varphi(r) = \sum_{hk\ell} F_{hk\ell} \exp\left[2\pi i(hx + ky + \ell z)\right] ,$$

whose maxima must correspond to the centers of the atoms. Then the first structure model can be obtained from this series. Structure factors are calculated by (6.8) and the intensities of the diffracted intensities $|F_{calc}|^2$ are compared with the experimental data $|F_{obs}|^2$. The overall reliability factor R, defined by the expression

$$R = \frac{\sum |F_{obs}| - |F_{calc}|}{\sum |F_{obs}|}$$

summed over all h,k,ℓ values, gives a criterion for the correctness of the model. If R = 0 then the observed and calculated structures are identical. If the value of R is very large (say, 50-60%), the structure is usually wrong, and evidently the whole determination should be restarted. If the R factor is less than 50% we can proceed to the refinement of the suggested structure, using the least squares method to adjust the calculated and observed values of F by small changes in the atom positions (x_k, y_k, z_k) and Debye-Waller temperature factors. A reliability factor of about 20% means that the atoms are placed in the model in correct positions, and the researcher decides, using all possible additional data, whether the obtained solution is in fact correct.

By using various forms of corrections in the expression for the structure factor and increasing the number of parameters through which F is expressed, the R-factor can be reduced to 4 - 5%. But after reaching the value ~20% the coordinates of the atomic centers usually stop changing, and reducing the R-factor to ~4% we pursue other ends.

For crystals with large unit cells the number of diffracted intensities can be several thousand. To solve such a structure problem sufficiently quickly, using such an enormous amount of data, one must have powerful computers.

The less ordered the structure is, the less information can be obtained from the coherent X-ray scattering. X-ray diffraction patterns, even from well-organized ("crystalline") polymers, contain only tens and not thousands of reflections. The structure determination then is impossible without the use of additional information. Of great help is the method of atom-atom potentials which will be considered in Chap.12.

X-ray studies of a system with a small degree of order can solve first of all problems other than the determination of atomic coordinates. One can determine the degree of order, the dimensions of crystallites and unit-cell parameters, and the dimensions and orientation of the elements of the supermolecular structure. These are considered in special monographs, for example, in [6.14,15].

7. Intermetallic Compounds

At least three quarters of the chemical elements, both natural and artificial, can be classified as *metals*. There are limiting cases (e.g., noble gases and halogens) where an element can be certainly considered as a *nonmetal*. But it is not clear whether polonium is a metal or a nonmetal, even though it is in the same column of the periodic table as the nonmetals oxygen, sulphur, and selenium. Boron, carbon, and nitrogen are elements from which are formed the millions of known (nonmetallic) organic compounds. Yet, carbides and nitrides which also contain carbon and nitrogen atoms (admittedly not in the same valence state as in organic compounds) are usually studied in metal physics.
The main criterion for dividing metals from nonmetals is the magnitude of the *electron exchange* between atoms, and hence the presence of *conduction electrons* in metallic solids. Even so tin is a *borderline case*: of the two dramatically different polymorphic modifications of tin only one (β-Sn) can be classified as a metal. Thus the elements on the borderline between metals and nonmetals can be called *metalloids*: boron, silicon, germanium, antimony, tellurium, and polonium.

The majority of metallic crystals are very simple. More than fifty metals possess at least one polymorphic modification consisting of close-packed spheres. The interaction between the particles building such crystals is surely not *directed*. For them the representation of atoms by spheres is fairly justified.

As Fig.7.1 shows, the expansion coefficients for most of these metals are the same. Close packings, such as face-centered cubic and hexagonal lattices with axial ratio $c/a = 1.63$, can exist at different temperatures. Thus, in order to compare sphere sizes, one should extrapolate all the structures to absolute zero: thus one can obtain meaningful values of *metallic radii*.

Though the number of metal elements having close packed structures is significant, it would be wrong to consider the *central* (i.e. spherically symmetric) *particle interaction* as a major characteristic of all metals. As was mentioned

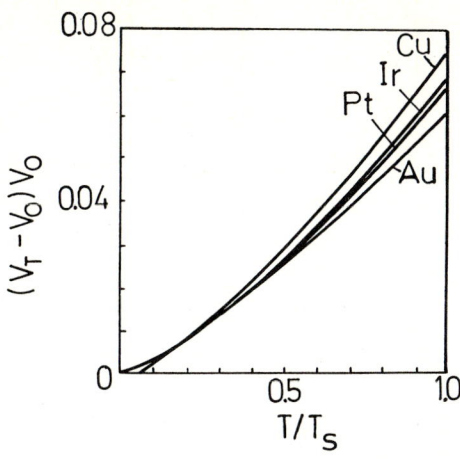

Fig.7.1. Temperature expansion coefficients of certain metals

in Sect.3.2.2, there is no doubt as to the directional character of the interatomic interaction, even for body-centered-cubic or hexagonal packing metals with the ratio c/a ≠ 1.63 (thirty of them!) On the other hand, close packing of spheres is found in the crystals of noble gases.

Finally, among the elements with metallic conductivity and other typical characteristics of metals one can find complicated low-symmetry crystal lattices where the unit cell contains several dozens of atoms (e.g. uranium, manganese). In such cases these low-symmetry metals seem not to be a single-component system, but rather an *alloy* consisting of several "types" of atoms of the same element. This assumption seems very natural as practically each atom in the unit cell can be in a different valence state. This idea is also reinforced by the fact that the atomic arrangements in these complicated structures of chemical elements coincide with the structures of many binary alloys.

The occurrence of electron exchange and the noncentral character of particle interaction significantly complicate the theory of metals and do not allow, at present, the structure of metals to be predicted from the electronic structure of atoms.

Nevertheless, valuable insights can be obtained by assuming that there is a pairwise interaction between metal atoms, both in single-component and in multi-component systems. Such an approximation works well only if this postulate is used in a general form. If one needs numerical results, i.e., data on the form of the interaction potentials between a pair of atoms, the so-called *pseudo-potentials* are used [7.1]. The curves of these potentials have alternating maxima and minima whose heights decrease with the distance from the center of action. In simple cases metal structures can be explained post factum, showing the possibility of choosing such potentials whose minima would occur at the radii of the coordination spheres.

Now it is obvious how difficult it is to explain rigorously the mutual solubility of metals, the formation of intermetallic compounds, and, worse, to obtain a phase diagram for a binary metal system. Attempts to make such predictions on the basis of two or three known parameters, e.g., atomic radii, are doomed to failure.

Now we shall consider metal systems, beginning with the structures of *intermetallic compounds*. This term will be applied only to fully ordered crystals. Unlike a chemical compound, an intermetallic compound loses its atomic "organization" in the crystal during fusion. In accordance with the definition of a component, the melt composition can be arbitrarily enriched with or depleted of any of the components.

We confine ourselves to the description of typical *binary alloys*. A unit cell of an intermetallic-compound crystal contains either one or only a few formula units $A_m B_n$. In most cases the integers m and n are small. Yet, there are some intermetallic compounds with unexpectedly high values of m and n. The aim of this chapter is to show how the formation of intermetallic compounds can be "understood", proceeding from the fundamental principles of crystal chemistry, i.e., from the tendency of the particles to form a close-packed structure and attain the highest possible symmetry.

As was said in the foreword, we do not discuss those explanations of structures and properties of metallic solids which are supplied by the electron band theory. The electron band theory of solids seems to be ineffective in predicting structures.

7.1 Classification Schemes for Intermetallic Compounds

The classifications of metallic compounds suggested by various authors differ considerably since they are based on different principles. One can proceed from phase diagrams. On the one hand, there are the compounds which have sharp melting point maxima on the melting curve at a simple stoichiometric ratio: according to the terminology suggested by KURNAKOV [7.2] such compounds are *daltonides*. On the other hand, phase diagrams very often have domains of existence of solid solutions which as a rule have structures different from those of the pure components. The domains of existence for intermediate solid solutions can reach the melting curve at a local maximum at an "irrational" mole fraction as shown in Fig.7.2 or they can be "hidden" among two-phase domains (Fig.7.3). However, if at a certain rational composition and a certain temperature an ordering occurs for such an intermediate solid solution, the obtained intermetallic compound is sometimes called an *intermediate phase*.

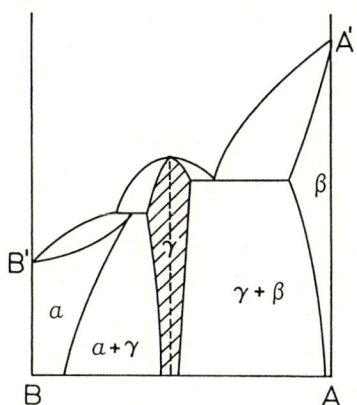 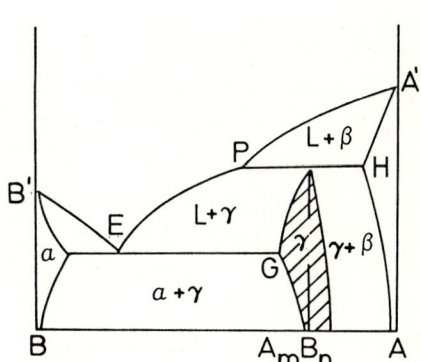

Fig.7.2. The case of melting intermediate phase

Fig.7.3. The case of not melting intermediate phase

More often, however, the term "intermediate phase" is related broadly to the whole region of intermediate solutions irrespective of the degree of structural ordering. The term will be used here presicely in this broad sense.

The classification of intermetallic compounds based on how many electrons are given by the metal to the electron gas has received widespread acceptance. More than fifty years ago HUME-ROTHERY [7.3] noted in some binary crystals the constancy of the average number of valence electrons per atom. Such crystals are numerous and therefore energetically favored. The most favorable valence electron ratios are 3/2, 21/13, and 7/4. Examples of such systems are, respectively, CuZn (Cu atom has one valence electron and Zn atom two valence electrons), Cu_9Ga_4 (Ga is trivalent), and Ag_5Al_3 (Ag is monovalent and Al is trivalent). It should be noted that these numbers can be obtained from the basic postulates of the quantum theory of the metallic state [7.4]. The idea of the theory is very simple: the system is stable if the first Brillouin zone can be filled with electrons. Proceeding from this idea, we can divide the intermetallic compounds into two groups: electron compounds (which correspond to the indicated electron concentrations) and all the rest.

A crystallographer would tend to classify intermetallic compounds by the arrangement of atomic centers. This approach would direct us to single out the binary alloys which are built on the basis of close-packed atoms, especially those with two-layered (hexagonal) and three-layered (cubic, face-centered) packing. If the sizes of the component atoms differ but slightly (up to 10%), the structure is determined by the arrangement of the alternating A and B atoms in the layers of packed spheres. If the difference in atomic sizes is large, structures will form which are called *interstitial phases* by metallographers. Here the packing is dominated by the atoms of one species, while the symmetry of the structure is determined by the distribution of the atoms of the second species over the octahedral and tetrahedral voids formed by the first species. However, such important compounds as carbides, hydrides, nitrides, and borides, which form interstitial phases, are considered by many investigators separately from intermetallic compounds.

If we agree upon what can be called a *"small" distortion*, we can single out structures which deviate only slightly from close packing.

The compounds built on the basis of the body-centered-cubic lattice are often considered as a separate group, the remaining compounds forming a group which can be called "other structures". A comprehensive geometric classification of these compounds is given by PEARSON [7.5]. It seems obvious that chemical principles are absolutely unsuitable for classification, because such compounds as Ce_2Ni_7, KNa_2, and $ReBe_{22}$ cannot be explained by valence theory.

No matter which classification is accepted, we must remember that it is empirical, and can help to predict the structure with a certain degree of probability only if there is a close analogy between the compound under consideration and other well-known compounds.

All the classification schemes discussed above are unsatisfactory. The compounds with "capricious" component ratios often do not obey the Hume-Rothery rule. Nor can we be sure that the atom gives up its valence electrons with equal readiness in all cases. Attempts to represent the crystals by spheres often fail: the spheres either do not touch each other at all or must overlap considerably [7.6].

7.2 Geometrical Analysis

The most natural classification of the binary alloys divides them into groups by the composition ratios. Of course, compounds differ widely in composition, but those with compositions AB, AB_2, AB_3, AB_5, and a few other simple compounds constitute about two thirds of the known compounds. Moreover, the compounds of each of these formulas form only a few structural types. However, it is true that the number of special cases is rather large.

A geometric analysis of intermetallic compounds shows that the main reason why a compound is energetically more favored than the two-phase mixture of the same composition is the possibility of forming a denser packing. Therefore, the simplest and perhaps most valuable characteristic of the intermetallic compound $A_m B_n$ is the *increase in density parameter* Δ given by [see (3.17)]:

$$\Delta = \frac{mV_A + nV_B}{V_{A_m B_n}} - 1 \quad .$$

Among the intermetallic compounds, there are a lot of compounds with rather complicated formulas. It seems, at first glance, that in these situations the crystal structure cannot be explained from a purely geometrical standpoint. We shall see, however, that here too the main principle of crystal chemistry works sufficiently well. We shall demonstrate by several examples that the formation of atomic clusters of approximately spherical shape is favored. These clusters can contain 12, 13, or 26 atoms. The crystal then is a close packing of such "superatoms" or "pseudoatoms". The formation of clusters and the building of atoms from such "superatoms" can be given a simple geometrical interpretation.

Now let us consider the most abundant structural types of intermetallic compounds. The data on unit-cell dimensions of metals and intermetallic compounds are from [7.7] and the lists of the compounds from [7.8].

7.3 AB Compounds

There are about a thousand compounds with the composition AB which are neither oxides nor halides. The majority of them belong to one of the six structure types: AuCu, CsCl, NaTl, ZnS, NiAs, and NaCl.

7.3.1 AuCu Type

The space group is P4/mmm. There are two formula units per tetragonal unit-cell. The unique atoms are positioned at the vertex and at the center of the base of a tetragonal prism; they are not symmetry-related. Each A (gold) atom has four neighbors of the same species (A) and eight B neighbors; each atom B (copper) is surrounded by 4 B and 8 A neighbors. The radii of the two coordination spheres, characterizing the surrounding by "like" and "unlike" atoms, are very close in length.

This circumstance explains why these substances are often superstructures. If the order in the arrangement of A and B atoms is disturbed, the lattice becomes face-centered cubic. Ordered structures of the AuCu type can be considered, in the first approximation, as close-packed structures with coordination number 12 (eight atoms of one kind and four atoms of the other). Whether such an approximation is justified depends on the differences in the AA, AB, and BB distances, or equivalently on the deviation of the c/a ratio from unity, or on the deviation of the AuCu-type unit cell from cubic symmetry. Usually, these deviations do not exceed 2-3%. In BiLi c/a reaches the maximum deviation 0.9.

Now let us consider the parameter Δ which has been calculated for thirty AuCu-type compounds. The parameter Δ is negative only for two Pd-compounds (with Fe and Cr). In all the other instances the structure of the intermetallic compounds is denser by 5-10% than the structures of the components. Of interest is the sharp increase in density (by 30%) of BiNa and CdPd. It is not accidental. The Bi structure is very porous and there are strong valence bonds. In CdPd, the Cd atoms, unlike the other atoms, are not spherical.

7.3.2 CsCl Type

This is the simplest of all the structures with the composition AB. (The model compound is a halide). It has a primitive cubic unit cell. The vertices of the unit cell are occupied by the atoms of the species (A), the centers of the cubes by the atoms of the other species (B). It is immaterial which atom is chosen as an origin. The space group is Pm3m. There is one formula unit per unit cell.

The nearest-neighbor arrangement of atoms can be described by two coordination spheres. The A atoms are surrounded by 8 B atoms, the AB distance being the shortest. Both A and B atoms have six "like" neighbors at a distance $r_{AA} = r_{BB} = 1.15\, r_{AB}$.

About 300 substances belong to this structural type. Nearly all the elements of the periodic table (or at least one element from every column) participate in forming CsCl-type structures. This structural type occurs for various ratios of the radii of A and B atoms. No matter what system of atomic radii is used, there are always extreme instances where the sum of the radii is either greater or less than half the body diagonal of the cube.

Therefore the packing coefficients calculated on the basis of a spherical model can hardly reflect the situation correctly. If one decides to use a geometrical model, the model of cut spheres should be used.

The increase in density Δ calculated for 240 compounds turned out to be negative only in 5 or 6 instances. The values of these negative coefficients are so small that they can be ascribed to experimental errors. In the majority of cases the positive coefficients are within the range 5-10%. Four Ga and four Hg compounds have very high values of the coefficient Δ, up to 30%.

7.3.3 NaTl Type

These are cubic crystals with 8 formula units AB per unit cell. The space group is Fd3m. Each atom has 8 neighbors, 4 of them being of the same species. The distances AA, AB, and BB are equal in accordance with the symmetry of the atomic arrangement which is rather simple: each atom is at the center of a cube (one eighth of the unit-cell); the vertices are occupied by four atoms of one species and four atoms of the other species, and form two tetrahedra, respectively.

The number of compounds belonging to this structure type is relatively small. The values Δ are 15-20%. No negative Δ values have been reported.

7.3.4 Zinc Blende (ZnS) Type

There are four formula units AB per cubic unit cell. The space group is F43m. Atoms of both species have the same surroundings. Atom A has 4 neighboring B atoms; atom B is surrounded by 4 A atoms which are equidistant from other A atoms. Both A and B atoms have 12 neighboring atoms of the same species, but the distances AA and BB exceed the distance AB by a factor 1.63.

Hence, such a structure can be similar to close-packed spheres whose tetrahedral voids are occupied by the atoms of the other species if the spheres are of appropriate sizes. The geometry of the structure is determined by the distance AB, which is one fourth of the body diagonal of the cubic cell.

There are dozens of ZnS-type structures. It seems that all of them are formed by the metals of group II, III, IV, and V (Be, Al, Si, As, etc.) of the periodic table and by the metalloids of the same groups.

7.3.5 NiAs Type

Two formula units AB occupy a hexagonal unit cell of the space group $P6_3/nmc$. Though the atoms A and B are in two-fold positions without any degrees of freedom, the distances between the atoms of the same and of different species observed for the various representatives of this family can differ, since the structure depends on two unit-cell parameters. Nevertheless, no matter how the c/a ratio changes, the coordination numbers describing the surroundings of atoms A by atoms B are constant and equal to 6. Atoms A are along the c axis at a z = 1/2. If the parameter a of the unit cell is greater than c/2, then atom A has only two nearest neighbors of the same species. If a is close to c/2, then the coordination number AA equals 8; and if a is much smaller than c/2, the coordination number AA equals 6 and the nearest neighbors are situated in the plane perpendicular to the six-fold axis. The same is true for the surrounding of atom B by other B atoms, i.e., the coordination number depends on the c/a value. The model of cut spheres takes account of these differences. The cuts should be made in a different way depending on the c/a value.

Of the twenty compounds for which Δ has been calculated, only AuSn has a negative Δ, -0.02. In all the other instances Δ is positive. The densest packing was observed for NiSb and CoSb.

7.3.6 NaCl Type

The simplicity of the NaCl structure can be compared only with that of CsCl. The face-centered-cubic lattice of NaCl (halite) is built of atoms A and B, forming their own sublattices shifted relative to each other by half a period along the lattice axis. The distance AB unambiguously determines the structure geometry. Each A atom is surrounded by 6 B atoms (and vice versa) and by 12 A atoms, which are at a distance AA 1.41 times larger than the distances AB.

If the ratio of the radii of the spheres representing the atoms has an appropriate value, the NaCl structure type (space group Fm3m, four formula units per unit-cell) can be represented as close-packed spheres whose octahedral voids are filled with the atoms of smaller size. Such a model works quite well for carbides, nitrides, and hydrides, whose numerous representatives belong to the NaCl structure type. In most cases, the most appropriate structure model is that of the cut spheres.

The values of Δ have been obtained for 46 metal compounds. Two compounds of As (with In and Pt) have rather high negative Δ values, namely, -0.14 and -0.17. The fraction of the structures with negative Δ (10%) is somewhat more than for other structure types.

7.4 Compounds AB_2

There are many intermetallic compounds with composition AB_2. As with the AB compounds only a very small number of structure types need to be considered. LAVES [7.9] was the first to emphasize the special significance of the structure types $MgCu_2$ and $MgZn_2$. These two structure types embrace about one fifth of all the known metal compounds.

As against AB compounds, most AB_2 compounds cannot be explained by close-packed spheres models. Neither is there any obvious relationship between the valence states of the elements and the formation of AB_2 compounds.

The parameter Δ is more than unity for nearly all the compounds of the four major AB_2 types ($MgCu_2$, AlB_2, $CuAl_2$, and $MgZn_2$), which will be considered below.

The structure types described in Sect.7.4 are characteristic for compounds where both components are metals. These types are not possible with elements from group VA, VIA, and VIIA of the periodic table (nitrogen, oxygen, or halogens).

7.4.1 $MgCu_2$ Type

AB_2 compounds of the $MgCu_2$ type crystallize with a cubic unit cell that belongs to the space group Fd3m, with 8 A atoms and 16 B atoms in special crystallographic positions without any degrees of freedom. This unit cell consists of two identical face-centered subcells of comparatively complicated structure, shifted relative to each other by about a quarter of the body diagonal. Each A atom is surrounded by 12 B atoms, and each B atom by six A atoms. The distances AB and BA are equal by symmetry. One interatomic distance fully determines the structural geometry of all the members of this family. The minimum AA distance is 1.04 times the minimum AB distance; the minimum BB distance is 0.85 times the AB distance.

The parameter Δ has been calculated for more than 150 structures. This structure type is densely packed: negative values of Δ are few, and they are most probably the results of experimental errors.

7.4.2 AlB_2 Type

The aluminium boride (AlB_2) structure has one formula unit per hexagonal unit cell and belongs to the space group P6/mmm. The A atoms are positioned at the vertices of the unit cell; the B atoms (at a two-fold special position without any degrees of freedom) form regular hexagons in planes located midway between the planes of the A atoms. Though the structure is very simple, the above information does not determine (as for the $MgCu_2$ structure) the local geometry since the ratio between the distances AA, BB, and AB, and the coordination depend on the axial ratio c/a.

This structure type is known for ten borides, whose c/a ratios are greater than unity and are usually between 1.05 and 1.10. For Hg compounds c/a is equal to 0.6 - 0.7.

The increase in density is the same as for other structure types: the fraction of negative Δ-values is insignificant. The density increases by 23% when the compound $SrGa_2$ is formed.

7.4.3 $CuAl_2$ Type

The space group is I4/mcm. The body-centered tetragonal unit cell contains four AB_2 formula units. Atoms A occupy a four-fold position without any degrees of freedom, that is, the atoms are situated at vertices and centers of two unit cells shifted by half a period along the axis c. Atoms B are at eight-fold positions with one degree of freedom and lie in the planes that are parallel to the unit-cell base and pass midway between the A atoms. Four B atoms form a square. The neighboring squares are rotated relative to one another. Different values of the axial ratio c/a and of the free parameter determining the position of atoms B provide even greater freedom than in the AlB_2 structures.

Among all the numerous compounds of this type for which the parameter Δ has been calculated only one compound, $NiHf_2$, has a negative increase in density, $\Delta = -0.11$. It would be of interest to remeasure the unit-cell parameters for this compound. Significant increases in density have been observed for $TiSb_2$, VSb_2, and $FeGe_2$.

7.4.4 $MgZn_2$ Type

This type has a hexagonal unit cell and belongs to the space group $P6_3/mmc$. There are 4 atoms A at sites with one degree of freedom. Atoms B are of two kinds: two B atoms occupy a two-fold position without any degrees of freedom [it is expedient to take one of them as an origin; then the second will have the coordinates (0, 0, 1/2)], and six B atoms occupy six-fold positions. These six atoms are distributed over two planes perpendicular to the c axis and spaced by 1/4 and 3/4 of a period from the zero plane.

This type of structure permits an even more diverse arrangement of atoms within the family, since the interatomic distances depend on the ratio c/a and on the value of two free parameters.

The increase in density has been calculated for about a hundred representatives of this group with no negative Δ values.

7.5 Compounds AB_3

Compounds that have a simple stoichiometric formula and a small number of atoms per unit cell can easily reach small free energy values. However, it must be admitted that this statement, which is made only because the known structures have small free energies, cannot be proven in general (i.e., for metals, organic or inorganic compounds). The simplicity of any structure has no relation to the principle of close-packing, since more complicated systems can possess higher density. It is more probable that simple structures are advantageous in terms of the lower relative thermal motion of their particles. But even so, it is difficult to say whether the simplicity is due to smaller vibrational energies, or to a higher vibrational entropy. In all the instances of large unit cells and many atoms, or of mixtures with complicated stoichiometric formulas, special "explanations" are needed for the existence of such "unusual" structures.

There is a great diversity of formulas describing binary metal alloys. The boundary between simple and complicated structures is arbitrary. To the list of simple formulas we add only AB_3. We confine ourselves to the analysis of Δ for three cubic lattices ($SiCr_3$, BiF_3, and $AuCu_3$ structure types) in which the atomic arrangement is uniquely determined by the edge length.

7.5.1 $SiCr_3$ Type

This type has a cubic lattice belonging to the space group Pm3n, with two AB_3 formula units per unit cell. The A atoms occupy the vertices and center of the cube. Two B atoms are positioned on the cube faces at a distance of a/4 from one edge and of a/2 from the other edge. There are no degrees of freedom, and therefore, the geometry is determined by only one parameter, which as usual is chosen to be the distance AB. Atom A is surrounded by 8 A neighbors, the distance AA being 1.55 times larger than AB. Atom B has two nearest A neighbors at a distance 0.89 and eight next-nearest A neighbors at a distance 1.09 times AB.

The results of Δ calculations are similar to those obtained for the AB and AB$_2$ compounds. Only five compounds of this type reveal a decrease in density by 1 - 2%, which could be due to experimental errors. All the positive Δ values are smaller than 0.02. Maximum values are observed for the compounds of As, Sb, and Hg.

7.5.2 Type BiF$_3$

This lattice belongs to the space group Fm3m. There are 4 AB$_3$ formula units per unit cell. The structure consists of four face-centered sublattices shifted by 1/4, 2/4, and 3/4 along the body diagonal. One of the cells (say, with the origin at the zero node) is occupied by A atoms, and three other cells by B atoms. There are no free parameters. The distance AB unambiguously determines the geometry of the structure. Atom A is tetrahedrally surrounded by four B atoms. The second coordination sphere of atom A is filled with 12 A atoms. Though the structure is very simple there are two kinds of B atoms. Those B atoms which are located in the lattices shifted by a quarter of the (one A and 3 B) diagonal relative to the A lattice have four nearest neighbors at the vertices of a tetrahedron. The second kind of B atom, shifted by 1/2 of the diagonal relative to the A lattice, has a tetrahedral surrounding built by four atoms of the same kind.

7.5.3 AuCu$_3$ Type

This is another simple cubic unit-cell; the atomic coordinates have no free parameters. The space group is Pm3m, and there is only one AB$_3$ formula unit per unit cell. Atom A is at the cube vertex, whereas B atoms occupy the centers of the faces. If the atoms could be represented by spheres of the same radii, they would form close packed structures. Atom A is surrounded by 12 B atoms; the B atom is surrounded by four A atoms. Atom B has 8 B neighbors at a distance of 1.0 times the minimum AB distance, and atom A has 6 A neighbors at a distance of 1.41 times AB.

If we randomly mix atoms A and B, a solid solution forms with a face-centered lattice of higher symmetry (space group Fm3m). Among the compounds of this type are superstructures that form from such solid solutions at lower temperatures.

The intermetallic compounds of this structure type are common. The parameter Δ has been calculated for about 150 compounds. No negative values have been obtained. A considerable increase in density takes place for some compounds of Ge, Ga, and Bi.

7.6 Number 13

One must mention that the *number 13* is very typical for the formulas of many intermetallic compounds. First of all, there is a large family of compounds with the formula AB_{13}. (There are AB_n compounds where n is any number up to at least 20, but these are less common than AB_{13} compounds.)

Secondly, there are many A_mB_n intermetallic compounds, where m + n equals 13, 26, 39, or 52. The number 13 can be obtained by various ways: 1 plus 12, 2 plus 11, 3 plus 10, 4 plus 9, 5 plus 8, or 6 plus 7. The composition A_5B_8 characterizes, for example, γ-brass: Cu_5Zn_8. Many representatives of the family have structures similar to that of brass and are often called the *phases of γ-brass*.

The detailed exposition of the crystal chemistry of all intermetallic compounds is beyond our aim. The distinctive features of intermetallic compounds with the complicated stoichiometric ratios of the components can be shown in examples of compounds with the number 13.

7.6.1 Compounds AB_{13}

A number of compounds with this formula crystallize in the same way as $NaZn_{13}$. Let us consider this structure in more detail and discuss the increase in density.

The unit cell is cubic with 8 AB_{13} formula units, i.e., it contains 112 atoms. The space group is Fm3c. The B atoms occupy two positions, I and II (one of them is an eight-fold position with no degree of freedom). This means that there are two face-centered sublattices shifted by half the body diagonal. Atoms B(I) are positioned at the nodes, centers of the edges, centers of the faces, and in the center of the unit cell. Thus the unit cell is divided into eight small cubic subcells. Atoms of Na also occupy an eight-fold position with no degrees of freedom and are located in the centers of the small cubes. If there were no atoms B(II) (which would be in a 96-fold position), the structure would consist of 8 unit cells of the type CsCl, thus forming one large cube.

However, we do have the CsCl lattice, but only in a special sense. Let us consider the surroundings of an atom B which occupies the vertex of a small cube. Its neighbors are twelve B(II) atoms with coordinates: $0yz$, $z0y$, $yz0$, $0y\bar{z}$, $0\bar{y}z$, $0\bar{y}\bar{z}$, $\bar{z}0y$, $\bar{z}0\bar{y}$, $z0\bar{y}$, $\bar{y}z0$, $y\bar{z}0$, and $\bar{y}\bar{z}0$. As the lattice is cubic, the distances from the origin to all these atoms are exactly the same, by symmetry.

The radius of the coordination sphere on which lie the centers of these nearest neighbors is $R = a(y^2 + z^2)^{1/2}$.

Thus, there is a rather compact group of 13 B atoms around each of the vertices of a small cube. The structure can be considered as that built by large spherical particles B_{13}. Such "*superatoms*" and the A atoms form a CsCl-type structure with the cube edge equal to a/2.

The value of the parameter y and z and the edge a are obtained for four intermetallic compounds in which B is a Zn atom. Lattice parameters have been measured only for the compounds with Ba, Ca, K, Na, and Sr. They are, respectively: 12.33, 12.13, 12.36, 12.29. and 12.21 Å. Of interest is not only that these values are practically equal, but also the equality of parameters y and z. They are 0.18 and 0.12 for all the structures. The Zn_{13} superatoms are built in the same way (they are icosahedra) and completely determine the structure.

By substituting the value of the cube edge into the expression for R, one obtains the radius of the coordination sphere 2.7 Å. Of course, the superatom should be delimited by an undulating surface and not by a sphere. The maximum radius of the "supersphere" is obtained by adding the radius of the Zn-atom (approximately 1.4 Å) to the radius of the coordination sphere. The result is 4.1 Å. The average radius of the superatom is 3.4 Å.

If the superatom were a regular sphere of such a radius the atom A would occupy a distance of only 3.7 Å on the cube body diagonal (whose length equals 21 Å). Not only this is sufficient for an A atom, it is even too big for it. The superatom is oriented in such a way that the A atoms fit into the "hollows" on the superatom's surface. The distance from A atoms to the nearest Zn atoms can be readily calculated to be about 3.5 Å. The features of this structure type are determined by the interaction of Zn atoms. Superatoms whose centers lie at a distance of a/2 from one another (i.e., 5.1 - 5.2 Å) are packed rather densely. The repulsion prevents atoms from approaching and decreasing the volume per A atoms. The atomic volume of Zn is 15.2 Å^3. The 13 Zn atoms in the Zn crystal have a total volume of 197 Å^3. If these 13 atoms are gathered into a spherical unit in such a way that the density remains the same, the radius of such a sphere becomes 3.6 Å, but the superatom in the structure AB_{13} has a smaller radius. Hence, the formation of such a superatom is favorable in terms of packing density.

Is there any gain in density? The A atoms fit loosely into the voids formed by the superatoms. Let us calculate the increase in density. The volumes for the unit cells of four compounds in which B is Zn are practically the same, about

1860 $Å^3$. One formula unit has the volume of 232 $Å^3$. If A = Na, the parameter $\Delta = (197 + 39)/232 - 1 = 0.02$, and if A = Ba, $\Delta = (197 + 69)/232 - 1 = 0.12$. Thus, there is an essential gain in density.

The unit cells of several compounds of this structure type with B = Be have been measured. The lattice parameters for the compounds where A is Ca, Cr, Mg, Pt, Th, or U vary within the range 10.17 - 10.40 Å.

Let us now consider the radius of Be superatoms. The radius of the coordination sphere of Be(II) atoms surrounding a Be(I) atom is 2.2 Å. The atomic radius of Be can be taken to equal 1.1 Å. Then the average radius of Be_{13} superatoms is 2.8 Å. The atomic volume of Be is 8.11 $Å^3$. Thirteen atoms, packed with the same density as close-packed spheres, would require a volume of 105 $Å^3$, i.e., the sphere would have a radius equal to 2.92 Å. Here, as in the previous instance, the superatom consisting of 13 particles located at icosahedron vertices provides a denser arrangement of particles.

Despite the fact that the large close-packed particles form large pores into which A atoms fit loosely, there is no loss in density. Calculation shows that no loss in density occurs even for the smallest A atoms, and for large atoms Δ is about 5 - 10%.

Also known are the unit-cell dimensions of two crystals isomorphous to those under consideration, having the compositions $RbCd_{13}$ and KCd_{13}. They have a cube edge of 13.8 Å and a coordination-sphere radius of 2.95 Å. The atomic radius of Cd is approximately 1.6 Å, the radius of the Cd superatom is equal to 3.75 Å, and the atomic volume of Cd is 21.6 $Å^3$. Thirteen Cd atoms can be placed within a sphere of radius 4 Å with the density of close-packed spheres. Again, the picture is the same: Thirteen Cd atoms are packed into a superatom with a density exceeding that of a Cd atom by nearly 10%.

Compounds of Cd possess the largest pores. If they were occupied by atoms with the size of Na, Δ would be negative. Therefore such compounds do not form. For the large atoms Rb and K, Δ is $\{[281 + 93 \text{ (or } 75)]/330\} - 1 = 0.13$ or 0.08. Thus we have proved once more the importance of the close-packing principle. Atoms of Zn, Be, and Cd also form superatoms consisting of 13 particles. (It should be noted that the atoms of these elements cannot be approximated by spheres as they crystallize in a hexagonal system with a c/a ratio different from the ideal value of 1.63. If the geometric model is insisted on, these atoms should be depicted by cut spheres or ellipsoids.) Compounds of these metals have a density which exceeds the density of the corresponding pure metals. Neighboring B_{13} superatoms, as well as Zn_{13}, do not share atoms. These superatoms form close-packed structures in the crystals of AB_{13} and are bordered by a wavy surface whose maximum radius is a sum of the coordination sphere and metal radii, and whose minimum radius equals that of the coordination sphere. The large pores of this packing are filled with atoms which have a size such that the system built up by A and B atoms would not have a negative Δ.

7.6.2 Compounds $A_m B_n$ with $m = 1$ and $n = 12$

The compounds UB_{12} and ZrB_{12} have a face-centered unit cell of the space group Fm3m. Four U-atoms occupy the vertices and the centers of the faces. The B atoms occupy a 48-fold position.

Let us determine whether this structure can be represented as the close packing of superatoms. The radius R of such a spherical superatom is given by $(4/3)\pi R^3 = (1/4)0.74 a^3$, where the lattice parameter $a = 7.477$ and R is therefore 2.65 Å. A U atom which occupies the vertices of the unit cell is surrounded by 24 (not 12) B atoms with the coordinates 0, x, 1/2-x. The distance from this U atom to its 24 nearest neighbors is 2.79 Å. Hence, no UB_{12} clusters can be singled out though formally the structure can be considered as close-packed spheres of radius ~2.7 Å. In contrast to the previous example, the 24 B atoms on the surface of the sphere belong simultaneously to several contacting superatoms. In the UB_{12} structure the arrangement of the atoms is defined by the valence bonds between B atoms, which form a three-dimensional framework whose free sites are occupied by U atoms.

The parameter Δ is barely positive. Let us consider the tetragonal crystal $ThMn_{12}$ by again representing the structure as the packing of approximately spherical superatoms. One formula unit occupies a volume of 189 Å^3. If the crystal is made up of close-packed spheres, the radius of such a sphere would be equal to 3.22 Å. The unit cell of $ThMn_{12}$ is body-centered. Atoms of Th occupy the vertices and the center of the unit cell and are surrounded by 12 nearest-neighbor Mn atoms at a distance of 3.15 Å. The second coordination sphere is at a distance of only 0.2 Å from the first. The increase in density Δ is practically zero.

The 13-atom clusters can also be found in the structure WAl_{12}. The body-centered cubic unit cell of the crystal has an edge equal to 7.58 Å and belongs to the space group Im3. After placing W atoms at the vertices and centers of the cubical cell, we can place aluminium atoms in such a way that they form regular icosahedra. The radius of the coordination sphere is then 2.73 Å. The maximum radius of the superatom is obtained by adding the aluminium radius of 1.42 Å to this value. Since the average radius of a superatom is assumed to be 3.45 Å, the volume of the sphere is 171 Å^3, whereas 12 atomic volumes of Al and one atomic volume of W add up to 215 $Å^3$. Thus, the superatom is very dense. The unit-cell volume equals 436 $Å^3$, while the group WAl_{12} occupies 218 $Å^3$. The packing coefficient of the body-centered unit cell formed by two superatoms with a volume of 171 $Å^3$ is too high (0.78). But this is not suprising as two aluminium atoms from the neighboring clusters are separated by a distance of 2.8 Å that is smaller than the diameter of the aluminium atom. Though increase in density is rather high ($\Delta = 0.25$) the superatoms do not overlap in this structure, and it can be unambiguously treated as the packing of clusters of approximately spherical shape.

7.6.3 Compounds A_mB_n with m = 2 and n = 11

The compound Mg_2Zn_{11} forms a primitive cubic lattice. There are three formula units per unit cell; the cell edge is 8.53 Å. The space group is Pma3. One third of the unit-cell volume equals 207 $Å^3$. No increase in density has been observed for this structure. The coordination number 12 is typical of Mg and Zn crystals. The atoms of all species in the structure have 12 nearest neighbors. One Zn atom is very tightly surrounded by its 12 neighbors (the radius of the coordination sphere is 2.575 Å), while the coordination sphere radii of the two Mg atoms are larger than those in the crystals of the pure components.

7.6.4 Compounds A_mB_n with m = 3 and n = 10

There are many cubic crystals with a unit cell containing *52 atoms*. These structures belonging to space groups Im3m, I$\bar{4}$3m, and P$\bar{4}$3m are often called phases of γ-brass. The compound Fe_3Zn_{10} belongs to this class. Its space group is Im3m. Fe atoms occupy a 12-fold position, and Zn atoms occupy a 16-fold and a 24-fold position. Each of the three positions has only one parameter ("degree of freedom"). The lattice constant is 8.96 Å. The volume per one formula unit is 180 $Å^3$, but the volume occupied by three Fe and ten Zn atoms in the structures of the components equals 187 $Å^3$. Consequently, the structure is dense, the density increase Δ being 4%.

This structure, as well as the other structures of the γ-phase of brass, can be built by two clusters of 26 atoms each. Such a grouping is advantageous if the symmetric cluster with no atom in the center is preferred. The centers of the clusters occupy the vertices and centers of the cubes in all the phases of γ-brass. In the case of Fe_3Zn_{10} the clusters are translation identical and possess a m3m symmetry. The center of such a cluster coincides with the "empty" center of the cube built by Zn atoms. Atoms are at a distance of 3 Å from one another so that the central part of the cluster is very porous. This cube is placed into an octahedron of Fe atoms. The outer part represents a cubo-octahedron (12 atoms) of Zn atoms. It is impossible to single out superatoms consisting of 13 atoms. However, the coordination numbers of all the atoms are 11, 12, or 13.

The crystals of Mn_3Al_{10} are different. Their hexagonal unit-cell (a = 7.543, c = 7.898 Å) contains two formula units. The space group is C6/mmc. Atoms are distributed over four independent positions with three independent atoms having coordination numbers 12, the fourth atom having coordination number 13. There are no superatoms in the structure, but the most important condition of the formation of intermetallic compounds is again fulfilled: the structure of the compound is more dense than the structures of the components. In fact, the parameter Δ has the value of 5%.

7.6.5 Compounds A_mB_n with m = 4 and n = 9

A classical representative of the group is the crystal Al_4Cu_9. Here 52 atoms are distributed over the sites of a primitive cubic lattice belonging to the space group $P\bar{4}3m$, the lattice parameter being 8.67 Å. It is possible to single out clusters of 26 atoms. As shown in Fig.7.4, the tetrahedron at the center of one of the clusters consists of Cu atoms (big circles) whereas that in the center of the other cluster is formed by Al atoms (small circles). As in all other crystals of the γ-brass phase, the center of one cluster occupies the vertex of the cube, and the center of the other cluster occupies the cube

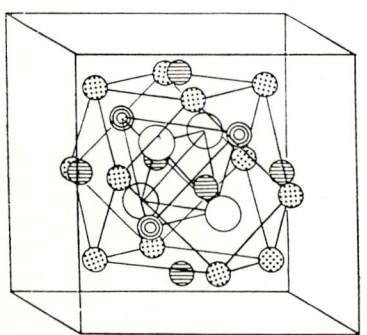 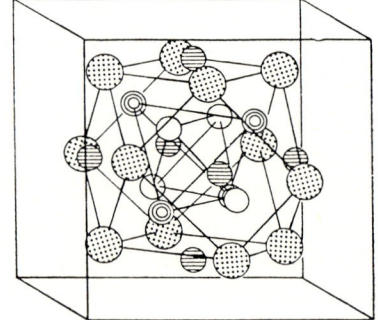

<u>Fig.7.4.</u> Al_4Cu_9 crystal structure

center. This structure type is slightly different from the other structures of the γ-brass phase: the two clusters are not identical. The atoms are distributed over the four independent positions in each of the clusters, with coordination numbers 13, 12, 12, and 11. The gain in the density Δ in this instance is about 7%.

7.6.6 Compounds A_mB_n with m = 5 and n = 8

These compounds have the structure type of γ-brass itself, Cu_5Zn_8. The lattice parameter for Cu_5Zn_8 is 8.84 Å; 52 atoms occupy a cubic unit cell that belongs to the space group $I\bar{4}3m$. Two 26-atom clusters are identical. In the center of the cluster, there is a tetrahedron of Zn atoms which is surrounded by a tetrahedron of Cu atoms, next there are octahedra of Cu atoms, and finally, a cubooctahedron of Zn atoms. The copper "belt" is "squeezed" between two Zn spheres. The increase in density Δ is 7%. Thus *structures with 52 atoms* per unit cell are not unexpected from the standpoint of close packing. The number 13 promotes the possibility for all the atoms to surround themselves with neighbors with maximum density. All the structures of the phases of γ-brass are essentially denser than the structures of the components. The crystals possess high symmetry.

Sometimes crystals of this type have the ratio of the number of free electrons to the number of atoms close to 1.65. And this value, according to the simple band theory of metals, corresponds to the condition of filling of the Brillouin zones [7.10]. It is difficult to assess the importance of this fact.

7.6.7 Compounds A_mB_n with m = 7 and n = 6

All four substances under consideration (Fe_7W_6, Fe_7Mo_6, Co_7Mo_6, and Co_7W_6) belong to the same structure type. There is one A_7B_6 formula unit per unit cell. The space group is $R\bar{3}m$. One atom is at the unit-cell vertex while six A atoms occupy a six-fold position, and three B atoms two-fold positions. Thirteen atoms form an icosahedron.

In structures of this type one can single out the group A_7B_6 not only as a cluster, but with some reservations as a superatom.

This can be demonstrated with the example Fe_7W_6. The unit-cell parameters in the hexagonal setting are a = 4.806 and c = 25.84 Å. the volume per formula unit is 172 Å^3, which leads to a value for Δ of about 4%. If the spheres are packed in a unit cell with the coefficient 0.74, the volume per sphere is 172 Å^3, and the sphere radius is 3.15 Å. Experience shows that the distance from the central Fe atom to six other Fe atoms is 2.41 Å, and to six W atoms 2.76 Å. The mean radius of the first coordination sphere is 2.48 Å.

The effective radius of the close-packed spheres of radius 3.15 Å is only 0.67 Å larger than the radius of the coordination sphere. Though this value is essentially smaller than the radius of a Fe atom, superatoms can be singled out in the structures of the described type. Such superatoms fill the lattice with normal density.

7.7 Close-Packing Principle for Intermetallic Compounds

In this chapter the selective consideration of some intermetallic compounds has been carried out from a certain point of view. We wish to clarify the role of the *close-packing principle* in the formation of intermetallic compounds. We have chosen several classes of compounds, some of them the so-called Laves phases, and others being the so-called electronic compounds.

The increase in density has been calculated for about a thousand substances. This parameter shows how significantly the ratio of the sum of atomic volumes of the components to the molecular volume of the compound deviates from unity. The coefficient does not suggest any structural model. It can be calculated on the basis of the densities of the compound and its component only, even without crystallographic data on the compound. In only twenty cases out of a thousand is the value of Δ appreciably negative. In seventy cases Δ, though being negative, differed from unity only by 2 - 3%. In all the other cases, increase in density was positive, and for half the substances it was within 5 - 15%. For about 200 compounds it was higher than 15%.

Thus it can be asserted with considerable confidence that an intermetallic compound forms only if the atoms in the compound can be packed more densely than in the crystals of the pure components. In other words, the potential energy of interaction must play a major part in the expression for the crystalline free energy. The vibrational energy and the electronic distribution also play a role, but a less important one. It seems that only in rare instances do these constituent parts of free energy dominate the potential energy.

By treating the *tendency to denser atomic arrangements* in the intermetallic compounds in this way, we are, in essence, associating the density of the packing with a presumed minimum of potential energy. As will be shown in the following chapters, for *organic compounds* this statement can be proved by corresponding calculations. For *intermetallic compounds*, it is so far only a hypothesis. But it is justified, first and foremost, by the increased density observed for these compounds. It is also confirmed by the tendency of intermetallic compounds to form icosahedral-type packing with coordination number 12, and by the behavior of components whose atoms reveal, to a certain degree, valence bonding. By ignoring central coordination tendencies and allowing the atoms to form spherical clusters we can explain the formation of compounds with very high density. This can explain the high value of Δ observed for many compounds of Bi, Ga, Ge, etc..

Another important argument in favor of the close-packing principle is that in numerous binary systems 5 - 7 intermetallic compounds of different compositions can form all with different electronic distribution and yet all with fairly dense structures.

In conclusion, we believe that the approximate model of a metallic compound as a system of close-packed particles of spherical (or close to spherical) shape works quite satisfactorily [7.6].

8. Solid Solutions of Metals

We consider only three simplest types of structures: A1 (face-centered cubic), A2 (body-centered cubic), and A3 (hexagonal), including not only the simplest packing of spheres (c/a = 1.63), but also some other structures belonging to the same space group $P6_3/mmc$, but with c/a ≠ 1.63. Thus, we will discuss the solid solutions formed by at least two thirds of all metals.

Unfortunately, we must admit that the role of the theory in predicting possible solid solutions of metals is minor. In fact, there is only one negative rule derived from experiment: there is no solubility if the atomic volumes of the components differ by more than 50%.

Solid solutions have been investigated for a very long time. There is a vast body of experimental data on the dependence of unit-cell parameters and properties of solid solutions on concentration. Of course, numerous attempts have been made to elaborate theories which would predict the mutual solubility of metals, but at best it has been possible to create models of only limited applicability, that explain the solubility of metals only a posteriori.

We refer the reader to other monographs [8.1-7] and present only the most important data for the A1, A2, and A3 structures. All data are taken from [8.8,9].

8.1 Unlimited Solubility

Unlimited solubility is possible only when both elements have the same type of structure. Of course, one should also bear in mind polymorphism. If one of the elements has several modifications, then only the high-temperature modification can form a continuous series of solid solutions. If both elements have several modifications, the formation of solid solutions is possible between the high-temperature modification of one element and the high-temperature modification of the other element; or other polymorphs of equal structure.

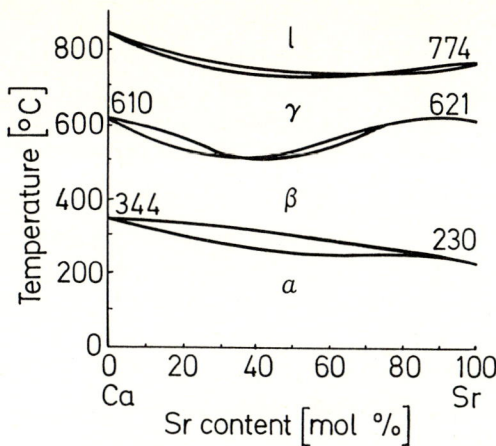

Fig.8.1. Ca-Sr phase diagram

A very special phase diagram is depicted in Fig.8.1. Calcium and strontium have three modifications each, α(A1), β(A3), and γ(A2), in the same temperature sequence. There are three types of solid solutions with the structures A1, A2, and A3.

It is obvious that the α-modification of one element cannot form a continuous series of solid solutions with the β- or γ-modification of the other. Moreover, "second" modifications can form a continuous series of solid solution only if the high-temperature modifications do so. There are only 6 elements with the A3 structure at low temperature and the A2 structure at high temperature, and 6 other elements with the A1 structure at low temperature and the A2 structure at high temperature. It is interesting, though somewhat unexpected, that there are no polymorphic transitions from the A3 close-packed structure to the A1 closed-packed structure with either increasing or decreasing temperature. One can also note that the modifications A1 and A3 do not transform into another simple structure type at low temperature. Yet, there is one exception, cobalt. Below 723 K the close-packed A3 structure replaces the close-packed A1 structure.

Finally, if two elements can form a continuous series of solid solutions it does not mean that they really do form it.

8.1.1 Elements with the fcc Lattice

There are 10 elements with an fcc lattice which exhibit no polymorphism. To these, we may add Co whose hcp polymorph appears at too low temperatures. Thus $11 \times 10/2 = 55$ binary systems are to be considered.

Ten systems can be excluded since the specific volumes of their atoms differ by more than 50%. Of the remaining systems, a continuous solubility takes place for 22 systems. There is no detectable solubility in 7 cases, and in the remaining 16 cases only partial solubility has been observed. Most peculiar are the differences in the solubility of the Au-Ni system (continuous solubility) and the Au-Co system (partial solubility) and the lack of solubility both in the liquid and solid states for the Ag-Ni and Ag-Co systems.

8.1.2 Elements with the HCP Lattice

There are only 19 elements which have polymorphs with the HCP lattice. Most of the systems formed by these elements have not yet been studied. Titanium could form continuous solid solutions with five elements whose atoms differ only slightly in size from Ti. Nevertheless these systems show no solubility. This fact may perhaps be attributed to polymorphism of Ti. Magnesium forms continuous solid solutions with cadmium and rhenium with osmium. The system Zn-Os has been studied thoroughly; though specific volumes of atoms differ only slightly, no solubility has been observed. Sodium and strontium could form continuous solid solutions, but in fact even partial solubility has not been reported.

8.1.3 Elements with the bcc Lattice

In principle, 22 elements with bcc lattice can form continuous series of solid solutions. Only one third of the systems have been studied well enough. It is highly probable, though, that the incomplete solubility data prove that there is no detectable solubility. Let us consider the results obtained for systems whose component atoms are close in size.

Iron forms continuous solid solutions only with chromium and vanadium even though Mn, Mo, Ta, Ti, and Zr could also be good partners. In the case of Zr, Ti, and Mo only partial solubility has been observed; in the remaining cases there is no solubility at all. There is no solubility between Pt and Va, Pt and W, or Mo and Mn. But in the case of the bcc phase, Ti obeys geometrical rules. Its partners in the formation of continuous solid solutions are Cr, Fe, Hf, Mn, Mo, Nd, Pt, Tl, and Th. Seven of these elements are completely soluble, the rest only partially soluble.

Of course, in some instances the absence of mutual solubility can be explained by the electronic structure. But in almost all cases, examples can be found where the components are mutually soluble and the electronic structure is similar.

For the majority of observed continuous series of solid solutions, the curves of unit-cell parameters versus composition have been obtained. These curves can always be roughly approximated by straight lines since there is no solubility when unit-cell parameters differ by more than 50%. The so-called rules of VEGARD (for cube edges a [8.10]) or of ZEN (for volumes V [8.11]) can be regarded as the first terms in a series expansion of functions $a(c)$ or $\bar{\bar{V}}(c)$, where c is the composition. If measurements are sufficiently accurate, one can obtain curves of various types. The specific character of the functional dependence of unit-cell parameters on the composition is determined by the features of the interaction potentials between atoms of components. These in turn can result in different types of short-range order. With a change in composition a porous lattice can change into a denser one, and the segregation of atoms can form or disappear. It is impossible to give a simple explanation of $\bar{\bar{V}}(c)$-curves, and therefore theories dealing with atom "compression" and simplified models of lattice distortions [8.12,13] are not satisfactory.

Detailed studies of short-range order and lattice distortions of solid solutions are necessary. Solid solutions strongly differ from one another, and only specific trends within limited families of solid solutions can be formulated.

8.2 Terminal Solid Solutions

Consideration of phase diagrams shows that there are only a few simple rules which can predict the mutual solubility of metals. DARKEN and GURRY [8.14] have suggested an approach for predicting solubility proceeding from the known sizes and electronegativities of the atoms. WABER et al. [8.15] have applied this method to analyze solid solutions for 850 systems. Their conclusion is rather pessimistic: the method has worked only for 61.7% of the cases, which is not much better as a flip of a coin.

The problem is very difficult, and attempts to create a theory using only two or three parameters seem to be very naive for the purpose of interpreting very slight percentage changes in lattice parameters. For example, there are indications of "unusual" behavior in the lattice-parameter curve of Au-Ag solid solutions. The "unusual" behavior of the curve consists in the fact that it has a minimum. Yet the whole range of the lattice-parameter change is less than 0.01 Å. No doubt, one can interpret such a fine effect only by considering short-range order; lattice distortions, changes in the vibrational spectrum, etc. The atomic volumes of Au and Ag differ by 0.5%. A sufficiently good approach consists in approximating the experimental curve by a straight line.

Though attempts to give a comprehensive interpretation to any $\overline{\overline{V}}(c)$ curve for solid solutions are doomed to failure, it is clear that in general the *geometrical model* can be of some use. Thus no solubility exists if the volumes of the atoms differ by more than 50%.

Take for instance a plot of the unit-cell parameters of the solid solutions formed when Ag atoms (specific volume 17 Å^3) are substituted by larger atoms. With larger guest atom volumes lattice parameters of the solid solution increase. The increase occurs in the following sequence: Mg (atomic volume 23 Å^3), As (21 Å^3), Cd (22 Å^3), In (26 Å^3), Sn (27 Å^3), Ti (18 Å^3), and Sb and Pb (30 Å^3). In general, the result is satisfactory. Of course, two exceptions — Mg and especially Ti — are slightly disturbing. But these are just fine details which the simple geometrical model does not pretend to be able to explain.

When Ag atoms are replaced by atoms of smaller volume the same trend is observed. The slope becomes increasingly negative in the following sequence: Al (16 Å^3), Zn (15 Å^3), and Cu (12 Å^3). Though this is quite satisfactory, unfortunately, these three solid solutions are followed by those with Pt, whose atomic volume is 15 Å^3, i.e., it is not smaller than that of Cu.

The situation is even worse with Ga. Its atomic volume is larger than that of Ag, yet the lattice period of the solid solution decreases. Let us note that Ga also caused exceptions from general rules in intermetallic compounds. This is probably connected with the noncentral character of the Ga-Ga and Ga-heteroatom interactions.

Cases of limited solubility should not be analyzed unless the whole phase diagram is also studied comparing the free energy of a solid solution with that of an intermetallic compound or of a solid solution formed using the space lattice of the other component of the system.

8.3 Solid Solutions on the Basis of Intermetallic Compounds

Binary systems with simple phase diagrams are relatively rare for metal alloys. Usually, the phase diagram is divided into some single- or two-phase regions. Often components can form three or four, and sometimes even seven or eight, *intermetallic compounds*. Some of them give maxima on the phase diagram; in such cases the diagram can be divided into separate parts. Other intermetallic compounds are formed according to a peritectic reaction and cannot be easily observed. Sometimes, only two-phase mixtures will form from components and intermetallic compounds or from several intermetallic compounds. In other instances intermetallic compounds form solid solutions that span smaller or larger regions of the phase diagrams. In the latter case the term *intermediate phase* is appropriate.

Usually intermetallic compounds of stoichiometric composition (*ordered crystals*) are within the concentration range of an intermediate phase. But solid solutions can also crystallize with the structure of the corresponding unstable intermetallic compound when the composition is outside the region of existence of the intermediate phase. Thus, for example, solid solutions with the structure of the intermetallic compound BeCu exist within the range 20 - 47 atom % Be.

The homogeneity region of a solid solution with the structure of an intermetallic compound usually spans several per cent of the relative molefractions. But phase diagrams of the brass type are not rare. This classical phase diagram is shown in Fig.8.2.

Here the α and η phases are the regions of existence of solid solutions with the structure of the pure elements Cu and Zn, respectively. In addition, there are three wide regions of existence of intermediate phases β, γ, and ε. The high-temperature δ phase exists in a narrow region of the phase diagram; the β phase is a solid solution with the structure of the intermetallic compound CuZn crystallizing in a bcc lattice with a = 2.98 Å. The structure has the same packing density as the α phase. It can be suggested that for the packing of Cu and Zn atoms whose shape only slightly differs from a sphere, a bcc lattice is more appropriate than a fcc one. The structure of γ brass

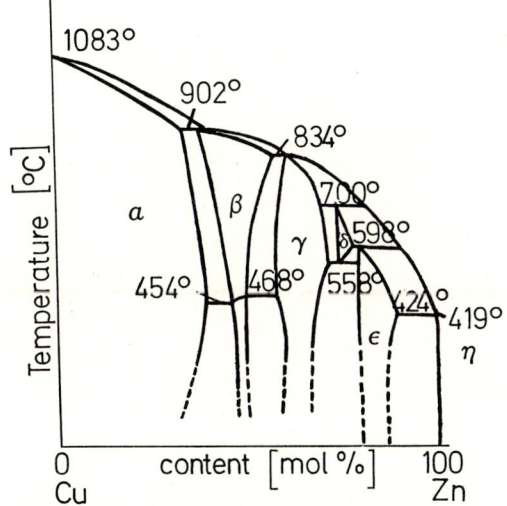

Fig.8.2. Cu-Zn phase diagram

(Cu_5Zn_8, 62.5 at. %) was discussed in Sect.7.6, and especially in Sect.7.6.6. As was indicated, the packing of 52 atoms in a cubic unit cell has significant advantages.

For the ε phase of the composition $Cu_{21}Zn_{39}$, a hexagonal lattice has been found. Unit-cell dimensions are a = 2.73 Å, c = 4.286 Å. The ratio c/a decreases from 1.63 (in pure Zn) to 1.567 in $Cu_{21}Zn_{39}$. The increase in density is 5%. The atomic arrangement is not known. Probably, atoms form a disordered crystal on the basis of an intermetallic compound whose composition is beyond the existence region of the ε phase.

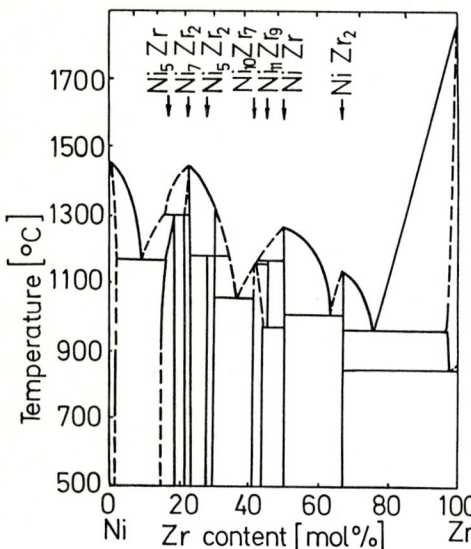

Fig.8.3. Ni-Zr phase diagram

Let us consider now the phase diagram Ni-Zr, in which intermetallic compounds form solid solutions only within a narrow band (Fig.8.3), if at all. In brass (Fig.8.2), the intermetallic compounds have no maxima on the melting curve. In contrast, of the seven known intermetallic compounds in the Ni-Zr system (Fig.8.3) three have melting-point maxima, and the other four are formed by peritectic reactions. We have calculated the increase in density Δ for all the Ni_xZr_y compounds whose unit-cell dimensions have been measured: the Δ values are all positive and lie within the limits 3 - 5%.

8.4 Ordering

For a disordered solid solution in which A particles "prefer" to be surrounded by B particles (rather than by A particles), we can presume that with a temperature decrease the degree of ordering will increase (the entropy term in the expression of free energy becomes less significant) and that at a certain temperature a phase transition disorder-order will occur.

Phase transitions predicted by Gibbs' thermodynamics are called *phase transitions of the first kind* (Ehrenfest's terminology). They occur at a certain temperature and are accompanied by a sudden change in volume $\Delta \bar{\bar{V}}$ and discontinuities in other physical properties. The phase transition of the first kind is actually the growth of a crystal of a new phase from the old one. The rules of such a transformation do not differ significantly from the rules of crystallization from liquids or vapors. The temperature of a phase transformation, i.e., the equilibrium temperature of two modifications, must be determined very carefully, since considerable *supercooling* or *superheating* can occur due to the difficulties in restructuring.

Very often, as a solid is cooled, the transition to an ordered state does not occur because the lattice does not restructure quickly enough: supercooling ensues, i.e., the temperature of the phase transition is "jumped over".

In 1914 KURNAKOV et al. [8.16] observed experimentally the phenomenon which led to Ehrenfest's theory of phase transitions *of the second kind*. These are the transitions with zero volume change, $\Delta \bar{\bar{V}} = 0$ across the phase transition. A new lattice, that forms in such a transition, is a low-temperature *superstructure* with respect to the lattice of the high-temperature disordered crystal.

There is no supercooling or superheating in phase transitions of the second kind. The formation of a new phase is not the growth of a new crystal, but a continuous restructuring of the lattice. At a certain critical temperature T_c the last particles occupy their "own" positions in the lattice and the disorder-order transition is complete.

Innumerable superstructures can be suggested for alloys of different compositions. Indeed, by increasing the unit-cell dimensions we find more and more ways in which we can distribute A and B atoms over the lattice points which were previously occupied by "average" atoms. Therefore the theories of LANDAU [8.17] and LIFSHITS [8.18] are of great importance. They have shown that there must be a strict relationship between the symmetry of the lattice of a disordered solid solution and the *symmetry of a superlattice* formed as a result of phase transition of the second kind.

Solving the system of equations (4.37) by the KHACHATURYAN method of concentration waves [8.19], one can obtain all the superstructures possible for the given structure of a disordered solid solution. For simple lattices, bcc and fcc, for example, the number of possible superstructures and of possible compositions is small. Thus, for bcc lattices the complete ordering in substitutional solutions can be realized only at the compositions AB, A_3B, and A_7B; the theory predicts only two AB structures [cubic of the CsCl type and cubic with 16 atoms per unit-cell (such a structure is known for ZnTl)], three A_3B structures (two tetragonal and one cubic with 4 formula units per unit cell), and one A_7B structure. The coordinates of crystallographic positions of all these complex structures can also be derived theoretically.

Similar calculations have also been made for fcc lattices, and solutions in which ordering occurs only at one crystallographic position in the lattice (the case of an interstitial solution). Khachaturyan's method is elegant and important for the theory of alloys, and, in principle, it can be extended to more complicated lattices.

However, this theory predicts possible superstructures, but does not say which of these structures exists in nature, since the final solution of (4.37) contains three energy parameters (if there are two σ parameters). These parameters can be calculated only if the law of particle interaction is known. For metallic systems this can be done using *pseudopotentials*, while for organic solid solutions, as we shall see in Chap.13, the interparticle potential is well known. All these parameters can be measured in X-ray diffuse-scattering experiments (Sect.12.9.3).

8.4.1 Investigation of the Order-Disorder Transition by X-Ray Diffuse Scattering

In the simplest cases the X-ray diffuse scattering measurements and the calculation of free energy by Khachaturyan's method of concentration-waves yield insights valuable on the transition of solid solutions from a disordered to an ordered state. From accurate measurements of the X-ray diffuse scattering at several points of the reciprocal lattice one obtains the energy parameters for (4.40): these are the Fourier transforms of the pairwise interaction potentials.

The solution of (4.40) give the dependence of the free energy on the temperature and concentration of solid solution. Thus we can obtain the phase diagram regions for disordered solid solutions that form superstructures as the temperature is lowered.

Few such studies have been carried out. We shall describe below the character and the temperature of the phase transformation in Cu_3Au alloys, and the derivation of the phase diagram for Fe-Al alloys.

8.4.2 Disorder in Cu_3Au

The disordered solid solution of Cu in Au has an fcc structure. When the temperature decreases, the alloy containing three Cu atoms per Au atom becomes completely ordered: Au atoms occupy the unit-cell vertices, whereas Cu atoms are at the face centers.

The probability $p(r)$ of a Cu atom occupying one of the nodes r of an fcc lattice in the case of partial long-range order can be represented (Sect. 3.3.4):

$$n(\underline{r}) = \frac{1}{4} + \frac{1}{4} \sigma \left[\exp(2\pi i x) + \exp(2\pi i y) + \exp(2\pi i z) \right] \quad . \tag{8.1}$$

This probability can have only two values: $1/4 - \sigma/4$ for 3/4 of the fcc-lattice nodes and $1/4 + 3\sigma/4$ for all the remaining positions.

Using (4.46,49), we obtain the free energy of the alloy:

$$F(\sigma) = \frac{1}{2} \Phi(0) \frac{1}{16} + \frac{1}{2} \Phi(2\pi a^*) \frac{3}{16} \sigma^2 + \frac{1}{16} kT \{(1 + 3\sigma) \ln[(1 + 3\sigma)/4]$$

$$+ 3(1 - \sigma) \ln[3(1 - \sigma)/4] + 3(1 - \sigma) \ln[(1 - \sigma)/4]$$

$$+ 3(3 + \sigma) \ln[(3 + \sigma)/4]\} \quad . \tag{8.2}$$

The condition for a free energy minimum ($\partial F/\partial \sigma = 0$) let us plot the degree of the long-range order σ versus kT/Φ (Fig.8.4). The theory gives the values of maximum and minimum free energy. The part of the curve corresponding to the minimum energy has a physical meaning up to kT_0/Φ when the free energies of ordered and disordered phases become equal. This is a point where a first-order phase transition occurs. The transformation temperature T_0 can be found by equating the free energy of a partially ordered phase with $\sigma \neq 0$, to (8.2) written for $\sigma = 0$. We find

$$T_0 = -0.205 \, \Phi(2\pi a^*)/\bar{\bar{K}} \quad . \tag{8.3}$$

Thus the problem is not quite solved since the obtained temperature is expressed in terms of the energy parameter Φ, which cannot be calculated direct-

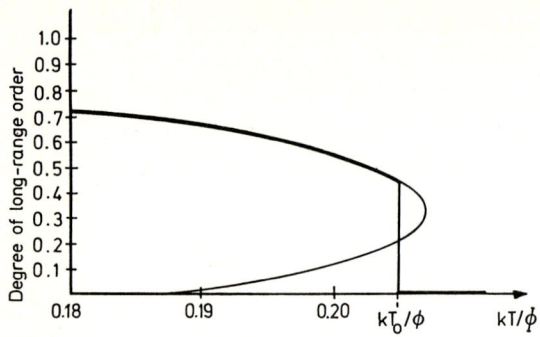

Fig.8.4. The degree of long range order σ versus kT/Φ for Cu_3Au alloy

ly. Hence the measurements of diffuse X-ray scattering can be of great help. The energy parameter Φ is related unambiguously as in (6.28), to the intensity of diffuse scattering at point 001 of reciprocal space.

MOSS [8.20] has measured the absolute scattering intensities for Cu_3Au at two temperatures. They are

$$I_D(a^*)/(f_{Cu} - f_{Au})^2 c_{Au}(1 - c_{Au}) = \begin{cases} 14.8 \text{ at } 678 \text{ K} \\ 6.7 \text{ at } 723 \text{ K} \end{cases},$$

Using $c_{Au} = 1/4$, we find the following values for the parameter under consideration Φ/k: -3370 K and -3280 K respectively. Substituting their average into (8.3), we find a theoretical temperature of the first-order phase transition $T_0 = 682$ K, while the experimental value is 663 K in remarkable agreement with Khachaturyan's theory.

8.4.3 Fe-Al Phase Diagram

The study of the Fe-Al phase diagram by SEMENOVSKAYA [8.21] (the first work of its kind) combined X-ray measurements with the method of concentration waves, and led to encouraging results.

The Fe-Al phase diagram is rather complicated. Below the liquids curve there are regions of disordered solid solutions with elements of long-range order typical of the AB structure and also regions of solid solutions with a cubic structure with sixteen atoms per unit cell. Within the latter region, there is a completely Fe_3Al structure. The boundary between ferromagnetic and paramagnetic alloys passes through the part of the diagram adjoining pure iron.

Semenovskaya tried to obtain the complete phase diagram of the system. We shall consider only the structure part of her work, and will not discuss the ferromagnetism.

First of all, let us write the expression for a static concentration wave for iron atoms in the Fe_3Al lattice. It has the form

$$n(\underline{r}) = c + x \exp(i\underline{k}_1\underline{r}) + y \sin(\underline{k}_2\underline{r}) \quad . \tag{8.4}$$

Here c is the concentration of iron, and x and y are the parameters of long-range order related to parameter σ of (3.13) by $x = -\sigma_1/4$ and $y = \sigma_2/2$.

The vectors $\underline{k}_1$ and $\underline{k}_2$ are directed toward points (1 1 1) and (1/2 1/2 1/2) of the reciprocal lattice of the bcc crystal of a disordered solution. In this lattice the function $n(\underline{r})$ has the three values $c+x$, $c-x-y$, and $c-x+y$, at 1/2, 1/4, and 1/4 of the nodes, respectively.

Using (4.36 - 40), we obtain the following expression for the free energy:

$$\begin{aligned}F = \frac{1}{2} \Big\{ &\Phi(0)c^2 + \Phi(\underline{k}_1)x^2 + \Phi(\underline{k}_2)\frac{y^2}{2} + kT \Big[(c + x) \ln(c + x) \\ &+ (1 - c - x) \ln(1 - c - x) + \frac{1}{2}\big((c - x - y) \ln(c - x - y) \\ &+ (1 - c + x + y) \ln(1 - c + x + y) + (c - x + y) \ln(c - x + y) \\ &+ (1 - c + x + y) \ln(1 - c + x + y)\big)\Big]\Big\} \quad . \end{aligned} \tag{8.5}$$

To obtain a numerical solution, we must know the values of three energy parameters of the Fourier transforms of the interaction potentials at the origin, $\Phi(0)$, and at two other points of the reciprocal lattice. If these are known, then by equating the derivatives of the free energy with respect to the parameters x and y to zero, we can find their equilibrium values for any temperature and concentration and hence determine the kind of phase diagram. This suffices for paramagnetic alloys. For ferromagnetic alloys we should add to the expression of free energy a further parameter, the magnetic free energy term, due to Heisenberg's spin exchange.

Stringent requirements are imposed on experiment, as described in detail by SEMENOVSKAYA and UMIDOV [8.22]. Five crystals were grown from a disordered solid solution with 9, 12, 16, 19, and 22% Al content, respectively. The crystals were quenched so that the structure of the high-temperature state was preserved at room temperature. The quenching temperature was established experimentally.

The X-ray diffuse scattering of monochromatic radiation was measured for the (120), (201), (021), and (210) bcc reciprocal lattice points to obtain $\Phi(\underline{k}_1)$, for the points (1/2 3/2 1/2), (3/2 1/2 1/2), and (1/2 3/2 1/2) to obtain $\Phi(\underline{k}_2)$, and at eight points in the vicinity of the nodes of the main reciprocal lattice in the direction [010] and [100], and near the node (220) in the direction [110] to determine the energy parameter $\Phi(0)$. The intensity was converted to absolute units (electron units per atom) by calibration with X-ray scattering data for amorphous quartz. Compton thermal and air scattering corrections were made.

The theoretical curves for the phase boundaries are similar to those observed in experiments. The maximum temperature and concentration differences in these diagrams do not exceed 10%. The result can be considered quite satisfactory given the approximations in the free energy formula.

8.4.4 Determination of the Degree of Long-Range Order by X-Ray Diffraction

Using the simplest methods of X-ray structure analysis one can determine the degree of long-range order in alloys by observing changes in the intensity of the lines in X-ray patterns of polycrystalline specimens. Of course, one could perform such studies even in the early days of X-ray structure analysis thanks to a wide choice of alloys that have simple cubic or hexagonal lattices in which ordering occurs with a change in temperature. Indexing of X-ray patterns in such cases posed no difficulties.

Let us consider ordering in AuCu alloys. We have a series of X-ray patterns obtained at different temperatures. How does the intensity of diffraction lines change with variations in the degree of long-range order σ (3.7)? This can be seen from the expression (6.21) for the structure factor. In the case of partial long-range order we have

$$F = \frac{1}{2}[f_{Au} + f_{Cu} + \sigma(f_{Au} - f_{Cu})]\left(e^{\pi i(h+\ell)} + e^{\pi i(k+\ell)}\right)$$
$$+ \frac{1}{2}[f_{Au} + f_{Cu} + \sigma(f_{Cu} - f_{Au})]\left(1 + e^{\pi i(h+k)}\right) \quad . \tag{8.6}$$

If the indices h, k, and ℓ are all even or all odd, then $F = 2(f_{Au} + f_{Cu})$, i.e., the intensity of the lines does not depend on the degree of order σ; otherwise $F = 2\sigma(f_{Cu} - f_{Au})$.

It is obvious that the degree of long-range order σ can be determined quite easily by measuring the intensity ratio of two lines, one major and one due to the superstructure. The ratio of the absolute values of structure factors with mixed even and odd indices, and with all even (or odd) indices obtained from experiment equals

$$\sigma = \frac{f_{Cu} - f_{Au}}{f_{Cu} + f_{Au}} \quad . \tag{8.7}$$

Knowing the values of f we can readily obtain values for σ.

A face-centered lattice is also formed by the centers of atoms in the alloys Cu_3Au, Cu_3Pt, Ni_3Fe, etc. In the case of complete order, Au atoms are at (0,0,0) and Cu atoms are at (1/2, 1/2, 0), (0, 1/2, 1/2), and (1/2, 0, 1/2).

The structure factor (6.21) for the general case of partial order has the form $f = (f_{Au} + 3f_{Cu})/4$, m = 3, n = 1:

$$F = \left[\frac{1}{4} f_{Au} + \frac{3}{4} f_{Cu} + \frac{3}{4} \sigma(f_{Cu} - f_{Au}) + \frac{1}{4} f_{Au} + \frac{3}{4} f_{Cu} + \frac{1}{4} \sigma(f_{Cu} - f_{Au})\right]$$

$$\times \left(e^{\pi i(h+k)} + e^{\pi i(h+\ell)} + e^{\pi i(k+\ell)}\right) \quad . \tag{8.8}$$

If the indices h, k, ℓ are all odd or all even, we have $F = f_{Au} + 3f_{Cu}$, i.e., the intensity of these (major) lines does not depend on the degree of long-range order. If h, k, ℓ are mixed, $F = \sigma(f_{Au} - f_{Cu})$; in the case of complete disorder these (superstructure) lines are absent, and in the case of complete order they acquire an intensity proportional to the square of the difference of the atomic scattering factors. In Cu_3Au alloys the cubic lattice is retained at all values of σ.

8.4.5 Short-Range Order

Short-range order in solid solutions can be determined from the measurements of X-ray or neutron diffuse scattering. Proceeding from the general formula (Sect.6.5.1) and making some simple transformations, we can represent the intensity of diffuse scattering in the form of a three-dimensional series:

$$I_D = K \sum_{\ell mn} \alpha_{\ell mn} \exp[-2\pi i(\ell\xi + m\eta + n\zeta)] \quad . \tag{8.9}$$

Here, K is a coefficient proportional to the product of concentrations of A and B atoms and to the square of the difference between their scattering factors $K \sim c_A c_B (f_A - f_B)^2$, ℓ, m, n are integer coordinates of the nodes of the crystal lattice, and ξ, η, ζ are coordinates of the points in reciprocal space. It is assumed that all the neighbors of the A and B atoms can be distributed over the coordination spheres. The number of atoms of any kind should be measured in all the coordination spheres. In such a formulation of the problem the function I_D has high symmetry; all the coefficients $\alpha_{\ell mn}$ obtained by changes of indices and signs are equal.

Experiment permits obtaining the coefficients α which determine the short-range order. If there are only Z_i sites (of which n_i are occupied by atoms A) on the ith spherical surface with center in the B atom, then

$$\alpha = 1 - n_i/c_A Z_i \quad . \tag{8.10}$$

The scattering intensity can be measured from experiment and represented in reciprocal space as a function of ξ, η, ζ. Knowing this function, we can calculate all the coefficients $\alpha_{\ell mn}$ using the conventional formula

$$\alpha_{\ell mn} = \int_0^1 \int_0^1 \int_0^1 I_D(\xi, \eta, \zeta) \exp[2\pi i(\ell\xi + m\eta + n\zeta)] \, d\xi \, d\eta \, d\zeta \quad , \tag{8.11}$$

and therefore find all the numbers n_i characterizing short-range order. The coefficients $\alpha_{\ell mn}$ have all the properties of the short-range order parameters. In the case of complete disorder $n_i/Z_i = c_A$ and all the coefficients $\alpha = 0$. In the case of complete order the absolute values of α are maximized.

Let us consider the application of the theory to $AuCu_3$ alloys. To describe an integer ℓ, m, n, we must enumerate the nodes, using as a unit, half an edge of a face-centered-cubic unit cell. Then the nodes in the centers of the faces nearest to 000 will have indices of the type 110, the vertices of the cube, of the type 200, etc. To clarify the meaning of the coefficients $\alpha_{\ell mn}$, let us determine their values in the case of complete order. The first coordination sphere — 1/m nodes of the type 100 — consists of 12 atoms. If we proceed from the Au atom at 000, all the 12 neighbors at 110 will be Cu atoms; $n_i = 12$, $Z_i = 12$, and c_A, the fraction of Cu atoms, equals 3/4; therefore $\alpha_{110} = -1/3$. The same result will be obtained if we proceed from Cu atoms at 000, since then $n_i = 4$ and c_A, the fraction of Au atoms, equals 1/4. The second coordination sphere 200 contains six atoms and n_i/Z_i; the fraction of the other species (relative to that at 000) equals zero. Hence $\alpha_{200} = +1$. It is easy to see that these two values will alternate, i.e., for coordination spheres with i even we obtain -1/3 and for the spheres with i odd we have +1. Knowing α for partial short-range order, we can easily find the character of the deviation in the distribution of atoms from both limiting cases.

It has been found, by the method described below, that for $AuCu_3$ $\alpha_{110} = -0.15$ at 406 C. Therefore, the number of Cu atoms closest to a Au atom n_i can be obtained from the expression $1 - 4n_i/3 \cdot 12 = -0.15$, whence $n_i = 10.35$. That is, 1.65 fewer Cu atoms are next to Au atoms than for a completely ordered lattice, or, 1.35 more Cu atoms are present around each Au atom than one would expect for a completely disordered lattice. The number of Au atoms nearest to Cu atoms can be found from the equation $1 - 4n/12 = -0.15$, resulting $n = 3.45$, which is 0.55 atoms fewer than in the case of complete order and 0.45 more than in the case of complete disorder.

The tendency of a given atom to surround itself by atoms of different kinds is revealed in the fact that the first shell around Au has an excess (compared with complete disorder) of Cu atoms, whereas the neighboring shells (2nd, 3rd, and 4th) then tend to have an excess of Au atoms.

The measurement of short-range order for the Cu_3Au alloy carried out by Cowley [8.23] was the first work of this kind. For the past thirty years the principles of such measurements have not changed, though some improvements have been made in the apparatus which have made experimental work somewhat easier. Cowley measured the intensity of scattering by the (100) face of a Cu_3Au single crystal. The crystal was kept above the critical temperature, so that there was no long-range order.

Measurements of the scattered-beam intensity were carried out for 250 unique reciprocal lattice points. In (8.11) integration was replaced by summation. The intensity due to thermal vibrations and to incoherent (compton) scattering was subtracted from the observed intensity, by extrapolation and by measurements at different temperatures.

In this pioneering work and also in later studies the static distortions of the lattice were neglected. Yet these distortions are not that small: as a rule, they are about 0.1 Å.

At present there is a significant amount of material about short-range order in alloys. We can refer the reader to a monograph and a paper by IVERONOVA and KATSNELSON [8.24,25]. The latter work gives an interesting list of the signs of the parameter of short-range order for the first coordination sphere. The data on the signs of the parameter of short-range order are essential. If $\alpha > 0$, we can expect decomposition with a temperature decrease, if $\alpha < 0$ ordering is possible.

It is useful to know what can happen if the crystal is observed for a sufficiently long time or if it is subjected to some external force that can accelerate restructuring, such as exposure to neutron radiation.

Sometimes one can detect phase transformations for systems where they were not observed before. Thus, relatively recently, decomposition in the Cr-Mo system has been found and ordering in Se-Te alloys. Yet the latter result is somewhat doubtful as Se and Te have chain structures.

Yet the methods for studying short-range order cannot be considered satisfactory. A priori, it is clear that systems can have a different character of order even when their X-ray or neutron diffuse scattering patterns are identical. Indeed, one can expect for solids the same kind of average short-range order as for liquids. Yet, we cannot exclude from consideration the alloys that have composition domains. Moreover, all indications are that they really exist. Here, two cases are possible. In one case, domains have a high degree of ordering in their "centers", the order decreasing towards the "edges" of the domains. In the second case, the domains themselves are ordered systems, but neighboring domains are shifted relative to one another, so that the long-range order is violated. It is quite probable that there is a specific order at which the crystal of the component A is surrounded by an orderly "fringe" of B particles. This specific "interblock" solubility has been discovered in organic solid solutions (Sect.11.4.2). To distinguish such variants of short-range order on the basis of X-ray scattering is indeed very difficult.

9. Inorganic Solid Solutions

We shall consider solid solutions of the simplest salts, oxides, and chalcogenides. We shall briefly discuss carbides, nitrides, and hydrides of metals. Inorganic molecular compounds usually form crystals, which can be regarded as regular packings of ions of different signs. These compounds are usually treated by inorganic crystal chemistry, and so their consideration falls outside the scope of this book. But some inorganic complexes, such as gas hydrates, are of interest for us, since they are closely related to the problems dealt with in this book. However, such complexes, which are not ionic packings, should be considered together with similar organic complexes, as discussed in Chap.14.

The theory of *solid solutions of inorganic compounds* is less developed than that of metal alloys. One of the sections will be devoted to the possibility of predicting solubility. We do not intend to consider here the vast experimental material existing on inorganic solid solutions. We are only interested in general laws for the formation of mixed crystals, which will be illustrated by some very simple examples.

9.1 Isomorphous Substitutions in Alkali Halide Salts

We shall confine ourselves to investigating the simplest case of isomorphous substitution: the substitution of ions. Moreover, we shall consider only the simplest mixtures, i.e., the *mixtures of alkali halide* salts, and deduce the simplest rules of isomorphous substitution of monovalent ions or cations. More detailed data on the substitution of solid solutions can be found in the monograph by URUSOV [9.1].

Most alkali-halide salts form crystals with face-centered-cubic lattices. Some compounds of Cs and Rb are polymorphic and have two modifications each. The low-temperature form has a B2 lattice. The data on solubility [9.2,3] are presented in Tables 9.1 - 9.

The first five tables contain data on isomorphous mixtures with different anions. The experimental data are not complete. In the cases where the data on solubility are not available, there is a dash in the corresponding square in the table. When mutual solubility is complete, a "1" is listed, when it is limited, "lim" is listed; when solubility ranges are known, the atom fraction range x of the atom given in the column heading is listed. The squares on the diagonal give the value of the cubic-lattice parameter for the pure alkali halide. If the square is occupied by two numbers, the upper value corresponds to a fcc lattice (B1 structural type) and the lower to the B2 structural type (CsCl type).

Table 9.1. Solubility data for LiHal systems

LiF	LiCl	LiBr	LiI	
B1 4.02	lim	lim	lim	LiF
	B1 5.14	1	lim	LiCl
		B1 5.50		LiBr
			B1 601	LiI

Table 9.1 supplies the data on Li salts. A continuous series of solid solutions has been observed only for the system LiCl-LiBr. The unit-cell parameters of pure LiCl and LiBr differ by 7%. Unfortunately, the type of phase diagram for the LiBr-LiI system is unknown. In this instance, the difference in parameters is insignificant, and the formation of a continuous series of solid solutions can also be expected. In the remaining cases, only limited solubility has been observed.

The gaps in solubility (on the eutectic line) are rather small and do not exceed 5 - 10%. Such gaps are displaced towards higher concentrations of the anion of smaller size. The effect of asymmetry of the atom-atom interaction is obvious: the tightness is less tolerable. Therefore the solubility limit is reached sooner if a small particle is replaced by a larger one than in the reverse case.

The situation is similar for the systems whose components have a common Na cation (Table 9.2). Only in the case of the NaCl-NaBr system whose lattice parameters differ insignificantly (6%) a continuous series of solid solutions takes place. When the heat of mixing was measured, it was observed that the curve obtained was symmetric, and the maximum value of the heat of mixing for

Table 9.2. Solubility data for NaHal systems

NaF	NaCl	NaBr	NaI	
B1 4.63	to 0.65		lim to 0.55 I	NaF
	B1 5.64	1	lim from 0.63 to 0.6	NaCl
		B1 5.97		NaBr
			B1 6.47	NaI

the equimolecular mixture was of about 300 cal/mole. The configurational entropy is higher than that at room temperature. Decomposition, if any, probably takes place at negative temperatures. The replacement of F anions by I anions has not been observed. At the same time up to 55% of I ions can be replaced by F ions.

Table 9.3. Solubility data for KHal systems

KF	KCl	KBr	KI	
B1 5.34	lim			KF
	B1 6.36	1	lim	KCl
		B1 6.55	1	KBr
			B1 7.07	KI

In systems with common K cations (Table 9.3) or Rb cations (Table 9.4), there are two systems with continuous solubility. The differences in lattice parameters are within 3-7%. Nevertheless, even such small lattice distortions lead to a significant increase in lattice energy. The system RbI-RbBr decomposes at 73°C.

Two cases of continuous solubility have been observed in systems with a common Cs cation (Table 9.5). The substitution of Cl and I anions for the Br anion results in small changes in lattice parameters. CsBr, CsCl, and CsI have polymorphism. Here, two series of continuous solid solutions on the basis of fcc and B2 lattices are possible. A difference in lattice parameters not exceeding 10% does not permit the replacement of ions throughout the entire range of concentrations.

Table 9.4. Solubility data for RbHal systems

RbF	RbCl	RbBr	RbI	
B1 5.64	lim	lim	lim	RbF
	B1 6.55	1		RbCl
		B1 6.87	1 $T_c = 73°C$	RbBr
			B1 7.34 B2 4.34	RbI

Table 9.5. Solubility data for CsHal systems

CsF	CsCl	CsBr	CsI	
B1 6.02	lim	lim	lim	CsF
	B1 6.97 B2 4.12	1	I to 20% Cl	CsCl
		B1 7.23 B2 4.50	1	CsBr
			B1 7.66 B2 4.57	CsI

Now let us examine systems with a common anion. The regularities will be the same. In the mixture of fluorides continuous solutions appear when Rb is replaced by K or Cs. The difference in the lattice parameter in this case is 5 - 6%. Continuous solubility disappears when the difference in parameters becomes 13% (Table 9.6).

CHANH [9.4] determined the phase diagram of the NaF-KF system with X-ray data. The larger K ion does not replace the Na ion in this system, and only an insignificant amount of substitution takes place on the eutectic line (7%). Solubility occurs only at temperatures higher than 460°C. The difference in lattice parameters of the two crystals is about 15%.

Table 9.6. Solubility data for MeF systems

LiF	NaF	KF	RbF	CsF	
B1 4.02	lim	lim	lim	lim	LiF
	B1 4.63	lim	lim	lim	NaF
		B1 5.34	1	lim	KF
			B1 5.64	1	RbF
				B1 6.02	CsF

Table 9.7 Solubility data for MeCl systems

LiCl	NaCl	KCL	RbCl	CsCl	
B1 5.14		lim	lim	lim	LiCl
	B1 5.64	1 decomp.	lim	lim	NaCl
		B1 6.36	1	1	KCl
			B1 6.55 B2 3.75	1	RbCl
				B1 6.94 B2 4.70	CsCl

Mixtures of chlorides (Table 9.7) have been investigated by many authors, most recently by Chanh. He investigated the NaCl-KCl system very thoroughly [9.5-7]. The melting points of NaCl and KCl are 800 and 768°C respectively. At a temperature of 650°C the liquidus and solidus curves merge at a point corresponding to the equimolecular concentration. Below these curves there lies a bell-shaped curve describing the decomposition of the solid solutions. The curve has a maximum at 60% NaCl, i.e., is displaced in the direction of the ions of smaller size. The maximum of the decomposition curve corresponds to 500°C, according to thermal measurements, and to 400°C, according to X-ray data. According to [9.4] the maximum heat of mixing is 630 cal/mole. This quantity was measured repeatedly, and the curves showing the heat of mixing as a function of concentration obtained by different authors differ significantly. According to many authors, the maximum heat of mixing is closer to 1000 than to 630 cal/mol.

The difference in lattice parameters of KCl and NaCl is 12%. It is obvious that for such a difference the small lattice distortions due to the substitution of Na atoms for K atoms can be balanced by configurational entropy only at high temperatures.

In all the works cited above the authors also investigated in detail the *kinetics of decomposition* of solid solutions, studying the continuous displacement and change in intensity of X-ray diffraction lines. The *activation energy* was measured for mixed crystals of different compositions, since the Arrhenius *law* is fulfilled. As should be excepted, it is lower when a solid solution contains more sodium (cations of smaller size). The activation energy is high, i.e., close to 20 kcal/mole.

We believe that the studies discussed above cannot determine the *decomposition mechanism*. For this, one must study the changes in short-range order as a function of temperature. For a system of three different atoms, such studies are difficult, yet possible. To study the decomposition, one should work in a very narrow temperature range: below 170°C the decomposition proceeds very slowly, whereas above 200°C it goes too fast.

There are still questions about the LiCl-NaCl system (Table 9.7). Literature data are inconsistent: some of them confirm continuous solubility, whereas others exclude it.

Cesium and rubidium chlorides are polymorphous. A continuous series of solid solutions are formed only by high-temperature modifications of an fcc lattice. Lattice parameters have been measured throughout the entire concentration range [9.8]. The unit-cell volume changes are linear. This solid solution possesses all the features of an ideal solution. The difference in lattice parameters is about 4%.

Cesium chloride forms a continuous series of solid solutions with potassium chloride also. In this case the difference in parameters reaches 9%. The same important regularity holds for all the chloride mixtures with limited solubility: small cations are better dissolved among large cations than large cations among small ones.

Table 9.8 is a summary of the experimental data on bromide mixtures. Though it is believed that CsBr and KBr form a continuous series of solid solutions, it is doubtful that this is really so. CHANH [9.8] has obtained the CsBr-RbBr phase diagram by the X-ray method. Since stable CsBr crystals are of a B2 structure type [9.8], a continuous series of solid solutions with other

Table 9.8. Solubility data for MeBr systems

LiBr	NaBr	KBr	RbBr	CsBr	
B1 5.50	1	lim	lim	lim	LiBr
	B1 5.97	1		lim	NaBr
		B1 6.55	1		KBr
			B1 6.87	to 15% to 75%	RbBr
					CsBr

bromides (of the B1 structure type) is not possible. The phase diagram shows that in the RbBr fcc lattice more than 75% of the cations are replaced at the eutectic temperature. In the CsBr crystal about 15% of the cations are replaced. As we have here two different structure types, the rule stating that large particles are less soluble than small ones does not necessarily hold.

For the NaBr-KBr system the heat of mixing has been measured. For the equimolar composition it is 800 cal/mole. The decomposition probably occurs at low temperatures, but this has not been documented.

Phase diagrams of NaI-KI [9.9] and KI-RbI [9.10] have been determined by the X-ray method. The first system decomposes at 243°C. The maximum of the decomposition curve is displaced, as it should be expected, in the direction of Na. The difference in lattice parameters reaches 9%. The composition dependence of the lattice-parameter is strictly linear. The same behavior is typical of KI-RbI solid solutions.

The other data presented in Table 9.9 have been obtained by less perfect methods. Nevertheless, these data do not contradict the general trends, and therefore we do not doubt their validity. In particular, Na substitutes for Rb more often than Rb for Na.

The experimental data on the mixtures of alkali-halide salts lead us to some general conclusions. Continuous solubility is observed when the difference in cubic-unit-cell parameters does not exceed 10%. Even in the case of very small differences in lattice parameters, solid solutions decompose at low temperatures. The decomposition curve is asymmetric: its maximum is shifted in the direction of ions of smaller size. In the case of limited

Table 9.9. Solubility data for MeI systems

LiI	NaI	KI	RbI	CsI	
B1 6.01		lim			LiI
	B1 6.47	1 decomp.	to 15%	lim	NaI
		B1 7.07	1	1	KI
			B1 7.66 B2 4.54	1	RbI
				B1 7.66 B2 4.57	CsI

solubility (when components are of the same structure type) the limits of solubility are essentially different: smaller ions dissolve easier than larger ones. The rules of substitution of anions are the same as for cations. The curves of the heats of mixture are symmetric. Their maxima are close to the equimolar composition. The maximum values of the heat of mixture are within the range of 300 - 800 cal/mole.

9.2 Effect of the Difference in Ionic Size on the Solubility Limits in Systems of Inorganic Compounds

The condition according to which solid solutions cannot form if atomic sizes differ by more than 15% is called by metallographers Hume-Rothery's *rule*. The same rule applied to ions replacing each other is known in inorganic chemistry as the Goldschmidt *rule*.

As has been noted in Sect.3.3.6 the Hume-Rothery *rule* is obeyed when metallic solid solutions do form, but unfortunately, it is not predictive. It is true that the difference in atomic sizes exceeding 15% excludes solubility. However, there are many cases where the atoms of the same size do not form solid solutions. Because of electron exchange between the particles or, in other words, because of the admixture of valence states, the gain in energy in the process of separate crystallization of the components cannot be overcome by a gain in entropy. It is very difficult to calculate the electronic component of the free energy.

Nevertheless, the predictability of the close-packing approach is much better for ionic inorganic systems than for metals. The geometric factor is again of the key importance. The lattice energy can be represented more accurately if we use an additive function of pairwise interactions. Atom-atom potentials are depicted by a curve sloping in the direction of larger interatomic distances and falling steeply toward small distances. Therefore, the same major rules of solubility stated for organic crystal chemistry are valid for inorganic chemistry (Sect.10.1.1). It is important, first of all, that the substitution of all ions making up one formula unit does not result in the formation of shortened interatomic distances or voids. At the same time, because of the asymmetry of the atom-atom potential curve, the formation of voids is always more tolerable than the formation of shortened distances.

This is the reason why the Goldschmidt rule should not be understood literally. The substitution of an ion by another ion gives different results for small and for large molecules (i.e. structural unit). Here, the difference in the specific volumes of mixed substances is important and not the difference in ionic sizes. This is discussed in the monograph [9.1], which underlines for miscibility the importance of the size of the whole structural unit.

This important idea can be illustrated by the example of the substitution of Ca ions (radius 0.99 Å) for Mg ions (radius 0.65 Å). The difference in ionic radii is significantly higher than 15%. Nevertheless, in the system MgO-CaO limited solubility has been observed. As it should be expected, MgO dissolves better than CaO. The respective values are 10.8% and 3.1% (at very high temperatures). So, there is no need to reject immediately a 15% criterion. The parameters of the cubic unit cells differ precisely by 15%!

In the system MgS-CaS the solubility reaches tens of per cent on both sides of the phase diagram. This difference is explained in the literature in a rather misty way by the difference in the covalence of metal-sulphur and metal-oxygen bonds. Of course, such a difference exists, but the real cause of greater solubility is that sulphur atoms are larger than oxygen atoms. The parameters of cubic unit cells of sulfides differ by only 9%.

It is quite natural that the large difference in ionic sizes has only a small effect on the unit-cell dimensions of substances such as $Ca_7[Si_4O_{11}]_2 F_2$ and $Mg_5Ca_2[Si_4O_{11}]_2 F_2$. These substances form a continuous series of solid solutions. The literature provides hundreds of such examples and the above-considered rules are fulfilled with practically no exceptions.

Now we want to say a few words about the effect of pressure on isomorphic substitutions in inorganic compounds. This problem is of interest, first of all for geochemists. As it should be expected, the solubility limit is lower when large ions replace smaller ones, and higher when the formation of a solid solution results in a decrease of the unit-cell volume. Such a behavior is natural when a unit-cell parameter varies linearly with concentration. When the lattice is porous and such a deviation from linearity is encountered rather than an increase in density, then the pressure, which makes the packing denser, reduces the solubility limit.

Many systems have been investigated which form continuous series of solid solutions at high temperatures and decompose at lower temperatures. In most cases the critical temperature (the maximum on the bell-shaped curve of decomposition) increases with pressure. Thus, for example, the mixture of the minerals $CaMgSi_2O_6$ and $NaAlSi_2O_6$ forms a continuous series of solid solutions up to the pressure of 30 kbar, but decomposition occurs at 40 kbar at $1425°C$. For the mixture of Mg_2SiO_4 and $CaMgSiO_4$ the critical temperature is $2100°C$ at 2 kbar and $2300°C$ at 10 kbar. Such examples are numerous. If a system possesses a limited solubility, an increase in pressure "elongates" the eutectic line. The effect of pressure on the critical temperature of decomposition of mixed crystals of KCl and NaCl has been investigated in detail. In the range between atmospheric pressure and 20 kbar the maximum of the decomposition curve increases by $200°C$.

It is easy to take into account the pressure effect within the framework of simple theories considered below.

9.3 Theory of Ionic Substitutional Solid Solutions

9.3.1 General Remarks

It is quite obvious that a rigorous theory of solid solutions cannot be formed. It is true, nevertheless, that for inorganic solid solutions we can accept the model of additive pairwise interaction with more confidence than for metals. Yet, in the majority of cases we cannot estimate the effect of mixed valence states on the ionic interaction. Also, one cannot calculate the potential energy of interionic overlap repulsions, which is most important in the theory of solid solutions.

Thus we must limit ourselves to semiempirical theories and try to limit as much as possible the number of free parameters.

The effects of the lattice distortions, thermal vibrations, and short-range order on the free energy of solid solutions can be considered, if with some difficulty. However, since the energy of mixing of solid solutions does not exceed 1% of the lattice energy, and since the latter reaches high values due to the electrostatic interaction between the particles, we can use the model of regular solutions. Moreover, it seems admissible to consider the lattice energy instead of the internal energy.

We believe that the principal theoretical problems can be solved by means of detailed consideration of the simple, but frequently encountered example of isomorphous replacement of a monovalent ion by another ion. Many theoretical investigations are devoted to this problem; the replacement of a cation or an anion in alkali-halide compounds is often taken as an example.

Consider the system of the KCl-NaCl type. The replacement of a K ion by a Na ion or vice versa results in an increase of configurational entropy thus favoring the replacement. But the equilibrium distances between the atoms are changed, and this tends to prevent the substitution. These two competing factors usually determine the type of the phase diagram. If the loss of lattice energy is insignificant, then continuous series of solid solutions are formed which remain stable down to the lowest temperatures. If the configurational entropy exceeds, at some temperature, the losses in lattice energy, then a continuous series of solid solutions will exist at high temperatures, but will decompose at a lower temperature. In the phase diagram, we shall see a bell-shaped curve of decomposition below the solidus curve. If the losses of lattice energy can be compensated by a gain in entropy only when limited replacement takes place, a phase diagram with a eutectic or a peritectic point will form.

When considering isomorphic substitutions in the systems with common cations or anions, the framework of the lattice built by the shared ions is taken to be constant, i.e., the structure type is not changed. When a small particle is substituted for a large one, the lattice parameter smoothly decreases; in the reverse case it increases. One can be sure that the parameter varies almost linearly if only because of the fact that the difference in unit-cell parameters of the components seldom exceeds 10%.

The slow change in the electrostatic potential with a change in the interatomic distance as compared to the change in interaction of all nonbonded atoms makes it possible to believe that the electrostatic energy of the crystal mixture does not differ from the energy of the solution of the same composition. The value of the energy of mixing is determined, we believe, by the difference in interaction energies of K ions and Na ions with Cl ions, their nearest neighbors.

The use of a computer makes it possible to improve our model by taking into account the interaction of impurity ions with several coordination spheres.

It is easy to introduce any dependence of the unit-cell parameter on concentration. It is not difficult to take account of the short-range order. It is expedient to limit oneself to rough models, which are, in fact, very valuable. They are the major tools in the hands of a creative technologist, as they permit fast estimation of possible experimental results.

Denote the replaced ion by the index 1. R_1 is the distance between the ion and its nearest neighbors in the pure component I. The corresponding distance in the crystal of the second component is denoted by R_2. We assume a linear dependence of the distance R_x between the "average" ions on the concentration x is

$$R_x = R_1 + (R_2 - R_1)x \quad . \tag{9.1}$$

Interaction potentials of the host and impurity ions are denoted by φ_1 and φ_2, respectively, where x is the fraction of ions 2 which have replaced ions 1. The energy of mixing can be written in the form

$$\Delta U = x\varphi_2(R_x) + (1 - x) \varphi_1(R_x) \quad . \tag{9.2}$$

As we deal with the energy of mixing, the potentials are calculated as a function of the energy of interaction in the pure components, i.e., $\varphi_1(R_1)$ and $\varphi_2(R_2)$ are equal to zero.

For atom-atom potentials a universal one-parameter function can be taken. The equilibrium distances between the interacting atoms (which are parameters of the universal atom-atom potential) are not equal. In our example they are larger for Na-Cl. Substituting the expression for φ into (9.2), we obtain a bell-shaped function whose maximum is shifted toward the atom of smaller size, i.e., the Na atom. Two free parameters can be used to fit the calculated and experimental curves. The values of these two parameters can be obtained from the calculations of lattice energy of pure components.

9.3.2 Review of Suggested Theories

We follow here the review given in [9.1]. In the earlier theories the authors solved the problem according to the scheme discussed in the previous section. We can mention a number of investigations which differ only in the choice of atom-atom potentials. Thus, WASASTJERNA [9.11] separated from the expression for the energy of mixing the electrostatic term and wrote the heat of the formation of solid solution as

$$\Delta H = -\frac{M}{R} + x\frac{M}{R_1} + (1-x)\frac{M}{R_2} + x\,\Psi_1(R_x) + (1-x)\,\Psi_2(R_x) \quad . \tag{9.3}$$

By expanding this expression into a series and ignoring the third and higher powers of $(R_2 - R_1)/R$, Wasastjerna came to an expression with five parameters. WASASTJERNA and HOVI [9.11,12] showed that these parameters can be expressed by experimental data. Roughly, the heat of mixture can be represented in the form

$$\Delta H = x(1-x)\frac{M}{2R}\left(\frac{\Delta R}{R}\right)^2 \left[x\,\Theta_2 + (1-x)\Theta_1\right] \quad , \tag{9.4}$$

$$\Theta = \frac{3 + 4T\alpha}{M\beta/6R^4 - T\alpha} \quad , \tag{9.5}$$

where M is the Madelung constant obtained from the lattice energies of the components, β is the isothermal compressibility, and α is the thermal expansion coefficient. If all these values are known for both components, the theory can be compared with experiment.

Within the same approximation (linear dependence of interatomic distance on concentration, series expansion up to second order) attempts were made to take into account the short-range order and distortions of the solid solution lattice [9.13,14]. For certain systems the agreement is excellent; for others it is not satisfactory [9.15].

Another model describes the possible shift of ions from the equilibrium position and considers the probability of various types of lattice distortions; HEITALA [9.16], using the same methods as Wasastjerna and Hovi, obtained the following simple expression for the heat of formation of solid solution

$$\Delta H = 4Ax(1-x)\,[1 + B(x_1 - x_2)] \quad . \tag{9.6}$$

Here A and B are free parameters whose order of magnitude can be evaluated if some physical constants of the components are known. This formula contains a term which is responsible for the asymmetry of the bell-shaped curve of the energy of mixing, displacing the curve in the appropriate direction (i.e., towards the ion of smaller size).

Many theoretical investigations have been carried out to study the behavior of an impurity atom in very dilute solutions. The interest was in the character of lattice distortions around the impurity atom. This interest is connected with the growing concern with preparing highly pure materials.

All these calculations require the introduction of a large number of arbitrary constants. These laborious calculations which use various models of ion displacements and take into account the variations in the repulsion, dipole-dipole, dipole-quadrupole, and other energies do not usually give any especially valuable results which could not be obtained by a less complicated approach.

Of course, all the difficulties inherent in the theory of mixed crystals of metals are also encountered in inorganic crystal chemistry when substitutions of ions with different outer electron structure are considered. Potassium and silver, calcium and cadmium, sodium and copper are considered to have the same ionic radii. Nevertheless, in some inorganic compounds, in which these ions occur, their mutual solubility is very limited or absent.

Since the partial exchange of electrons by atoms can be estimated only by means of quantum-mechanical methods, which are not easy, one must resort to semiempirical theories. We describe here one of such attempts [9.17].

The heat of mixture is represented as a function of the energies of atomization, and not in terms of lattice energies of solid solutions of the components. The energy of atomization is written in the form

$$E = -M\varepsilon^2/R + \Psi(R) \quad . \tag{9.7}$$

Here ε is the effective charge of the atom, and $\Psi(R)$ is the short-range component of the energy. The author assumes a linear concentration dependence of the lattice parameter and the validity of the formula $\varepsilon = x\varepsilon_1 + (1-x)\varepsilon_2$; the calculations are carried out by Wasastjerna's scheme, i.e., by using a series expansion in terms of the small parameter $\Delta R/R$, neglecting terms greater than second order.

Derivatives in $\Psi(R)$ can be expressed, as it is done in Wasastjerna's theory, in terms of the experimental thermal expansivities and compressibilities of the components. Fortunately, the expression of the heat of mixture includes only the difference between the effective charges of the components. Thus the heat of mixture can be expressed in terms of experimental quantities and only one arbitrary parameter, the difference of the effective charges of component atoms. The bell-shaped curve of decomposition obtained in this calculation is asymmetric and shifted in the desired direction, i.e., towards the substitution of larger ions for the smaller ones. Despite many arbitrary assumptions, it is possible to achieve for many systems good agreement with experiment by a trial-and-error technique with only one parameter.

Many papers are devoted to the interaction energy between ions as functions of a great variety of parameters. However, we shall discuss only one of these papers. GROENWEGEN and HUISZOON [9.18] calculated the Debye-Waller factor for structures of the sodium chloride type. The ion-interaction model which was used in this work coincides basically with KELLERMAN's model [9.19]. The second coordination sphere is included for the short-range interaction. The interactions were assumed to be pairwise and additive, and were divided into Coulombic long-range forces and short-range interactions. The model parameters were the effective ionic charges and the first and second derivatives of the short-range pairwise potentials taken at the equilibrium interatomic distances. Groenwegen and Huiszoon calculated the parameters using the frequencies of the longitudinal and transverse optical phonons, three elastic constants, and the lattice parameters. On the grounds of certain quantumechanical calculations, it was possible to neglect the non-electrostate interaction of the positive ions. With all these assumptions, the model had five parameters. The study indicated that the Debye-Waller factors could be predicted quite successfully. Interesting theoretical investigations published recently are [9.20 - 22].

9.4 Interstitial Solid Solutions

The term "interstitial" seems not to be very felicitous. But it has been used for such a long time that it seems unwise to change it. An interstitial solid solution is a solution in which a foreign molecule does not substitute for a host molecule, but occupies another crystallographic position.

Similar to the substitutional solid solution, the interstitial solid solution can be ordered or disordered. As has been shown by KHACHATURYAN [9.23,24], the theory of order-disorder transformations can be used for both substitutional and interstitial solid solutions.

The distinctive feature of interstitial solid solutions is that they cannot form a continuous series of solid solutions. When an interstitial solid solution is formed the unit-cell volume either remains constant or increases.

Interstitial solid solutions can be formed beginning with a zero concentration of the impurity. Most carbides, nitrides, and metal hydrides are solid solutions of this kind. Atoms of metals usually form simple structures which can be considered as close-packed spheres or cut spheres (Sect.3.2). Carbon, nitrogen, and hydrogen atoms in a valence state unusual for them can easily occupy the voids of such packings. Nevertheless, very often an increase in impurity percentage is followed by a *lattice reconstruction* made necessary to maintain the packing density.

Since the lattice is sensitive to distortions, in interstitial structures not all the octahedral and tetrahedral voids can be occupied. Thus in carbides and nitrides the fraction of filled voids usually does not exceed 30%. Small hydrogen atoms can occupy up to 3/4 of the close-packing voids.

The nature of the bonding of the interstitial atoms and atoms forming the framework of the lattice is of no importance. Here, as usual, the geometric factor plays the key part or, in terms of energy, it is the short-range component of the atom-atom potential which is important.

We believe that the term "vacancy" should not be used in all the cases when a crystallographic position is not completely occupied by atoms, because it gives the erroneous impression that the system is thermodynamically unstable. In fact, the system would be unstable if by some artificial technique we could fill all the voids of the close-packed spheres in carbides, nitrides, or hydrides. This rarely happens. In most cases thermodynamically stable inorganic structures built by large cations and small anions have many vacancies. It is much better to say in such cases that not all the crystallographically permissible positions are filled with atoms.

Now, let us discuss the interstitial solid solutions in which the main lattice is built by anions, whereas the cations are in some way distributed over octahedral and tetrahedral voids. Such a solid solution whose range of existence begins with a zero concentration is rarely found. Thus, the existence of a continuous series of solid solutions of a metal and its oxide (or sulphide) is impossible. A *continuous solubility* of two inorganic substances of the composition AB or AB_3 is also impossible. But *intermediate phases* of variable composition are frequent. These solid solutions are usually called *nonstoichiometric chemical compounds* [9.25] or *berthollides*, probably for historical reasons, and are described as AB_x, ABC_x, or AB_yC_x. Since these nonstoichiometric compounds exist as solids and decompose during solution or fusion into two or more components, it is clear that we deal here with multicomponent systems. Thus, the above formulas can be justified by the fact that metals often can form several oxides, fluorides, or sulphides. Therefore if x or y changes within a small range, it is difficult to say which system, binary or ternary, is under consideration. In the way of example, the oxide TiO_x can be considered as an intermediate phase of a variable composition of the titanium-titanium dioxide system or as a phase of a system formed by two different titanium oxides.

Differences in terminology are of minor importance. It is sufficient to know that nonstoichiometric compounds existing within some concentration range are solid solutions. It is very important to interpret unambiguously the available structural results. Unfortunately, the literature on mixed inorganic systems contains many inconsistent data and dubious statements. This can be explained, first and foremost, by the difficulties in carrying out careful X-ray experiments for these systems: the *powder method* often leads to erroneous conclusions, whereas it is very difficult to grow and investigate a single crystal.

Despite the fact that oxide and chalcogenide crystals are the systems whose atoms are not bonded by a "pure" ionic bond, the close-packing approach applied to interstitial solid solutions permits one to understand, usually with no difficulties, the rules of solubility. We will discuss only some simple examples.

A great number of intermediate phases of the composition of MeO_x (O standing for oxygen, sulphur, or selenium, etc.) with the simple NaCl structure are known. The concentration range of these solid solutions is rather wide. For example, for TiO_x the values of x are within the interval from 0.7 to 1.25 (all the data are borrowed from [9.25]). The most widespread viewpoint is that atoms of Ti and O have "their own" places in the lattice. If $x < 1$, there are vacancies in the nodes of the face-centered lattice of oxygen atoms; if $x > 1$, there are vacancies in the nodes of the Ti lattice. Then at $x = 1$ we have an ordered structure such that one Me and O atom are absent per unit cell.

Of course, there is no reason to deny a priori the possibility of the formation of *superstructures* at low temperatures. But conclusions about the possible formation of a lattice in which both anion and cation positions are vacant seem to be very dubious. If in TiO_x solid solution an oxygen atom is present as a dinegative ion its diameter is equal to 2.8 Å. Unit-cell dimensions of solid solutions are known. The half-length of the face diagonal of a unit cell of a TiO crystal equals 2.9 Å. Thus, if titanium or any other metal in the crystals of a solid solution is in the form of a multicharged positive ion, its size is very small and the TiO_x structure can be considered as *close-packed spheres of oxygen anions* with randomly distributed metallic cations, at least at high temperatures. However, in TiO_x the use of the term "*vacancies*" in the case when not all the octahedral or tetrahedral positions of the close-packed spheres are occupied can be measleading. Similarly, it is incorrect to state that a significant fraction of the spheres forming the lattice framework in TiO_x is absent: this would result in significant distortions of the lattice, which in turn would lead to large energy losses.

Consider now a simple geometrical model of solid solutions formed by two binary inorganic compounds. A careful analysis shows that in all cases the geometrical factor is the major fact determining the possible solubility and the interval of concentrations of intermediate solid solutions. It is especially obvious when the structure of mixing crystals can be represented as close-packed anions. By starting with a cubic close-packed structure and filling the octahedral voids, or one set of tetrahedral voids, or two sets of tetrahedral voids, or all the octahedral voids and one set of tetrahedral voids, or all the tetrahedral and octahedral voids, we can obtain the NaCl, ZnS, CaF_2, CuMgSb, and BiF_3 structure types, respectively.

All these structure types differ only in the pattern by which voids are filled by cations. In a certain range determined by the relative sizes of anions and cations, we can predict the formation of solid solutions from the components of such structure types.

For example, intermediate solid solutions are formed within wide concentration ranges for the systems CaF_2-YF_3, SrF_2-LaF_3, and CaF_2-ThF_4. The volumes of the fluorides being mixed differ by 2 - 5%. In the system SrF_2-LaF_3, solubility starts at zero LaF_3 content. The lattice parameter, as it should be expected, increases: the specific volume of the second component is somewhat higher. However, the presence of 33% of the second component increases the lattice parameter by only 1%.

A similar situation is observed in other systems. In the PbF_2-BiF_3 system, solid solutions exist in the full range of the values of x from 2 to 3. The unit-cell parameters of the components differ by less than 1%.

Intermediate solid solutions of variable composition are also observed in the system La_2O_3-U_3O_8 for the value of x from 1.8 to 2.4. If the component formulas are rewritten as $LaO_{1.5}$ and $UO_{2.67}$, and volumes for these formula units are calculated, the obtained values are 55 and 47 $Å^3$, respectively. The difference in volumes is about 15%, or about 5% in linear dimensions.

Consider one more example, that of solid solutions with the structure intermediate between that of nickel arsenide (type B8) and the structure of type C6. Both structures can be considered as hexagonal close-packed anions whose octahedral voids are filled with Me atoms. The difference between B8 and C6 types consists in the following: if we draw planes normal to the hexagonal axis and passing through all the layers and anions A and through the centers of all the octahedral voids, it will turn out that in the B8 structure, cations are in the planes passing through the centers of the voids, while in the C6 structure, a layer of cations C alternates with an "empty" plane O. The scheme of layers is -A-C-A-C-A-C-A- in the first (B8) case and -A-O-A-C-A-O-A-C-A- in the second (C6) case (O denotes an "empty" layer).

Thus we can proceed from the structure A_2C to the structure AC, filling step by step all the octahedral voids with cations. Solid solutions of this type are formed by many chalcogenides. If we write the formula for a phase of variable composition as A_xC, then in the majority of cases the values of x are within the range of 1.5 to 2. Filling of voids usually starts with zero content of AC. Chalcogenides of the composition AC are rare (TiTe exists).

In all the above-considered examples the role of the packing factor is obvious. Solid solutions can be formed only when the density of the packing permits it, because the gain in entropy is not really high enough to compensate for the loss in interaction energy between the atoms.

It is also possible to see the key role of the short-range component of the potential. As in the case of substitutional solid solutions, the nature of chemical bonding of the atoms is less important. A simple theory for solid solutions which uses only two free energy terms (universal short-range atom-atom interaction and configurational entropy) seems to us to be most successful.

Only further investigations will show to what extent such a statement is justified. Inorganic crystal chemistry is such a vast and diverse field that we cannot be sure of the correctness of our assertions in all cases, since only a small part of the enormous volume of experimental data has been considered.

10. Conditions of Formation of Substitutional Organic Solid Solutions

The conditions of solubility in organic systems are specific enough to be considered separately. First, there is usually no electron transfer between the particles building the crystal. Second, they are nonspherical and sometimes of very intricate shape. In addition, particles can be asymmetric.

Here two approaches are possible. The first and foremost is the *geometrical approach*, in which we consider the shape of the host molecule and that of the guest molecule and use the packing density in the mixed crystal as well as the symmetry of the crystals of the pure components. Using such an approach, we can make only qualitative predictions: from geometrical analysis we can state only whether solubility is possible or not, and whether it is small or significant. As to continuous solubility, negative statements ("continuous solubility is impossible") can be made with fair ease, but positive statements must be hedged as probable statements.

The second approach is quantitative. As the method of atom-atom potentials can calculate the interaction energy of the molecules, we can at least determine a priori the maximum value of the substitution energy. This allows good predictions of solubility.

All the theories in which thermodynamic parameters are expressed in terms of pairwise interactions of particles can be used and checked for organic mixed crystals because one can carry out numerical calculations by the method of atom-atom potentials.

In this chapter we shall discuss the effectiveness of both methods for predicting the conditions of solubility and formation of *substitutional solid solutions*. In Chap.12 we will show how they can be applied to interpret experimental data.

The formation of *interstitial solid solutions*, where an admixture molecule occupies a crystallographic position which should be vacant in the host crystal, will be considered in Chap.13.

10.1 Geometrical Analysis of Substitution

Since in most organic crystals there is no electron transfer between particles, it is clear that the geometrical factor plays the key role and, with some reservations, even the decisive one.

10.1.1 Major Rule of Solubility

Trying to predict solubility, we should compare, first of all, the shapes of the host and impurity molecules. This procedure was discussed in Sect.3.3.6. The molecules, contoured by intermolecular radii, should be superimposed as well as possible, and then the volumes of superimposed and nonsuperimposed parts of the molecules should be calculated. Now, the coefficient of geometric similarity ε (Sect.3.3.6) is determined. Experience shows that values of ε differing from unity by more than 20% preclude solubility. Thus the similarity of the shapes of molecules is a necessary condition of solubility.

CHANH et al. [10.1] investigated the solubility of naphthalene and its β derivatives. If a hydrogen atom is replaced by the groups OH, NH_2, Cl, CH_3, and SH, the coefficients of geometric similarity are 0.95, 0.93, 0.89, 0.87, and 0.85, respectively. Solubility is observed in all these cases. We shall analyze these interesting results quantitatively in Sect.12.6. Here we should like to underline the following. The phase diagrams of naphthalene with its α derivative show that α derivatives cannot be dissolved in naphthalene. The coefficient of similarity ε for α substituted naphthalenes and naphthalene is practically the same as for β substituted naphthalenes.

The point is that the comparison of the shapes of molecules is only the initial step in the prediction of solubility. If ε differs significantly from unity, we can state with confidence that solubility is impossible. If ε is close to unity we can say nothing. The geometrical consideration of the problem consists in the following. Let us imagine a host molecule to be removed from the lattice and replaced by an impurity molecule. Then it should be determined how significantly intermolecular distances have changed. It is possible that the "protruding" parts of this substituent occupy a lattice site where the packing is loose. Then the substitution does not lead to the appearance of shortened intermolecular distances. In this case solubility is possible. Vice versa, if the "protruding" part of the substituent molecule occupies a site which was tightly packed before the replacement, the distortions will be great and solubility is impossible.

Such is the case of naphtalene replacement by the molecules of β- and α-chloronaphthalene. In the former case the replacement of a hydrogen atom by a halide results in a small contraction of intermolecular distances (by 0.2 Å). In the latter case the α hydrogen in a napthalene crystal is at a distance of 2.82 Å from the carbon atom of the neighboring molecule. This value is only slightly less (by 0.1 - 0.2 Å) than the equilibrium non-bonded distance. If this hydrogen atom is now replaced by a chlorine atom the intermolecular distance becomes less than the standard one by 0.6 Å.

Now we can formulate the basic *rule of solubility*. The molecule A dissolves in the crystal B only if the coefficient of the geometrical similarity of the molecules ε exceeds 0.8, and if a replacement of the molecule A by the molecule B does not disturb significantly the *molecular packing*.

But what is a significant disturbance of the packing? We can answer the question as follows. It is well known that in different crystals the contact distances vary within 5%. If, on substitution, the contraction of interatomic distances amounts to 10 - 15%, the solubility is doubtful. We must make one more important remark. An *increase in density* up on substitution is less tolerable than an *increase in porosity* (= decrease in density). This is quite understandable as the curve of the atom-atom potential is steeper in the direction of small distances and smoother towards large distances. Such a situation is observed in the systems acridine-anthracene [10.2], stilbene-dibenzyl [10.3], napthalene-β-chloronapthalene [10.4], and naphthalene-thionaphthol [10.1]. This regularity is especially prominent in the case of solid solutions of paraffins [10.5,6].

It seems that a decrease in interatomic distances of about 10% and a decrease in density of about 20% are still tolerable. Of course, the number of deviations from normal contact distances is very important.

The less the packing of the host molecules is disturbed by the "guest" molecule, the higher the solubility. However, such a statement can be made only with some reservations. First of all, the short-range order is very important. The formation of obvious short-range order with an increase in concentration can lead to the intolerable lattice distortion.

Most important, however, is that in the often-encountered case of eutectic systems with limited solubility, two solid solutions with different structures are in equilibrium. To know whether the solubility is high or small, we must know not only how the "guest" molecule A spoils the packing of the molecules B, but also to what extent the A packing is impaired when the impurity molecule B enters the A crystal. The geometrical analysis of a system such as acridine-anthracene shows that the disturbances of the anthracene lattice are small, whereas those of acridine lattice are significant. This explains the very high solubility of acridine in anthracene. In the dipyridyl-diphenyl system [10.7], which has the same type of phase diagram, the solubility is small on both sides of the phase diagram. The packing violations are also small and are of the same order on both sides of the diagram.

10.1.2 Exceptions to the Basic Rule of Solubility

It would be more accurate of us to speak of the two additions, rather than of two exceptions to the basic rule.

The first additional rule states that solubility is impossible if the impurity molecule would break the intermolecular *hydrogen* bonding network. As is known, the energy of the *hydrogen bond* is rather significant, and the gain in configurational entropy cannot compensate for the violation of the framework of bonds.

In organic compounds, hydrogen bonds can promote the formation of strongly interacting pairs of molecules (β-naphthol), islands of several molecules bound by these bonds, chains of molecules, ordinary or double layers (glycin), and crystals in which hydrogen bonds form a three-dimensional framework. In the latter case (resorcinol for instance) the melting point of the crystal is significantly higher than that of compounds with similar molecular weight, that lack such hydrogen bonds.

In some instances it is possible to imagine that the "guest" molecule enters the host crystal without breaking hydrogen bonds. For example, it is quite possible that fluororesorcinol dissolves in resorcinol. But the molecule of dichlorobenzene or benzene, most likely, will not dissolve in resorcinol (here, we take the risk of predicting the results without making experiments).

In many organic crystals the molecules are bound by hydrogen bonds in pairs. The replacement of such a pair by two molecules containing no hydroxyl or amino groups is, of course, accompanied by some loss of lattice energy. However, in such a case, the loss of energy is less significant than that due to a breakdown in the bonding network. Therefore in such a case small solubility is probably possible. The reverse situation is quite possible: undoubtedly cases exist where molecules with hydroxyl and amino groups can enter a crystal with no hydrogen bonds. Such molecules can enter the host crystal in pairs: then the packing is not violated, and there is a gain in energy due to the formation of local hydrogen bonds between pairs of molecules.

The second addition to the basic rule of solubility concerns molecules with permanent dipole moments. Here, we must notice that though some bonds have dipole moments, the molecular packing is not affected if the sum of the bond dipole moments within a molecule equals zero. Thus in the molecule of p-nitrobenzene the dipole moments due to the bonds between the 1,4 carbons and the nitro groups are oppositely oriented. The molecule as a whole does not possess a dipole moment, and its behavior in the packing of a one-component or mixed crystal is the same as that of a non-polar molecule. Moreover, as far as it can be judged by the meager experimental data and proceeding from general considerations, dipole moments do not affect the solubility provided that the molecules in the crystal are oriented so that two (or more) molecules form an island with no net dipole moment. Such a substitution does not affect the electrical part of the energy (examples are supplied in Sect.12.4).

It can be expected that solubility is not possible in only one case, namely, when the crystal as a whole possesses an electrical moment (*electret*). This is the (rather rare) case where all the dipole moments of all the molecules have the same direction. This situation has been encountered in studies of solid solutions of coumarine with naphthalene [10.8], as discussed in Sect.12.4. Even one molecule of naphthalene entering the coumarine crystal must change the energy of dipolar interaction: thus, solid solution does not form despite the close similarity of shapes and sizes of the mixed molecules.

Thus, whereas for metallic alloys only predictions of insolubility could be made, for organic systems "positive" predictions of mutual solubility can be made with fair accuracy.

10.2 Conditions of Continuous Solubility

Some additional conditions are necessary for the formation of a continuous series of solid solutions. These conditions are quite obvious for atomic crystals: the crystals must belong to the same space group, the atoms must occupy the same crystallographic positions, and the unit-cells must be similar in dimensions. These are also valid for molecular crystals. However, the crystal consists not of spherical particles but of molecules which can possess different symmetries and shapes, thus the conditions of continuous solubility become more complicated. Some interesting features appear which are worth special consideration.

10.2.1 Crystals of the Components Belonging to the Same Structure Type

The same space groups and the same number of molecules per unit cell are not sufficient conditions for the formation of a continuous series of solid solutions. Even when, in addition to the similarity of the symmetry, the shapes and sizes of the unit-cells are also similar, we still cannot infer that the components will form a continuous series of solid solutions. Even under these conditions partial solubility may be impossible. About half of all the known molecular crystals belong to the space group $P2_1/c$. Varying the arrangement of the molecules in this low-symmetry space group, we can create different packing modes but the unit-cells will be similar in shape and size.

For instance, continuous solubility does not occur, despite equal symmetry (same space group) and similar molecular shape, in two systems: the diphenyl-mercury-tolan system (considered in detail in Sect.12.9) and the anthracene-phenazine system. In these two instances the molecules are *tilted* with respect to the unit-cell axes in a *different way*; the molecular packings are essentially different and cannot transform continuously into one another.

This can be understood by considering the energy of molecular interaction in organic crystals, which can be represented by a multidimensional surface. Such surfaces have many minima: the deepest minimum corresponds to the real structure and is separated from shallower minima by potential barriers. Energy surfaces of the molecules of similar shapes are also similar, but the relative depths of the several minima can be different, and new minima at modified orientations may become energetically favorable. In such cases structures with essentially different molecule packings are realized: unit-cell parameters and the molecular orientation relative to the unit-cell axes are changed.

Continuous solubility can be realized between two substances if their energy surfaces are similar not only in general features, but also in the *location* of their *deepest minima*. Strictly speaking, the crystals are related to the same structure type if both structures "are in the same potential well".

In Fig.10.1 structures A and B are essentially different in terms of their multidimensional parameter p, though the energy surfaces are quite similar.

We have seen that for crystals to have the same symmetry (even with similarly shaped molecules) is not a sufficient condition for the formation of continuous solid solutions. But another fact seems to be more interesting and specific for molecular crystals: equal symmetry is not even a necessary condition for the formation of continuous solid solutions.

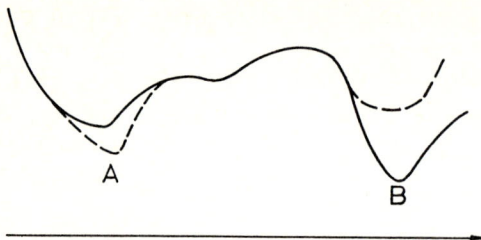
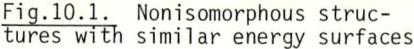

Fig.10.1. Nonisomorphous structures with similar energy surfaces

10.2.2 Continuous Loss of Symmetry

This specific situation can arise if the molecule of the substance A is asymmetric, whereas crystals B are built by symmetric molecules or are racemic or pseudoracemic. Asymmetric molecules are those which possess neither a center of inversion nor a symmetry plane. Such molecules can exist in two forms (right- and left-handed) and cannot be superimposed by translation operations. They can be transformed into one another only by an imaginary reflection through a mirror.

Such *asymmetric (chiral*, or handed) *molecules* can form symmetric crystals. In such crystals half of the molecules are left-handed, whereas the second half are right-handed. They are situated in pairs related by a center of inversion or by a symmetry plane. If all the unit-cells are identical, the crystal is called a racemate. Sometimes left- and right-handed molecules are distributed in the crystal randomly, and the center of inversion (or the plane of mirror symmetry) exists "on the average". Such a crystal is called a pseudoracemate.

An organic compound consisting only of left- or right-handed molecules crystallizes in the majority of cases in the asymmetric monoclinic space group $P2_1$ or the orthorhombic group $P2_12_12_1$.

If a host crystal built of asymmetric molecules A is penetrated by a "guest" molecule B similar to A in shape and dimensions, a substitutional solid solution is formed, the symmetry of A remaining the same. It is not so if the asymmetrical molecule A enters the symmetrical crystal B (for example, a racemate). Several left- (or right-) handed molecules cannot, in principle, imitate either the center of inversion or the symmetry plane. Yet, if it is allowed by the geometrical similarity, a substitutional solid solution can be formed. But when the first A molecules enter the crystal B, the latter immediately loses its symmetry (formally, already at zero doping by A). With an increase in the fraction of A molecules this loss of symmetry becomes more and more appreciable. Strictly speaking, the title of this section is misleading, since the loss of symmetry occurs immediately, and the crystal B loses its center of inversion or symmetry plane even in the presence of the smallest fraction of molecules A in the crystal. However, the physical properties of crystal B change with an increase in the concentration of molecules A in such a way as if the loss of symmetry occurred continuously.

If the left- and right-handed molecules are similar in geometrical form, we can obtain continuous series of solid solutions, the condition of identical symmetry of components being violated.

One can derive all the cases of the formation of such continuous series. More than twenty years ago several such possibilities were pointed out [Ref. 10.9, p. 234]. Thus, one can pass continuously from a crystal of the space group $P2_1/c$ with four molecules per unit-cell to a crystal belonging to the space group $P2_1$ with two molecules per unit-cell.

A molecule of naphthalene (space group $P2_1/c$, two molecules per unit-cell) can lose its center of inversion if two centrosymmetric naphthalene molecules are replaced by left- (right-) handed molecules having a similar shape. The crystal does not "feel" the entrance of the first few molecules, but with an increase in the number of "guest" molecules, a piezoelectric effect appears as well as additional X-ray reflections forbidden for the centrosymmetric space group. With further increase in the concentration these effects become more intense. However, we know of no practical examples for this case.

Let us consider another case. The crystal A consists of two left-handed molecules and belongs to the space group $P2_1$. The racemate consisting of right- and left-handed molecules has the space group $P2_1/c$ with four molecules per unit-cell. If the difference in shape of the left- and right-handed molecules is insignificant, solid solutions can form. Such systems do exist, and there are about several dozens of phase diagrams available in the literature which show continuous solubility between the racemate and left- (right-) handed modifications.

When the first left-handed molecule enters the crystal, the loss of symmetry occurs immediately, but it manifests itself only gradually. A subcell starts to form, and only when the number of molecules A reaches 100% does the unit-cell volume decrease by half. Thus if a solid solution forms between an optical enantiomer and a racemate, we observe two jumps from both sides of the phase diagram. From the side of the enantiomer, unit-cell dimensions change by a jump, and from the side of racemate, the symmetry also changes suddenly.

In a series of excellent studies in France (Sect.12.7) such binary systems of enantiomers and racemates have been investigated in detail by X-ray diffraction.

10.3 Energy Calculations

10.3.1 Approximate Estimation of Mixing Energy

From the general formula (4.6,7) it follows that the mixing energy can be described by

$$\Delta f_m = \Delta V + \Delta f_{dist} + T\Delta S_c + \Delta f_{vibr} + \Delta f_{ord} + \Delta f_{conf} \quad . \tag{10.1}$$

Here, the electronic energy term is omitted. The formation of solid solutions is impossible even if the mixing energy is positive at the smallest concentrations of x: if x = 0.01, then $T\Delta S_c$ = 0.03 kcal/mole; thus, there is no solubility if

$$\Delta V + \Delta f_{dist} + \Delta f_{vibr} + \Delta f_{ord} + \Delta f_{conf} > 0.03 \text{ kcal/mole} \quad . \tag{10.2}$$

As was mentioned in Sect.4.1 it is most convenient to consider separately two aspects of the effect of replacement on the potential energy of particle interaction: ΔV is given by $0.01(V_{AM} - V_{MM})$, where V_{AM} and V_{MM} are the energies of interaction of host molecules and impurity with the environment. These values assume that the lattice experienced no distortions. The deformation is taken into consideration through the term Δf_{dist}. It is obvious that lattice distortions disperse the stress resulting from the addition of a "guest" molecule. Hence $\Delta f_{dist} < 0$. As we wish to find whether solubility is possible, we should make the inequality stronger, rewriting it in the form

$$\Delta V + \Delta f_{vibr} + \Delta f_{conf} > 0.03 \text{ kcal/mole} \quad . \tag{10.3}$$

We have omitted here one more term Δf_{ord} from the general expression for free energy (4.6). Since we want to estimate whether weak solubility is possible, it can be assumed that the impurity molecule is surrounded by the host molecules.

Now only three terms are left in the formula. Two of them, ΔV and Δf_{conf} can easily be calculated with sufficient accuracy by the atom-atom-potential method. It is most difficult to account for Δf_{vibr}, the term describing the vibrational spectrum modification by the impurity molecule in the crystal. As indicated in Chap.4 there is no theoretical method of evaluating Δf_{vibr} though it is possible to estimate it qualitatively through the changes in X-ray diffracted intensities [see (4.52)]. Our studies of more than ten binary organic systems show that weak solutions give no changes in intensities: this implies that the potential energy of interaction ($\Delta V + f_{conf}$) is responsible for the solubility of the impurity molecule.

Interaction curves are more or less ordinary. Atom-atom potentials can be described sufficiently by a universal function containing only one parameter, the equilibrium distance between the interacting atoms. In such an approximation, the problem reduces to the difference in molecular shapes. This justifies the simple geometrical rule formulated in Sect.10.1.1.

Calculations show that in all cases where solubility occurs, the presence of a guest molecule either decreases the interaction energy, or increases it by an amount allowed by (10.3). We shall return to the calculations of ΔV when we discuss specific examples. Here, it is important to emphasize that (10.3) can predict with high probability the cases where solubility is impossible.

10.3.2 Possibility of Calculating Phase Diagrams

Phase diagram calculations require expressions for the free energy as a function of concentration and temperature. Chapter 4 has shown that, in general, such an expression cannot be found.

Yet, one can solve two particular problems. First, one can determine the *solubility limits* for a given type of phase diagram, provided that the crystal structure of the solid solution is known. Second, one can study the *transformations* occurring as a function of temperature within the region of existence of solid solutions by varying the theoretical arrangement of component molecules over different crystallographic positions.

The first problem is solved according to the scheme outlined in Sect.4.3: one can obtain solubility limits for organic systems by using atom-atom potentials. However, the calculations outlined in Sect.4.3 are approximate, and valid only at low temperatures. By removing the major simplifications and by taking into account different variants of the nearest neighbors (of both kinds), we can make the calculation valid for concentrated solutions. Furthermore, we can introduce into the free-energy formula the term Δf_{vibr} to account for thermal vibrations. This yields interesting results even within the framework of the quasi-harmonic approximation, i.e., when the characteristic temperature is made to depend on concentration by using two or three parameters.

It will be shown in Sect.12.1 that such a calculation predicts correctly the phases diagram of diphenyl-dipyridyl.

The second problem, i.e., studies of transformations in the region of solid solutions, will be discussed in Sect.12.9.3 with the example of the diphenyl mercury-tolan system.

10.4 Unit-Cell Dimensions of Solid Solutions

As we saw in Chapter 7, for metallic systems there is a wealth of experimental data on the changes in crystal unit-cell parameters as a function of impurity concentration. This is because it is easy to measure unit-cell parameters by X-ray diffraction, particularly for cubic or hexagonal polycrystalline substances, because such data can be readily interpreted in terms of a solubility model. Unit-cell parameters can be obtained with high accuracy by the Debye method.

Experimentally, the dependence of the unit-cell parameters of cubic-lattices on concentration usually confirms Vegard's law. Such behavior is considered to be normal. It is the linear dependence of the unit-cell parameters on the change in concentration that is unexpected. It was shown by MOTT and NABARRO [10.10] that for solid solutions this linear dependence of the unit-cell volume on concentration can be derived from the model of *elastic spheres*. This linear dependence holds when the moduli of compression of the spheres of two kinds are the same.

For continuous series of solid solutions of metals we often observe slightly convex curves. In the elastic spheres model the positive or negative deviations from linearity can be explained by the sign of the product of the difference between compression moduli times the difference between the sphere radii.

To characterize atoms by bulk values of compression moduli is somewhat artificial. Such a theory ignores the crystal's structure type and the nature of bonding between the atoms. For molecular crystals, the elastic spheres model is not valid because molecules are figures of complicated shape rather than spheres.

The general approach to organic crystals outlined in [10.9,11] may allow us to study the dependence of unit-cell volumes on concentration for solid-solution crystals.

We shall consider the meager data available for organic solid solutions and show how they can be interpreted by a *geometric model*. We shall also pay special attention to the case where impurity molecules in the matrix crystal do not change the unit-cell parameters and volume. The discovery of such cases brought the conclusion that there are two types of substitutional solubility in solid solutions of organic compounds, *true* and *interblock solubility*.

In solid solutions of organic substances a change in concentration very often leaves one or two lattice parameters unchanged. Then a change in volume, which is sometimes very significant, occurs due to the change of only one parameter. Such a behavior results in some important conclusions about the nature of substitution. We shall defer the discussion to when we consider concrete examples in Chap.12.

10.4.1 Lattice Loosening

The data on the unit-cell volumes for continuous solubility have been obtained by French scientists. Figure 10.2 shows the curves for the system of C_8 - C_{10} dicarboxylic acids [10.12]. Unlimited solubility is observed despite the significant difference in the molecular chain lengths. The coefficient of geometrical similarity ε (3.18) is 0.80. The authors had good reasons to believe that the formation of the solid solution is due to a specific "adaptation" of the molecular lengths. It can be assumed that six C_8 molecules (48 carbon atoms) can fit the space left by 5 C_{10} molecules (50 carbon atoms). The plot of volume versus concentration for the system of dicarboxylic acids C_6 - C_8 is shown in Fig.10.3 [10.13].

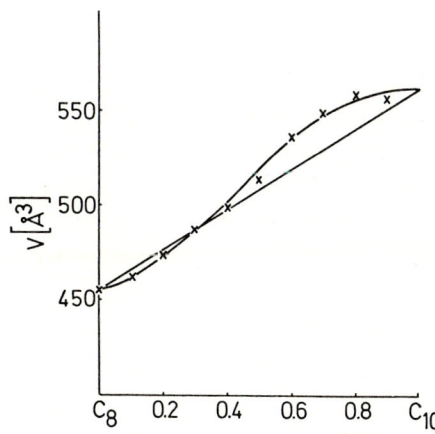

Fig.10.2. Unit-cell volumes versus composition for the system of dicarboxylic acids C_8 - C_{10}

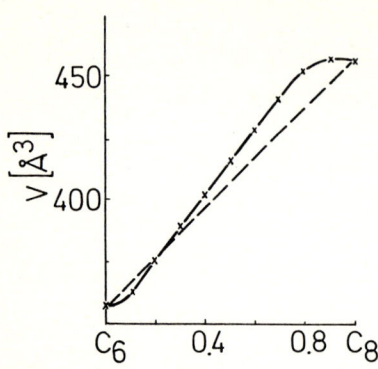

Fig.10.3. Unit-cell volumes versus composition for the system of dicarboxylic acids $C_6 - C_8$

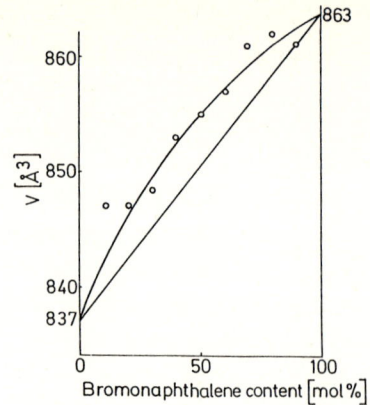

Fig.10.4. Unit cell of the low temperature forms of bromonaphthalene and chloronaphthalene

Continuous series of solid solutions are formed by many β-derivatives of naphthalene. In [10.14] the dependence of the unit-cell volume on the concentration of low-temperature forms of bromonaphthalene and chloronaphthalene has been measured. The plot of volume per unit cell (containing four molecules) is shown in Fig.10.4. The molecules of fluoronaphthalene form a crystal with orientational disorder which simulates a center of inversion. The space group is $P2_1/c$ with two molecules per unit cell. The same structure is observed for β-chloronaphthalene at high temperature. As shown in [10.15], the high-temperature modification of chloronaphthalene forms with fluoronaphthalene a continuous series of solid solutions. The plot of the unit-cell volume as a function of concentration is depicted in Fig.10.5.

The high-temperature forms of β-naphthol and of fluoronaphthalene have orientational disorder. The mixed crystals are isomorphous and form a continuous series of solid solutions. The volume curve is similar to the curve of dicarbolic acids, and can be constructed from the data of [10.16].

These examples show that there are two types of curves: convex curves and S-shaped curves which intersect a straight line at a point corresponding to some intermediate composition.

How can the difference in the shape of the curves v(c) be explained on the basis of a geometrical model? The following assertions seem to be quite obvious. The plot of volume versus concentration is linear when the *packing coefficient* is constant within the whole range of concentrations (and, consequently, is the same for all the components). If experimental points are

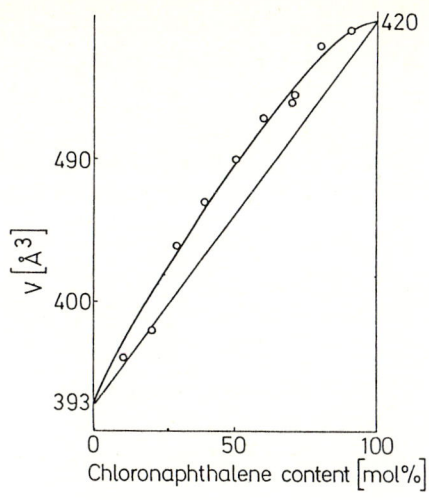

Fig.10.5. Unit-cell volume for the system of the high-temperature modifications of chloronaphthalene and fluoronaphthalene

above the straight line, it means that the structure becomes loose, i.e., the coefficient of packing on the corresponding part of the curve is smaller than that for the pure component. This means that if all the points of the curve lie above the straight line, the solid solutions of all the concentrations are "*looser*". By considering packing coefficients, it seems unlikely that all the points of unit-cell volume versus concentration curve can lie below the line connecting the unit-cell volumes of the components, as this would imply an increase in density at all concentrations. In fact, such curves have not been observed.

The decrease in the packing coefficient needs no special explanation. But the increase in the density of the arrangement of the molecules in a solid solution, even in a limited concentration range, must be explained.

For dicarboxylic acids the *increase* in the packing coefficient occurs when small molecules are replaced by larger ones. If we accept the hypothesis that molecules of different lengths "adapt" during the formation of the chain [10.12], the shape of the volume curve seems to be natural. At small concentrations of C_{10} molecules, five such molecules replace six C_8 molecules. This means that the volume for 48 carbon atoms is occupied by 50 atoms. At large concentrations of C_{10} molecules the reverse occurs, and the packing coefficient should decrease. Thus the S-shaped curve is explained.

If larger molecules of naphthol enter the naphthalene crystal, the increase in density is due most probably to a hydrogen bond between the two molecules: two naphthol molecules turned towards each other by their hydroxyl groups replace two naphthalene molecules. The density of packing increases. When the number of naphthol molecules is sufficiently large they "dictate" the type of packing. Then, it becomes more correct to say that two naphthalene molecules replace two naphthol molecules bound by a hydrogen bond, and the structure becomes looser.

Limited solubility should be considered from the same standpoint. The increase in density Δ given by (3.17) is a function of concentration. It can be expressed in terms of packing coefficients. The values of Δ, as well as the shape of the plot of volume versus concentration, are interpreted geometrically. Thus, the entry of a larger anthracene molecule into an acridine crystal drastically loosens the structure. The same occurs when the matrix is the crystal of para-dibromobenzene and the guest molecule is para-diiodobenzene. An increase in the packing coefficient is usually observed only when the guest molecules are small.

10.4.2 Interblock Solubility

Twenty years ago a conclusion was drawn that in organic systems two types of substitutional solutions are possible, *true* and *interblock* solutions [10.17]. In solutions of 1,8-dinitronaphthalene in 1,5-dinitronaphthalene and of β-chloronaphthalene in naphthalene one could grow perfect single crystals of solid solutions containing up to about 3% impurity. Precise X-ray measurements showed that the unit-cell parameters were the same as those of the crystals of the pure majority constituent.

To understand the lattice distortions caused by the impurity in both instances, the geometry of substitution was carefully examined [10.12].

In the structure of pure *naphthalene* there are closely packed layers in the ab plane ("ab layers") which are superimposed on one another. Hydrogen atoms in the β positions of the naphthalene molecule do not participate in the packing of the molecules into ab layers and when the layers are superimposed, the minimum distances between C and the two β H atoms from neighboring layers turn out to be larger than the sums of the relevant intermolecular radii (3.15 and 2.55 Å instead of 2.97 and 2.34 Å). Therefore it was suggested that the Cl atom in a β position can occupy the "porous" site of the structure. A geometrical analysis shows that at optimum substitution (the atomic coordinates of naphthalene nuclei coincide) a series of shortened distances appear (again, compared to typical non-bonded intermolecular radii). In particular, the Cl-C distance becomes 2.74 Å (shortened by 0.8 Å) and Cl-H = 2.34 Å (shortened by 0.6 Å): such "compressions" would require 9-10 kcal/mole. Therefore, instead we expect a significant increase in the length of the unit-cell edges, especially the axis c, and yet this is not observed.

Similarly, in the crystal structure of 1,5-*dinitronaphthalene* all the atoms of the molecules are in the same plane that coincides with the ab plane, thus forming a closely packed layer. The contacts between layers are realized only through the carbon atoms (c = 3.69 Å).

In the optimum method of substitution, a series of shortened distances appear, the shortest being in the ab layer: N - O = 1.87 Å (shortened by 1.2 Å) and O - O = 1.9 Å (shortened by 1.11 Å). Total energy losses in such a substitution are of about 15 kcal/mole. Again, an increase in unit-cell axes is thus predicted for solutions of 1,8-dinitronaphthalene in 1,5-dinitronaphthalene.

Yet, in other systems investigated, 2 - 3% impurity, which distorts the lattice to a lesser degree, did yield significant changes in lattice parameters. It will suffice to give one example. When 3% of anthracene enters an acridine crystal, one of the lattice parameters increases by 0.5 Å. Substitution results in compression which "cost" only 1 - 2 kcal/mole. Yet, in the two "anomalous" cases discussed above the parameters do not change, at least within an accuracy of ± 0.01 Å.

Thus, comparing experimental data with the geometrical analysis of optimum substitution and with estimates of the limiting values of substitutional energy, one arrives logically to two different *mechanisms of substitution*. The first mechanism yields the *true substitutional solid solutions*. Here, impurity molecules replace matrix molecules in the whole bulk of the mosaic block with a probability determined by the concentration. The distortions in packing arising in this process can be recorded by X-ray photography as noticeable changes in the dimensions of an "average" unit cell.

The second mechanism of substitution leads to a solid solution which we shall call an *interblock solution*. Here we assume that the molecules of impurity occupy the "defect" sites of the crystal or are located at the boundaries of mosaic blocks, "adapting" themselves in a certain way to the packing of matrix molecules. The existence of such solutions can be proved when the geometrical analysis of substitution predicts a noticeable increase in the volume of the unit cell, but this prediction is not confirmed by experiments.

Thus, we can suggest, for example, the following model for oriented distribution of β-chloronaphthalene impurity in naphthalene matrix: the naphthalene nuclei of the "guest" molecules in the block cells close to the boundaries are placed so that the atoms of Cl "protrude" from the blocks as a fringe. The estimates of the linear dimensions of the mosaic blocks in such a substitution model give reasonable results. The suggested model is also confirmed by the very *poor quality* of the grown crystals of solid solutions.

Table 10.1. The unit-cell parameters of p-dibromobenzene and a solid solution containing 4% durene

Parameters	Pure-p-dibromobenzene	96%-p-dibromobenzene 4% durene
a(A)	4.107 ± 0.003	4.106 ± 0.003
b(A)	15.484 ± 0.009	15.483 ± 0.009
c(A)	5.836 ± 0.0008	5.837 ± 0.0008
β(°)	112.693° ± 0.002	112.703° ± 0.002

The interblock type of solid solutions has also been observed in the durene-para-dibromobenzene system (Sect.12.8). The impurity durene (1,2,4,5-tetramethylbenzene) molecules whose volume exceeds that of the host molecules by 16% should bring significant changes in the unit-cell dimensions in the case of a true solution. Indeed, geometrical analysis proved that both possible methods of substitution (Sect.12.8.2) would lead to an increase in energies by more than 10 - 15 kcal/mole. The unit-cell parameters for the durene-p-dibromobenzene system are listed in Table 10.1. Since there are no significant changes in unit-cell parameters it seems that interblock solubility takes place. Impurity durene molecules replace host molecules only in the "defect" sites and at the boundaries of mosaic blocks. As the number of impurity molecules does not exceed 4 per 96 host molecules, such a substitution does not deteriorate the crystal quality.

The requirement of partial geometrical correspondence of the host and impurity molecules necessary for the formation of the interblock type of solid solution was exploited by BELIKOVA and BELYAEV [10.18] for the creation of new luminescent crystals: naphthalene and anthracene were used as a matrix, whereas anthranilic acid, methylanthranilat, and other substances were used as dopants. The effects of impurities on the optical properties of the host crystal were explained by the geometric similarity of host and impurity molecules and by the possibilities of forming an interblock solid solution [10.18].

The same partial correspondence is a necessary condition for the choice of matrix and impurity in the production of quasilinear impurity spectra (SHPOL'SKY effect) [10.19].

PAKHOMOV et al. [10.20] established two different rules for the distribution of impurity molecules in the host crystal from measurements of the diffusion rate of impurity molecules in a solid solution crystal. The results obtained confirm the existence of interblock and true types of solid solutions.

11. Ordering in Organic Solid Solutions

In crystals and other condensed media built of molecules, we generally encounter specific problems of ordering which do not exist in crystals built of atoms or ions. If an atom or an ion behaves as a spherical body, and vector properties of the atom (say, magnetic moments) are neglected, the mixed systems built of such particles can possess only *positional ordering*. The vector drawn from the unit-cell origin to the center of an atom determines its position unambiguously. The description of ordering reduces to describing the positions of points of two or more different types.

This is not so with molecules. Each molecule can be rigidly connected with its own coordinate system. It is convenient to choose the origin of such a system at the center of symmetry (if any) of the molecule or at its center of mass. Sometimes it is reasonable to use some atom of the molecule as the origin. The arrangement of the "centers" of molecules characterizes the positional ordering.

As any molecule has at least one atom bound valently to two other atoms, it is always possible to connect rigidly the coordinate system of the molecule with its position in the crystal. Assigning the angles between the coordinate system of the molecule and that of the crystal, we can characterize the molecular orientation. The orientations of molecules occupying the same crystallographic positions are, in general, different but related by symmetry elements. If molecules occupy several independent crystallographic positions, each of them has its own set of orientations. The guest molecule also has its own coordinate system. If in the formation of a substitutional solid solution the mutual arrangement of coordinate axes of impurity and host molecules is the same in all replacements, there is no orientational disorder in the system.

But there are cases where the replacement can occur in different ways. The host molecule removed from the matrix and the guest molecule occupying its place can have different orientations. If these variations in orientation are more or less random within the unit cell, there is an *orientational disorder* in the solid solution. The following three types of orientational disorder can occur.

1) The host molecules occupy only one crystallographic position. The guest molecule can replace the host molecule in two or more different orientations.

2) The guest molecule replaces the host molecule in only one way, but the host molecules occupy two or more independent crystallographic positions.

3) Orientational disorder is due to both causes.

Ordering in molecular crystals does not reduce to the orientational ordering. Organic molecules can exist in several conformations. The transition of a molecule from one conformation to another may involve overcoming a significant energy barrier. It is clear that conformational transitions of molecules in solids can be hindered since not only intramolecular but also intermolecular forces of interaction need to be overcome.

Thus the formation of solid solutions can yield one more type of disorder. We shall term it *conformational disorder*. There are several types of conformational disorder.

1) Molecules of the host crystal have only one conformation. Impurity molecules have two conformations, but their shapes differ only slightly. The host molecules can be replaced by guest molecules of either conformation.

2) The host crystal belongs to quasi-binary systems. This means that the host molecules possess two favorable conformations, each of which occupies its own crystallographic position. The guest molecules have only one conformation but are able to replace matrix molecules of either conformation.

3) The host crystal is a system with conformational disorder (one of the variants of one-component solid solution, Sect.11.1). The guest molecule is capable of substituting any of the host molecules. In such a case a two-component disordered system appears.

4) The disorder appears because both host and impurity molecules exist in two (or more) conformations and are capable of replacing one another.

Thus in molecular crystals we can encounter not one but three types of ordering: *positional, orientational,* and *conformational*. Correspondingly, three types of short- and long-range order can be distinguished.

All the considered cases of disorder and their combinations are possible, but only a few of them have been investigated.

11.1 One-Component Crystals with Orientational Disorder

Molecules with different orientations can overlap. This possibility can be especially often observed in the derivatives of aromatic substances. It is quite clear that there are four ways of superimposing the molecules of fluoronaphthalene (or some other monoderivative) so that the aromatic rings coincide.

There are numerous examples where the superposition of the same molecules in two or more orientations results only in a slight change in their contour. In Sect.3.3.6 we introduced the *parameter of geometrical similarity* $\varepsilon = 1 - \lceil /r$ (r describing the overlapping parts of the molecules and $\lceil$ their protruding parts) to characterize the differences in the shapes of molecules. Geometrical similarity can also be used to characterize a molecule having several orientations.

Let us assume that a molecule is removed from an ordered crystal. Now put back the same molecule into the vacant place but in another orientation. Some lattice distortion will naturally appear, and contact distances will change. If these changes are due to insignificant energy losses, they can be compensated for by the gain in entropy. When the geometrical similarity of two orientational states of the same molecule differs from unity only slightly (according to experimental data this difference can even amount to 5-7%), the formation of a one-component crystal with the orientational disorder of molecules becomes advantageous.

From a theoretical standpoint such a crystal does not differ from a two-component (binary) system. It makes no difference whether we consider the mixture of two different molecules or two identical molecules in different orientational states. Of course, the same is true of the conformational disorder. Such systems can be called *quasi-binary* or one-component solid solutions.

One may consider two cases of one-component molecular systems with orientational disorder, one where symmetry increases (next Section), and one where symmetry is decreased (Sect.11.1.2).

11.1.1 Disorder with Increase in Crystal Symmetry

It has been found that many crystals with a regular lattice (when distortions do not exceed 0.001 Å) possess orientational disorder, which increases their symmetry. The most interesting cases are when planar molecules lacking a center of inversion are oriented in opposite directions so that the crystal as a whole becomes centrosymmetric, i.e., contains a center of inversion.

A crystallographic proof of such a situation is not difficult. Some asymmetric molecules can form a crystal whose unit cell contains two molecules and has the symmetry of the centrosymmetric space group $P2_1/c$. But a two-fold position in such a crystal possesses a center of inversion. One is forced to admit a statistical disorder in the orientations. The "average" molecule is centrosymmetric, and the complete X-ray structure investigation yields the electron density of this "average" molecule, showing that exactly one half of the molecules have one orientation and the other half are oriented in the opposite direction. Examples are azulene whose molecule consists of one five-membered and one seven-membered ring ("quasi-naphthalene"), β-fluoronaphthalene, and chloronitrobenzene [11.1 - 4].

The geometric model explains these structures "overturning" molecules; we do not create any specific compressions or loosenings. In some instances calculations have been carried out by the method of atom-atom potentials, which showed that a short-range order was possible such that the losses in potential energy were covered by the gain in entropy ($\Delta S = 2 \ln 2 = 1.38 \text{ cal K}^{-1}\text{mol}^{-1}$) due to disorder.

In all these instances, in addition to the gain in entropy, there is a gain in symmetry. Generally speaking, nobody has proved that the higher the symmetry the better. Yet it is highly probable that the presence in the lattice of even an "average" center of symmetry results in a gain not only in configurational entropy but also in other free energy terms.

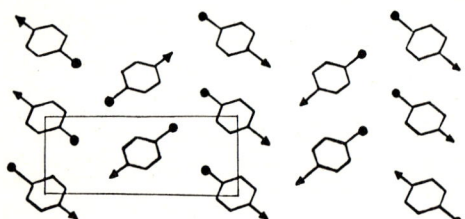

Fig.11.1. Planar molecules without an inversion center exhibit this element of symmetry "on the average"

The preservation of a center of inversion in the crystal does not lead (as was mentioned in Chap.3) to a decrease in packing densities. Therefore, schemes of the kind shown in Fig.11.1 occur frequently. Also, it often happens that noncentrosymmetric molecules have in the crystal a two-fold rotation or a mirror-plane symmetry. Thus, for example, the molecule of cis-azobenzene occupies a site with two-fold rotation axis symmetry. It can be predicted that the de-

rivatives of this molecule, in which the hydrogen atom is replaced by some
other small atom, will form a crystal where the two axis exists statistically.
For this to occur, the scatter in orientations should not lead to any essential
distortion of the lattice.

It would be very interesting to determine the structure of monoderivatives
of adamantane $(CH)_4(CH_2)_6$. Depending on the position of the substituent, the
molecules of monoderivatives must, we believe, be scattered over a large number of orientations to give the "statistically" high site symmetry — 43 m —
occupied by the molecule of adamantane in a cubic crystal.

Gains in entropy are noticeable at high temperature. It is quite possible
that certain structures are determined only at relatively high temperatures,
whereas at low temperature a disorder-order phase transition takes place. This
is exactly the case for β derivatives of naphthalene [11.5,6].

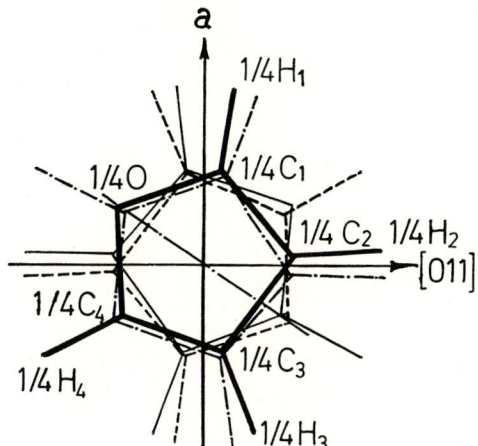

Fig.11.2. Orientational disorder in the furan crystal

Furan, (C_4H_4O), has a crystal structure with orientational disorder [11.7].
A furan molecule has the approximation shape of a regular pentagon where four
vertices are occupied by CH groups and one vertex by the oxygen atom. Furan
melts at $-86°C$. The high-temperature phase is stable between $-123°C$ and $-86°C$.
Its crystals have orthorhombic symmetry, space group Cmca with four molecules
per unit cell, and a unit-cell volume of 391 $Å^3$. The structure is characterized by orientational disorder as shown in Fig.11.2. All the four positions
are equally probable. The "average" molecule has m symmetry. In addition to
an energy gain which we cannot evaluate quantitatively, the crystal gains a
configurational entropy of R log 4 = 2.76 cal $mol^{-1}K^{-1}$ due to the disorder.

The low-temperature phase of furan, however, is characterized by an ordered
arrangement of molecules.

11.1.2 Disorder with No Increase in Crystal Symmetry

There are cases where orientational disorder of molecules does not lead to any change in crystal symmetry. The increase in symmetry, considered above, occurs because the fractional occupation of the molecules with different orientations is strictly dictated by the multiplicity of the crystallographic site symmetry. Thus in the case of centrosymmetric structures, exactly half of the molecules are directed "to the right", whereas the other half are oriented "to the left".

Let us consider the disorder in the crystals of hexa-substituted benzene derivatives. A molecule with two chlorines in para positions and bromines in all the other positions can have, at least at high temperatures, three different orientations in a crystal. An "average" molecule has a center of inversion, and the structure becomes isomorphous to that of hexabromobenzene, in which the molecule in the crystal preserves only the center of inversion out of all its molecular symmetry elements.

There is an essential distinction here from the examples given in Sect. 11.1.1: different weights can be assigned to different orientations. Here, a uniform distribution over various orientations leads to no gain in symmetry.

Investigation of the structure of 2-amino-4-methyl-6-chloropyrimidine [11.8] revealed orientational disorder. The substitution of the methyl group by a chlorine atom or vice versa would increase the symmetry of the molecule. The small difference in volume between the chlorine atom and the methyl group makes the disorder in the crystal favorable for these molecules. The space group is $P2_1/c$; there are four molecules per unit cell. The molecule occupies a general position, and disorder will not increase the symmetry. But the two orientations with different positions for the chlorine atoms and the methyl groups are not equivalent. Disorder occurs with unequal numbers of molecules in two orientations. This is obvious from the values of electron density at sites occupied either by chlorines or by methyl groups and also from the distances between the carbon atom and substituents. These distances are equal to 1.76 and 1.69 Å, respectively. The bond lengths are between these values and are not equal. It is possible to evaluate the fraction of the molecules occupying each of the positions and the energy of interaction of the molecule with its neighbors. Such a calculation confirms the nature of the molecular distribution over the two orientations.

11.2 Orientational Disorder in Binary Systems

Here, we shall confine ourselves to the geometrical approach to the problem. Let us recall once again that the condition of *geometrical similarity* is essential for the formation of solid solutions. In other words, the guest molecule must be of such a shape and dimension that it will fit the site in the host crystal. If this requirement is fulfilled, two types of impurity molecule packing in the matrix crystal are possible. The first type occurs when the guest molecule is so different from the host molecule that the replacement of host by guest *cannot bring coincidence* of atomic positions in the mixed lattice. There are reasons to believe, for instance, that benzene (C_6H_6) and thiophene (C_4H_4S) are mutually soluble. Such solutions have not been studied yet. The second type of impurity molecule packing occurs when there *is rough coincidence* of most atoms in the resulting mixed crystal.

It is clear that the substitution laws can be stated more firmly when the molecules of the matrix and impurity are geometrically similar. Examples for orientational disorder, both here and in Chap.12, will be simple substitutions of planar aromatic molecules (benzene or naphthalene) differing only by one or several hydrogens or radicals R surrounding the aromatic molecules. The picture can be idealized, i.e., we can assume that the nuclei of aromatic molecules coincide exactly with each another. In order to simplify the task and to concentrate on the orientational disorder, we shall neglect the slight nuclear displacements, i.e., neglect both lattice distortions and positional disorder. Such an approximation would be exact in the case of deuterium substitution.

11.2.1 Disorder Due to Admixture Molecules

We shall follow the classification of orientational disorder accepted at the beginning of this chapter. Consider the case when the molecules of the main crystal occupy one crystallographic position. Orientational disorder in such a system can appear when geometrical similarity permits the admixture molecule to replace the host molecule in different orientations. Here we shall repeat much of what was said earlier about one-component pseudosolutions.

What can be said about the situation where the host molecule occupies a general position? It is clear that replacements of the host molecule by the guest molecule in different orientations do not all have the same probability. Let the host crystal be that of 1,8-dichloronaphthalene and assume that the molecule occupies a general lattice position. Let the guest molecule be 1-

chloronaphthalene. The positions 1 and 8 of the host molecule are chemically equivalent. But if the molecule of 1,8-dichloronaphthalene is in a general lattice position, the atoms 1 and 8 are in different crystalline environments. We can state with fair confidence that orientational disorder appears and that the impurity molecule enters the crystal in two possible ways with unequal relative probabilities.

To show this effect, the host molecule need not occupy a general lattice position. In Chap.12 we shall consider in detail the dissolution of para-dibromobenzene in durene. The durene host molecules are at inversion centers of the host lattice. The impurity molecule must replace the durene molecules at the center of inversion, but such a substitution can occur in two different ways, inequivalent from the standpoint of intermolecular surrounding.

Assume that the geometry permits substitution of a molecule of the host crystal in one or several orientations, and that the host molecule occupies a position with a certain symmetry, say, the position with the center of inversion. If the guest molecule does not possess this symmetry element, then an important phenomenon will necessarily occur, which we shall term *symmetry simulation*.

For example, a naphthalene molecule preserves in the naphthalene crystal only the center of inversion. When the naphthalene molecule is replaced by any alpha or beta derivative, the guest molecule, if such a substitution really takes place, will occupy, with *equal probability*, two positions related by the center of inversion. This will occur because in both positions the substituent will find itself in the same intermolecular surrounding. There are no exceptions to this rule. The guest molecule will have as many orientations (with equal probability) as are required by the symmetry of the crystallographic position of the host molecule.

There are many examples of solid solutions in which the equally probable substitution at different orientations of the nonsymmetrical guest molecules simulates the high symmetry of the host crystal.

Another example of symmetry simulation occurs in the dissolution of meta-dichlorobenzene in hexachlorobenzene. Host molecules occupy in the crystal a position with inversion symmetry. The impurity molecules must imitate the center of inversion. For example, if chlorine atoms of some guest molecules are brought into coincidence with atoms 1 and 3 of the host molecule, then an equal number of guest molecules must enter the host lattice with their chlorine atoms at positions 4 and 6. But the guest molecule can also have a second orientation in the host crystal, namely chlorine atoms of the guest molecule can occupy positions 2 and 4. This substitution involves a different

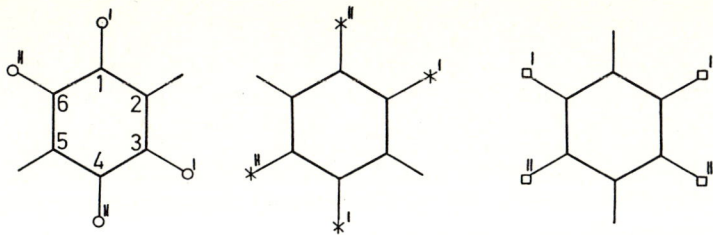

Fig.11.3. Six variants of substitution for the dissolution of meta-dichlorobenzene in hexachlorobenzene

intermolecular surrounding than the first one. But to simulate the center of inversion, the same number of molecules should occupy positions 1 and 5. Finally, a third variant of substitution is shown in Fig.11.3 by squares. Altogether, there are six variants of substitution or, rather, three pairs of variants. Different fractions of molecules can participate in the substitutions shown in the figure by black circles, crosses, and squares. But to simulate the center of inversion in each of the three variants of substitution the fractions of molecules indicated by one and two dashes must be equal. This rule of symmetry simulation cannot have exceptions.

11.2.2 Lattices with Crystallographically Inequivalent Positions

Now we shall consider a host crystal whose unit cell contains molecules in symmetry-independent positions. The existence of two or more crystallographically inequivalent positions, in which the molecular orientations relative to the crystal axes are different, forbids, in principle, solid solutions with a single orientation of impurity molecules. Of course, the variants of orientational disorder are more diverse in these cases.

Let us take as a matrix the crystal lattice of biphenylene, $C_{12}H_8$. There are 6 molecules in a crystal with a monoclinic unit cell, space group $P2_1/a$; they are distributed over two crystallographically inequivalent positions (Fig.11.4). The orientations of the independent molecules relative to the crystallographic axes are different. Two of the six molecules (denoted by 1 in Fig.11.4) occupy the centers of inversion whereas the other four molecules (denoted by 2) are in the general position. What should happen if we add, say, molecules of 3-R-biphenylene or 3, 10 -R-biphenylene? In the first case (Fig. 11.5), we shall have 8 different orientations, of which six are symmetry-independent. In the second case (Fig.11.6), the number of possible orientations

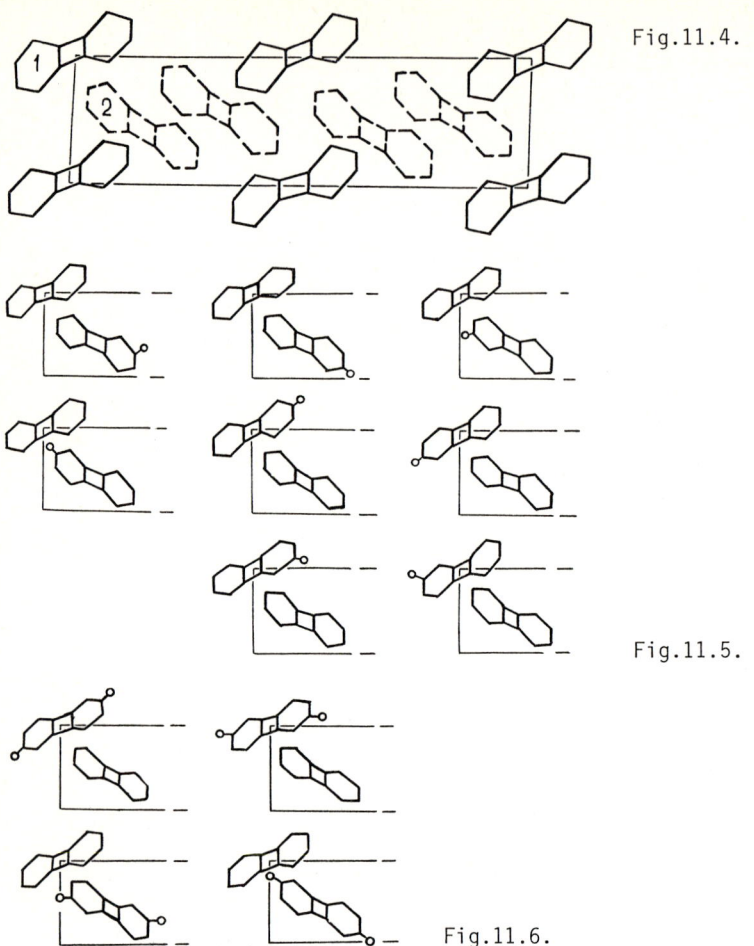

Fig.11.4.

Fig.11.5.

Fig.11.6.

Fig.11.4. (101) projection of the biphenylene crystal structure
Fig.11.5. Eight different orientations occurring by replacing biphenylene by 3-R-biphenylene
Fig.11.6. Four types of orientations occurring by replacing biphenylene by 3-R-biphenylene

reduces to four, when all four orientations of the impurity molecules are crystallographically inequivalent.

These examples should be sufficient to illustrate the general principles of orientational disorder. Though their investigation has only begun, we would like to emphasize once again that investigating the distribution of impurity molecules over different positions is probably the most refined method of studying molecular interactions and verifying free-energy algorithms.

12. Structures of Organic Solid Solutions

The structure of a solid solution can be studied only if the researcher succeeds in growing a single crystal. This is explained by the fact that organic crystals tend to crystallize in the monoclinic system and less frequently, in the orthorhombic system. We very rarely meet cubic, tetragonal, or hexagonal organic crystals; in those cases the investigation is certainly very convenient thanks to their simplicity. However, even this simplicity does not allow the elementary use of the X-ray powder method because of the large sizes of the unit cells.

In Sect.2.4.3 we showed how difficult it is to produce single crystals of a solid solution. Although solid organic substances have attracted the attention of physicists, crystallographers, and biologists in recent years, little attention has been given to the structure of solid solutions of organic substances: only a few dozen solid-solution single crystals have been subjected to X-ray analysis.

There is however a wealth of information on phase diagrams obtained by thermographic analysis. This is because organic solid solutions are easy to investigate by thermography (low melting temperatures and transparency).

At this point we should remember the warnings given in Chap.2 about the great number of erroneously defined solid solutions. Nevertheless, some general conclusions on the types of phase diagrams of organic substances can be made with sufficient certainty.

First of all — the reasons are quite clear — continuous solubility in the solid state is a rare thing. Because geometrical similarity is a stringent condition that is rarely fulfilled, most of the phase diagrams are eutectics with no partial solubility or, at least, no true solubility.

Polymorphism in the world of organic matter is found just as often as in other kinds of crystals. Correspondingly, the reader is referred to [12.1] for complicated phase diagrams.

Organic complexes (molecular compounds) are found rather often in nature. Unlike the metallic systems, we do not encounter complicated stoichiometric compositions with two exceptions: clathrate compounds (which are extremely specific complexes often with complex formulas) and hydrates and solvates (where substances crystallizing with five or seven molecules of water or other solvents are commonplace). Clathrates are discussed in Sect.13.2. If one disregards clathrates and solvates, organic complexes usually have a stoichiometry of 1-1, 2-1,.... Most complexes are formed by molecules which do not possess geometrical similarity. Of course, in such cases intermediate phases, which are often found in metals, cannot be formed.

Therefore, the phase diagrams of organic substances with potential complex formation are generally very simple. They are diagrams of the eutectic type without solubility and in the case of molecules close in shape diagrams of the eutectic or peritectic type, with limited solubility. Phase diagrams with continuous solubility are extremely rare since, in addition to the geometrical similarity of the molecules, their formation requires the isomorphism of crystalline structures.

Below we describe in detail experimental work wherein both the phase diagrams and solid-solution structures have been determined. A fairly thorough structural study was conducted only for the tolan-diphenyl mercury system, and for the durene-para-dibrombenzene system. In all the other cases the experimental results are incomplete. In addition to describing experiments, we present calculations to illustrate how well the existing imperfect theory can explain and predict the experimental results.

12.1 Diphenyl-2,2'-Dipyridyl   [12.2,3]

The molecules of these two substances are very close to each other in form and size (Fig.12.1). There may be several kinds of dipyridyl molecules differing by the substitution sites of the CH group by N. The figure shows that we used 2,2'-dipyridyl in this investigation.

The difference in the volumes of these molecules (155.4 and 137.1 Å for diphenyl and dipyridyl, respectively) appears only because of the "hollow" near the nitrogen atom. It was expected that in this system, solid solutions would form with a high concentration of impurity molecules. Continuous solubility seemed improbable because of the great difference in the orientation of the molecules, in spite of the fact that both crystals have the same structure.

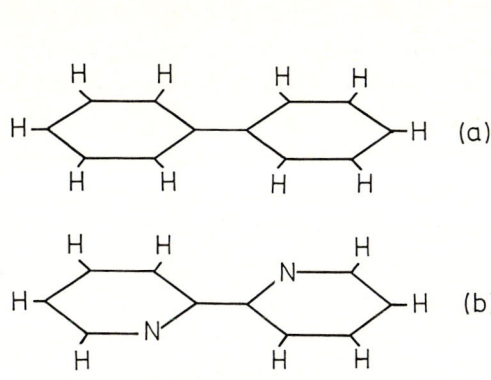

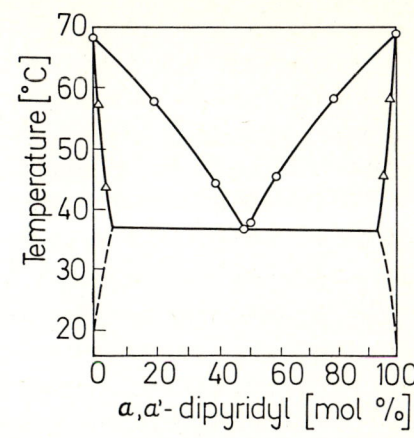

Fig.12.1. The molecules of (a) diphenyl and (b) 2,2'-dipyridyl

Fig.12.2. Phase diagram of the diphenyl-dipyridyl system

The phase diagram of the system is of the eutectic type (Fig.12.2 and Table 12.1). The eutectic point is at a concentration of 49% dipyridyl and a temperature of 37.5°C.

Table 12.1. Melting points of diphenyl-dipyridyl solid-solution crystals

Liquid Composition [mol. %] 2,2'-dipyridyl	Temperature [°C] of Liquidus Curve	Crystal Composition [mol. %] 2,2'-dipyridyl
0	68.6	0
20	57.9	1.4
40	44.3	5.0
50	37.2	8.0
50	37.2	95
60	45.7	97
80	58.4	98.5
100	69.6	100

Observation of the crystal growth has revealed that pure components and mixed crystals grow in plates substantially different from one another. For an α phase with a diphenyl structure, they are thin, soft, and deform readily; for a β phase with the dipyridyl structure, they are thick, hard, and fragile. The drastic difference in the appearance of the crystals has allowed each solid-solution crystal to be attributed to a definite phase, even without an X-ray study.

Solid solutions with a diphenyl structure contain from 0 to 8% dipyridyl, whereas in the dipyridyl phase the solid solutions contain from 0 to 5% diphenyl. So, the expectations of a high solubility were not realized.

The eutectic point was obtained by extrapolating the liquidus lines. The correctness of its determination is supported by the fact that thin plates and thick plates of two phases, hypoeutectic and hypereutectic, of solid solutions of a limited concentration were grown in a eutectic composition melt, using a seeding needle.

Crystals of solid solutions begin to decompose with time even at room temperature. This makes it possible to establish the limits of solubility in the solid state.

Unfortunately, the variation of unit-cell parameters with concentration was not measured. On the other hand, the energy of substitution was estimated with sufficient accuracy. It was shown that the difference in the variation of the substitution energy depends little (within the limits of a hundredth of a kcal/mole) on the composition of the first coordination sphere.

When substituting a diphenyl molecule for a dipyridyl molecule, a relatively rare case of decrease in the lattice energy (lattice distortions are supposed to be absent) was discovered. This can be easily interpreted by a geometrical approach: in a perfect dipyridyl structure the molecules are packed so that the hollows near the nitrogen atom remain empty. When a diphenyl molecule is substituted into the vacant site, the hollows begin to fill up. A denser packing is created. It is quite possible that the volume of the solid-solution cell is less than the one which is given by the linear law.

The diphenyl phase represents a solid solution with a positional disorder. The dipyridyl phase is a solid solution having not only a positional, but also an orientational disorder. In fact, the symmetry of diphenyl molecule is higher than that of dipyridyl. Upon substitution, the dipyridyl molecule can assume two different orientations. The orientations differing in the location of the nitrogen atoms of one molecule with respect to the surrounding atoms of the neighboring molecules, may appear to be unequally probable. Calculations of ΔV have shown, however, that the difference in the energy of the two substitution modes is insignificant. Moreover, the substitution energy is only slightly affected by the presence of one, two, or three dipyridyl molecules in the first coordination sphere. A value of ΔV = 1.2 kcal/mole can be accepted for both modes.

In Sect.4.3 we described a method roughly estimating solubility limits, the method being used to estimate the solubility of this concrete system. In order to obtain the solubility limits given by (4.28,29), we must obtain the quantities A, B, L, and M, in (4.28,29) through the substitution energy in both phases of the system.

If we denote the energy of replacement of the dipyridyl molecule by the diphenyl molecule by ΔV (it equals 0.25 kcal/mole) and the energy or replacement of the diphenyl molecule by the dipyridyl molecule by ΔV_1 and ΔV_2 (they both equal 1.2 kcal/mole), then the limits of solubility are:

$$x_A^\beta = \frac{\exp\left(-\frac{\Delta V + \Delta V_1}{\Theta}\right) - \exp\left(-\frac{\Delta V + \Delta V_1}{\Theta}\right)\left[1 + \exp\left(-\frac{\Delta V_2 - \Delta V_1}{\Theta}\right)\right]}{1 - \exp\left(-\frac{\Delta V + \Delta V_1}{\Theta}\right)\left[1 + \exp\left(-\frac{\Delta V_2 + \Delta V_1}{\Theta}\right)\right]},$$

$$x_B^\alpha = \frac{\exp\left(-\frac{\Delta V_2 + \Delta V_1}{\Theta}\right) - \exp\left(-\frac{\Delta V + \Delta V_1}{\Theta}\right)\left[1 + \exp\left(-\frac{\Delta V_2 - \Delta V_1}{\Theta}\right)\right]}{1 - \exp\left(-\frac{\Delta V + \Delta V_1}{\Theta}\right)\left[1 + \exp\left(-\frac{\Delta V_2 - \Delta V_1}{\Theta}\right)\right]}.$$

(12.1)

The numerical estimates for the solubility limits are shown in Fig.12.3.

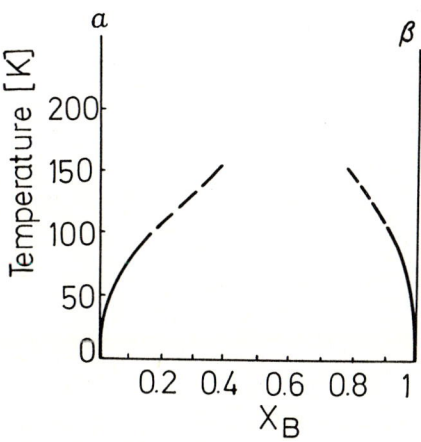

Fig.12.3. Theoretical diagram of state for the diphenyl-dipyridyl system

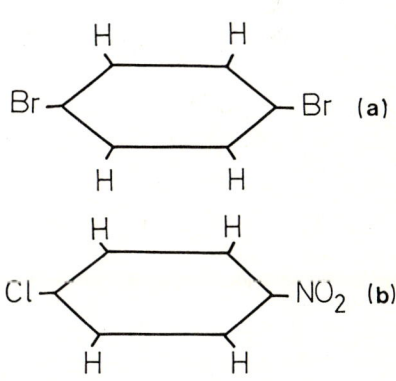

Fig.12.4. The molecules of (a) dibromobenzene and (b) para-chloronitrobenzene

As could be expected, the solubility curves deviate considerably from the experimental data in the high-temperature ranges (these temperatures are higher than the Debye temperature, which is about 100 K). Of significance at these temperatures is the vibrational part of the free energy Δf_{vibr} which we have neglected.

The theoretical estimates explain the asymmetry of the solubility curve found experimentally. The difference $x_A^\beta - x_B^\alpha$ in the equilibrium values of the solubility boundary composition is positive over the full temperature range corresponding to the solid state of the diphenyl-dipyridyl system; if we neglect the two inequivalent positions of the admixture molecules B_1 and B_2 in the α phase, this difference vanishes.

Thus the asymmetry of the solubility boundary in the low-temperature range is accounted for by the increase in the entropy of the α phase at the cost of disordering the impurity molecules.

12.2 Para-Dibromobenzene — Para-Chloronitrobenzene [12.4]

This system is of interest for studying the influence of polarity of one of the components on the formation of solid solutions: the para-dibromobenzene molecule is nonpolar, while para-chloronitrobenzene has a dipole moment of 2.83 D [12.5].

The molecules of these compounds are rather close in form and size (Fig.12.4): the molecular volume of $C_6H_4Br_2$ is 126.4 $Å^3$, and that of p-$C_6H_4Cl\,NO_2$ is 115.4 $Å^3$. Both compounds crystallize in the $P2_1/c$ space group with two molecules per cell, as should be expected for the centrosymmetrical p-$C_6H_4Br_2$ molecules; the acentric p-$C_6H_4Cl\,NO_2$ molecules produce the inversion center statistically, positioned on the average along the crystal in two antiparallel positions. As a result the molecular packing coefficient in this structure (0.67) is substantially less than in the first structure (0.76). The unit-cell parameters are:

p-$C_6H_4BR_2$: a = 4.10; b = 5.75; c = 15.36 Å, β = 112°38' .

p-$C_6H_4Cl\,NO_2$: a = 3.84; b = 6.80; c = 13.35 Å, β = 97°31' .

The diffraction data show that the packing structures differ greatly from each other, as do the unit-cell parameters. Hence one cannot expect a continuous series of solid solutions in this system.

The phase diagram of the para-dibromobenzene — para-chloronitrobenzene system was described long ago. In Fig.12.5 two eutectic points and a sloping maximum were fixed at approximately 50% of p-C_6H_4Cl NO_2 as shown by the dashed and dotted line. In [12.4] the method for growing single crystals was used to determine the phase diagrams. Observation of the crystal growth in a thermostat shows that pure para-dibromobenzene has two equiprobable growth forms: needles and plates which grow simultaneously from the melt on a seeding needle. In mixed melts containing about 10% para-chloronitrobenzene admixture, crystals of both forms appear with equal probability, and then the plates disappear. The possible existence of a peritectic point was disproved by X-ray examination; it showed that all the needles and plates were identical in structure.

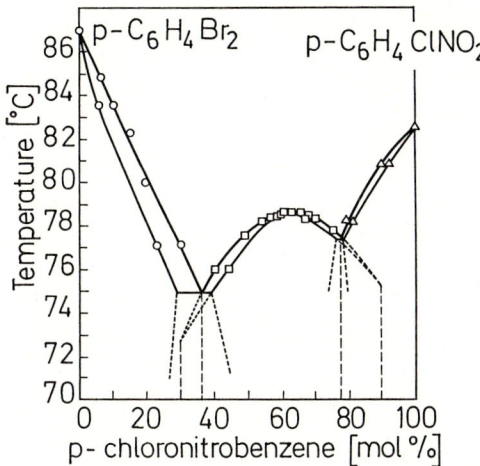

Fig.12.5. The phase diagram of p-dibromobenzene and p-chloronitrobenzene

The phase diagram points out the existence of three solid phases which are separated from one another by two eutectic points: the first phase, at 75.3 °C and 35.5% para-chloronitrobenzene, separates the α phase having a para-dibromobenzene structure from the β phase; the second, at 77.6 °C and 78% of para-chloronitrobenzene, separates the β phase from the γ phase having a para-chloronitrobenzene structure. The eutectic points E_1, E_2 were obtained by extrapolating the liquidus lines. No attempts were made to simultaneously grow crystals of both phases in melts of eutectic composition, as all the crystals have a similar appearance (needles) so that their separation without a cumbersome X-ray examination is a difficult matter.

In the β phase region the liquidus line produces a sloping maximum whose center corresponds to a 60 - 65% C_6H_4Cl NO_2 solid solution. This maximum does not correspond to a complex having a simple stoichiometric ratio of components.

The phase diagram described is a rare exception in the family of phase diagrams of organic substances; here, we encounter an *intermediate phase* which is a disordered solid solution with a structure distinct from that of the components. This is proved by X-ray diffraction study of crystals with different compositions. The crystal structures of all the three phases have been established.

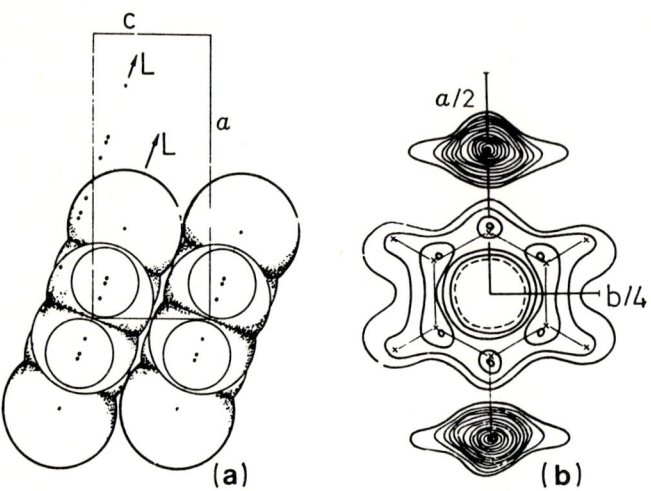

Fig.12.6. Crystal structure of the β phase (intermediate phase) in the p-dibromobenzene and p-chloronitrobenzene system: (a) molecular packing; (b) approximate Fourier synthesis

Single crystals of the β phase (having the structure of a pure para-dibromobenzene) contain from 0 to 28% para-chloronitrobenzene. Of course, in the α phase the molecules of admixture realize statistically the inversion center.

Single crystals of the α phase are stable in the concentration range from 44% to 74% $C_6H_4Cl\,NO_2$. The unit-cell parameters are a = 9.21, b = 9.61, c = 3.9 Å, β ≈ 90° (for 60% p-$C_6H_4CL\,NO_2$ composition) and the cell volume is 345.2 Å. X-ray diffraction requires that the β phase belong to one of the three X-ray space groups (Fig.12.6): C2/m, C2, or Cm. The X-ray data cannot in principle allow a precise or unique refinement as shown in Fig.12.6(b). Nevertheless, in each of the space groups there must be no less than two moleculels in a cell; let us assume that Z = 2, and k = 0.7 for a conventional packing coefficient for organic structures and calculate the intrinsic volume of the β phase "molecule":

$$v_{mol} = \frac{k \cdot v_{cell}}{Z} = \frac{0.7 \cdot 345.2}{2} = 121 \text{ Å}^3 \quad .$$

This value appears to be the average between the intrinsic volume of the molecules of pure components, i.e., β phase must contain 2 molecules per unit cell. Thus the β phase represents a solid solution with a positional and orientational disorder not differing in this respect from the α and γ phases. The "average" molecules are translationally identical in this structure.

The γ phase single crystals contain from 80 to 100% $C_6H_4Cl\,NO_2$, i.e., from 0 to 20% of para-dibromobenzene admixture molecules. More data on this interesting system can be found in [12.6,7].

12.3 Acenaphthene – α-Nitronaphthalene [12.8]

In this system one can study the arrangement of impurities in symmetry-inequivalent positions of the acenaphthene ($C_{12}H_{10}$) crystal lattice [12.9].

Acenaphthene and α-nitronaphthalene ($C_{10}H_7NO_2$) molecules (Fig.12.7) satisfy the geometrical conditions necessary for the formation of substitutional solid solutions. The α-nitronaphthalene molecule in Fig. 12.7(b) is in its optimal conformation when the nitro group plane is rotated by about 47° relative to the aromatic plane [12.10]. The impurity α-nitronaphthalene molecules may "adjust" slightly to the acenaphthene residing in the lattice, since the deviation of the nitro group rotation angle by 5 - 7° does not require much energy.

In the phase diagram displayed in Fig.12.8, the eutectic point corresponds to a composition of 70.5% α-nitronaphthalene at a temperature of 40.7°C. The diagram is characterized by poor mutual solubility of the components, especially in the hypoeutectic phase. In a melt containing 70% α-nitronaphthalene one can grow well-faceted acenaphthene phase crystals, the admixture content being not higher than 12%. The study of the structure of this phase is particularly interesting.

Figure 12.9 demonstrates the dependence of the a, b, and c unit-cell axes and of the volume per single "average" molecule on the concentration of the admixture molecules. A small difference in the geometrical sizes of the molecules of the substances being mixed results in slight changes in the cell volume. These slight changes, however, reliably show the solubility type in the acenaphthene phase: replacement of the matrix molecules by the α-nitronaphthalene admixture molecules naturally decreases the volume per single "average" molecule in the unit cell of the solution.

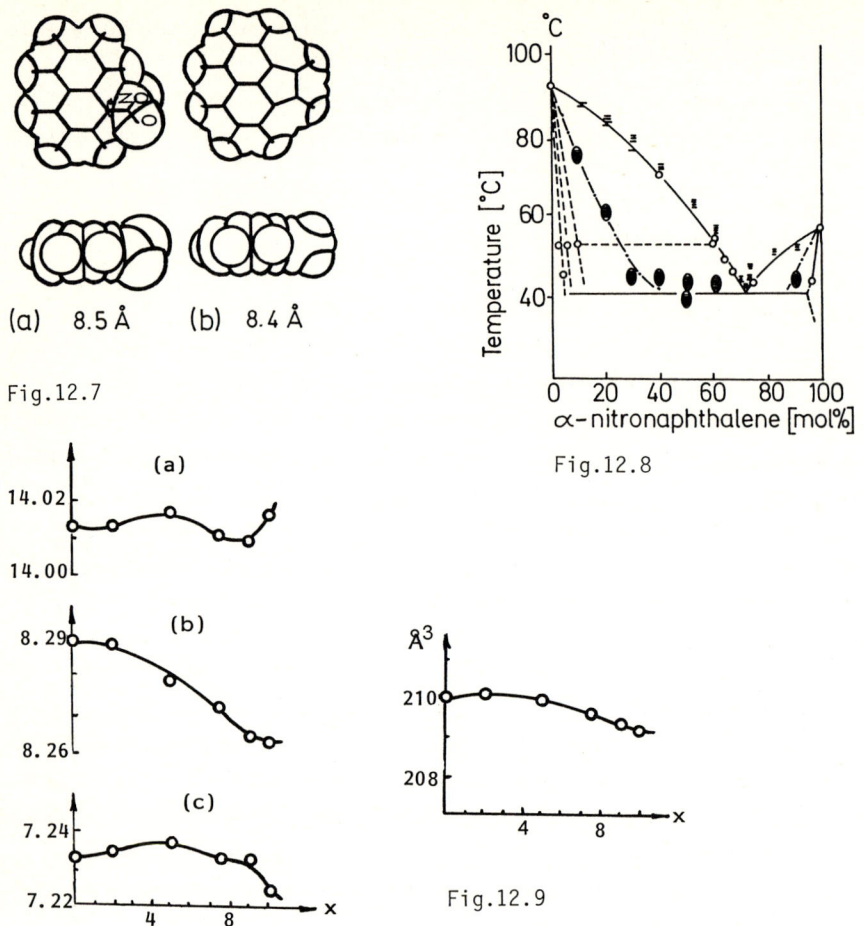

Fig.12.7. Acenaphthene (a) and α-nitronaphthalene (b) molecules
Fig.12.8. Acenaphthene — α-nitronaphthalene phase diagram
Fig.12.9. Unit-cell dimensions and volume of the acenaphthene — α-nitronaphthalene mixed crystal as function of α-nitronaphthalene content

The acenaphthene unit cell contains 4 molecules (space group $Pmc2_1$) which are located in special positions on the symmetry planes. Thus the molecules maintain in the crystal one of the two molecular symmetry planes, namely, the one perpendicular to the molecular plane (Fig.12.10).

In one layer, the molecules are positioned so that their plane coincides with the (001) lattice plane. In the second layer, the plane of the acenaphthene molecule is considerably tilted (by more than 20°) to this lattice

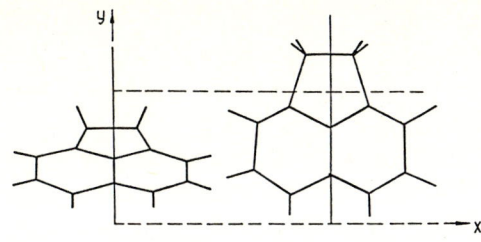

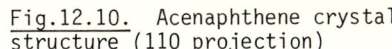

Fig.12.10. Acenaphthene crystal structure (110 projection)

plane. When replacing the lattice molecules, the α-nitronaphthalene molecules can replace the "parallel" and "tilted" positions with different probabilities. Of course, the substitution proceeds in such a manner that the nitro group of the admixture molecule is oriented in the same direction as the acenaphthene methylene groups. It is clear that the orientational disorder is not limited to the possibility of replacing the basic molecule in the two above-mentioned positions. The replacement occurs with symmetry simulation. Hence, there are four, and not two, different orientations for the α-nitronaphthalene guest molecules. The distribution between the two lattice sites — "parallel" and "tilted" — can be in principle of any kind, whereas the distribution of the two possibilities at each lattice site is of equal probability.

What is the difference, in the order of magnitude, in the observed electron densities during substitution? Half of the nitro group replaces the methylene group. The difference in the number of electrons per molecule is 3.5 electrons. If the "parallel" and "inclined" positions were equally occupied, the difference in the number of electrons would be 1.75.

In the work cited a crystal containing 10% nitronaphthalene was studied. With a uniform substitution the numbers of the electrons per molecule would vary by 0.175 electrons. If preference is given to one of the possible crystallographic positions, the number would vary by 0.350 electrons, not at an atomic position, but in some volume. Such a variation is difficult to reveal by X-ray analysis.

Nevertheless, a classical X-ray study of a crystal containing about 10% nitronaphthalene was carried out. The set of observed intensities was not very large; it contained 375 nonzero reflections.

Several variants were employed for refining the solid-solution structure by the least-squares technique on the basis of the nitro-group atoms' "weights" and the CH_2 group being substituted by it. In the first variant the impurity is distributed equally over the two independent positions (the discrepancy factor $R = 20.3\%$); in the second variant the admixture molecules are distributed at the site of the molecule in the parallel position only ($R = 18.2\%$);

in the third, the admixture is located at the sites of the "tilted" position (R = 20.2%). Then, an electron density series was obtained, the C and H atom coordinates of only the aromatic nuclei being used for calculating the phases of the structure amplitudes. The maxima corresponding to the CH_2 bridging groups, which were not used to calculate the phases, proved to be slightly lower than those of pure acenaphthene. In addition, the maximum of the molecules in parallel positions was not very pronounced. This finding agrees with the smallest of the three R factors, and allows one to suppose that the α-nitronaphthalene admixture molecules replace the matrix molecules predominantly in the parallel position.

With such small differences revealed by experiment, it is especially essential to supplement these observations by calculations using the atom-atom potentials technique. A calculation of the substitution energy was made using the approximation of an infinitely dissolved solid solution, i.e., it was assumed that the guest α-nitronaphthalene molecule was surrounded (in the calculated radius of 15 Å from each of its atoms) only by host acenaphthene molecules. The guest molecule was located in a geometrically optimal manner, alternating between the site of the first and second independent molecule of the acenaphthene matrix. The calculation took into account the possible variations in the angle between the nitro-group plane and the plane of the aromatic nucleus. The calculation showed that it was considerably more convenient for the impurity molecule to be accommodated in the "parallel" position. The results of the geometrical analysis are analogous. If a nitronaphthalene molecule is placed in the inclined position, one of the intermolecular distances, O-H, proves to be dramatically shortened.

Thus, in spite of the low sensitivity of the X-ray experiment, the confirmation of its results by calculation of the substitution energy allows one to conclude that the impurity occupies preferentially in the "parallel" lattice site.

12.4 Naphthalene – Coumarine [12.11]

Molecules of both substances are close in size and form (Fig.12.11). Unlike the nonpolar naphthalene molecule, a coumarine molecule ($C_9H_6O_2$) has an appreciable intrinsic dipole moment of 4.51 Debye [12.5]. This system was also investigated to determine whether the geometrical closeness of sizes and forms of the molecules are a sufficient condition to form solid solutions with a substantial concentration of guest molecules.

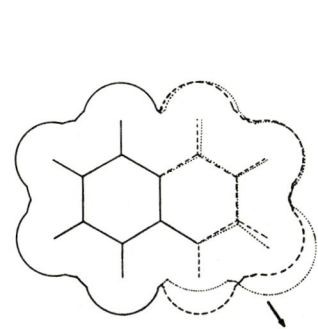
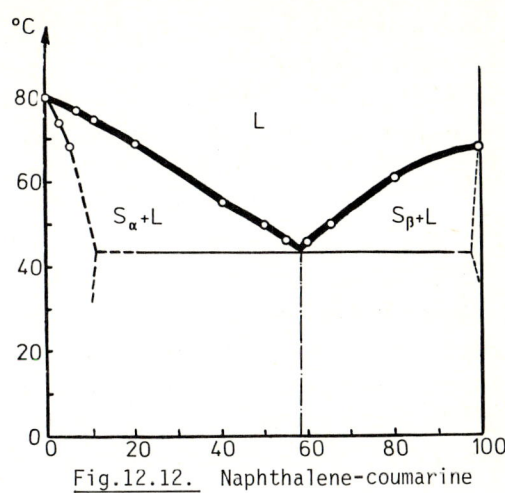

Fig.12.11. Superposition of naphthalene and coumarine molecules

Fig.12.12. Naphthalene-coumarine phase diagram

A continuous series of solid solutions in this system is impossible because the crystals of the pure substances have different symmetries: naphthalene is monoclinic $P2_1/c$ ($Z = 2$), and coumarine belongs to the acentric orthorhombic space group $Pca2_1$ ($Z = 4$).

Figure 12.12 displays the phase diagram of the system. The diagram is of the eutectic type. The eutectic point (temperature 44°C) was obtained by extrapolating the liquidus lines. It corresponds to 42% naphthalene and 58% coumarine. The solidus line was determined with less accuracy.

The problem with the investigation is that the investigators failed to grow solid-solution crystals in the naphthalene phase with a content in the melt between 30% coumarine and a eutectic composition. A droplike bulge appears on the seeding needle.

When the needle fuses inside the melt, the liquidus line can be reliably determined. But when it is removed from the melt, the mother liquor remaining in the pores and in the roughness of the bulge leads to concentration variations of up to 10%. The extrapolation of the solidus line established by observing the fusion end-point for single crystal of the solid solution enables us to assert that the maximum content of coumarine in the naphthalene phase is 10%.

From the coumarine side of the phase diagram, well-faceted crystals grow in the melts at all concentrations. The determination of the solidus-line points from these crystals has shown that the end of the temperature range in which they melt is not more than 0.1° lower than the melting point of pure coumarine, i.e., these crystals contain less than 1% naphthalene, which is on the borderline of detectability. Thus, the geometrical conditions for the formation of solid solutions are not fulfilled in this case.

The dissolution of naphthalene in coumarine demonstrates a rather rare case where the electrostatic interactions between the molecules turn out to be in the foreground. We spoke of this possibility in Sect.10.1.2. However, the presence of a large dipole moment of coumarine does not prevent the dissolution of nonpolar molecules like naphthalene in the coumarine crystal.

Here are some examples of systems with a relatively good solubility whose components are substances consisting of polar and nonpolar molecules. In the acridine-anthracene system [12.12] the acridine molecules ($C_{13}H_9N$) are polar ($\mu = 2.0$ D). The solubility is good from both sides of the phase diagram. We should pay attention to the fact that when the acridine molecules enter into the antracene ($C_{14}H_{10}$) crystals, they simulate the inversion center. On the other hand, the pure acridine crystal also has an inversion center. This implies that the crystal is constructed as if there were a pair of dipoles with opposite directions.

In the para-dibromobenzene — para-chloronitrobenzene system (Sect.12.2) the molecules of the latter component ($\mu = 2.53$ D) are polar, but in the crystal they occupy an inversion center. It is obvious that for the statistical realization of the inversion centers, the molecular dipoles residing in the neighboring lattice points must have an antiparallel orientation. In this system, a fine intermiscibility of the components is observed in all three phases.

In the naphthalene - β-chloronaphthalene system [12.13] the molecules of the latter component ($\mu = 1.6$ D) are polar, but this does not hinder a fine solubility of the nonpolar naphthalene molecules in the β-chloronaphthalene matrix (up to 44 molar %). Again we observe the same situation: the β-chloronaphthalene crystal possesses an inversion center.

In the acenaphthene — α-nitronaphthalene system the molecules of both components ($\mu = 1.50$ and 3.90 D, respectively) are polar. Nevertheless, the acenaphthene phase can contain up to 12% impurity molecules. Solubility on the part of nitronaphthalene is also observed. Examining the crystal structures we see that when the inversion center is absent; the dipoles of the molecules creating the crystal have different directions.

In the anthracene-carbazole system, the molecules of the latter component ($C_{12}H_9N$, $\mu = 2.10$ D) are polar; in the carbazole structure [12.14] this molecule maintains one of its symmetry planes (perpendicular to the molecular plane), occupying the corresponding special position (space group Pnam, Z = 4). The solubility in the system is satisfactory; the carbazole matrix can contain up to 30% nonpolar anthracene molecules.

Of the known phase diagrams of the solid solutions of naphthalene with its β-derivatives, we note the system where the guest molecules, β-naphthol (μ = 1.4 D), enter pairwise to simulate a statistical inversion center (Sect.12.6). We find the same situation in the anthrone-anthraquinone system [12.15].

In conclusion, the principal rule of solubility (Sect.11.1.1) remains in force in two cases: first, molecules with dipole moments which crystallize in centrosymmetrical groups that have an inversion center, do so either in pairs or by means of an orientational disorder that simulates the inversion center; and second, in space groups without an inversion center, polar molecules will orient in different directions because of some peculiarity of the crystalline symmetry.

The coumarine crystal presents a different picture. The structure of this interesting compound was established only quite recently, and the calculations were carried out in [12.16,17]. These studies have shown that coumarine crystals belong to the space group $Pca2_1$ with four molecules per cell that occupy one unique crystallographic position. Thanks to this, the dipole moments of all the molecules are oriented in one direction.

As previously asserted by the author [12.18,19], the contribution of the dipole-dipole interaction energy to the total energy does not exceed 1 - 2 kcal/mole, whereas the intermolecular energy for the crystals constructed from molecules of about the same size as those of coumarine is close to 20 kcal. This statement remains valid even when the molecular dipole moments are oriented in the same direction. To define the role of the dipole-dipole interaction in establishing the equilibrium structure of the coumarine crystal, we calculated the parameters corresponding to the lattice-energy minimum using the atom-atom approximation. In the process of minimization nine parameters were varied simultaneously: the size of the unit-cell edge, the Eulerian angles determining the orientation of the molecule, and the coordinates of the molecular center of gravity. The calculated parameter values of the structure proved to be very close to the experimental values, and the theoretical interaction energy value at the energy minimum, 16.0 kcal/mole, differed little from the theoretical energy at the configuration of the experimented structure, 15.4 kcal/mole. These estimates prove once again that the lattice energy of the equilibrium structure of a molecular crystal is determined primarily by the dispersion interaction.

The comparison of the heat of sublimation of coumarine, 17.8 kcal/mole, with the calculated theoretical lattice energy shows that the dipole-dipole interaction energy must be less than 2 kcal/mole.

GAVUZZO et al. [12.17] defined the coumarine structure by using, at the first stage of investigation, the method of finding the lattice-energy minimum. In doing so, they made an attempt to take account of the electrostatic interactions, but arrived at the conclusion that though the coumarine molecules do possess sufficiently large intrinsic dipole moments, the total contribution of the dipole energy to the lattice energy was negligibly small and did not influence in any way the final results.

However, the small electrostatic interactions of the coumarine molecules manifest themselves in the process of formation of solid solutions.

In the naphthalene phase of the coumarine-naphthalene system the solubility is quite natural: the polar coumarine guest molecules can replace statistically the centro-symmetrical host molecules. The simulation of the inversion center ensures the antiparallel arrangement of the dipoles in the neighboring lattice points.

Coumarine molecules create a very dense packing with an almost parallel arrangement of the dipole moments. The electrostatic interaction energy does not force the molecules out of the potential well formed by ordinary (dispersive) intermolecular interactions.

However, the formation of solid solutions in such a structure does not adhere to one geometry. Coumarine and naphthalene molecules are very close in volume and shape. However, the entrance of a naphthalene molecule into a coumarine crystal substantially decreases the electrostatic interaction energy, disrupting the chain of dipoles along on the screw axis 2_1. The gain in entropy cannot compensate for the losses in the electrostatic energy which reach tenths of kcal/mole even when an insignificant number of naphthalene molecules enter into the solution.

Thus, the coumarine phase in the naphthalene-coumarine system is an example (so far the only one) where the required condition of the proximity of the geometrical sizes of the molecules being mixed turns out to be insufficient for the formation of solid solution, even in the absence of hydrogen bonds.

The statement made by PENKALA [12.20] concerning the "negative effect of the difference in the dipole moment values on the reciprocal solubility of substances in a solid phase" is by itself insufficient. Only when we know how the polar molecule packs in a crystal can we make a definite prediction.

12.5 Dicarboxylic Acids [12.21,22]

One expects a significant mutual solubility of normal hydrocarbons and of their derivatives. Reference [12.19] discusses certain general rules obeyed by phase diagrams of the binary systems from these substances. With a similar symmetry of components and not too great a difference in the chain lengths, we can expect miscibility in all proportions. Unfortunately, X-ray studies of such systems have been extremely rare. This is due, first of all, to the fact that it is difficult to produce single crystals.

CHANH et al. [12.21,22] studied the phase diagrams and parameters of solid solutions of normal dicarboxylic acids. In Sect.10.4.1 we plotted curves of the cell volume as a function of concentration for the two systems studied by these authors.

Investigation was made by X-ray powder photography and by differential thermal analysis. The authors have established that the phase diagrams in the C_6-C_8, C_8-C_{10}, C_{10}-C_{12}, C_{12}-C_{14} systems have the same form. All systems under study decompose at a temperature of around $50°C$, and the bell-shaped decomposition curve is shifted toward the component consisting of larger molecules.

How does mutual substitution of molecules differing in length by 2.5 Å occur? Such a question can be posed not only in respect to dicarboxylic acids, but also in respect to any paraffin derivatives. One of the possible explanations (the one given in [12.21,22]) was presented in Sect.10.4.1. It assumes a pronounced short-range order wherein a large number of short molecules are adjusted to a smaller number of long ones. Another explanation assumes the helicity of longer molecules. This question can be solved if sufficient time and labor is spent on growing solid-solution single crystals and studying them by X-ray diffraction.

12.6 Naphthalene and Its β Derivatives [10.14,16, 11.2]

The investigation of the phase diagrams and the structures of solid solutions which are formed by naphthalene and its β derivatives and by β derivatives among themselves is of interest for studying intermolecular interactions. In a great number of papers [10.1,14-16, 11.5, 12.23,25-31], CHANH et al. studied mixed crystals of components, the structural data of which are collected in Table 12.2.

Naphthalene has one modification. The parameters of three modifications (Form I,II,III) have been measured for β chloronaphthalene. No intermediate

Table 12.2. The structural data for naphthalene and its β derivatives (high-temperature forms)

Substance	Form I	Form II	Form III
Naphthalene	a=8.259±0.003 Å b=5.980±0.002 Å c=8.668±0.004 Å β=122°36'±6' $P2_1/a$ Z = 2		
β-naphthol	a=8.295±0.0012 Å b=5.880±0.006 Å c=9.030±0.010 Å β=121°12'±12' $P2_1/a$ Z = 2		a=8.138±0.008 Å b=5.935±0.003 Å c=36.27±0.02 Å β= 119°58'±6' Ia Z = 8
β-thionaphthol	a=7.749±0.002 Å b=5.915±0.002 Å c=10.748±0.004 Å β=118°52'±5' $P2_1/a$ Z = 2		a=8.383±0.002 Å b=5.860±0.002 Å c=41.181±0.008 Å β=116°47'±5' $P2_1/a$ Z = 8
β-methylnaphthalene	a=7.800±0.002 Å b=5.983±0.002 Å c=10.805±0.004 Å β=119°54' $P2_1/a$ Z = 2		a at 100°C a=8.43±0.01 Å b=5.81±0.01 Å c=41.25±0.03 Å β=117°00±20' $P2_1/a$ Z = 8
β-naphthylamine			a=8.642±0.002 Å b=6.003±0.002 Å c=32.414±0.006 Å β=112°34'±5' $P2_1/a$ Z = 8
β-bromonaphthalene	at 50°C monoclinic $P2_1/c$; Z = 2 a=7.824 Å b=6.045 Å c=10.736 Å β=118.83°	monoclinic $P2_1/c$; Z = 4 a=7.693 Å b=5.926 Å c=22.706 Å β=123.52°	parameters unknown
β-chloronaphthalene	at 48°C monoclinic $P2_1/a$; Z = 2 a=7.743 Å b=5.955 Å c=10.675 Å β=119.95°	at 25°C monoclinic $P2_1/c$; Z = 4 a = 7.687 Å b=5.950 Å c=22.146 Å β=123.82°	at 180°C monoclinic Z = 8 a=7.685 Å b=5.945 Å c=40.33 Å β=114.52°
β-fluoronaphthalene	a at 25°C monoclinic $P2_1/a$; Z = 2 a=7.796 Å b=5.969 Å c=9.955 Å β=122.87°		

forms (Form II) which are stable at room temperature have been found for the other derivatives. One modification each has been found for β-fluoronaphthalene and β-naphthylamine.

All the high-temperature forms (Form I) produce crystals with two molecules per cell and the usual space group $P2_1/a$ (or $P2_1/c$). Form I has an orientational disorder imitating a center of symmetry.

The simple examination of the unit-cell parameters can lead to the following conclusions. Forms I of naphthalene and β-naphthal (we use the notations of the French authors) belong to one structure type. All the three naphthalene halide derivatives in Form I are ascribed to another type. The fluoronaphthalene structure was determined [12.12]. Figure 12.13 exhibits naphthalene and fluoronaphthalene molecule packing projections that differ sharply from each other. The possible arrangements of C-F bonds in Fig.12.13b are indicated by dashed lines.

All the naphthalene β-derivatives are supposed to possess similar energy surfaces which have two minima approximately of the same depth. One of them

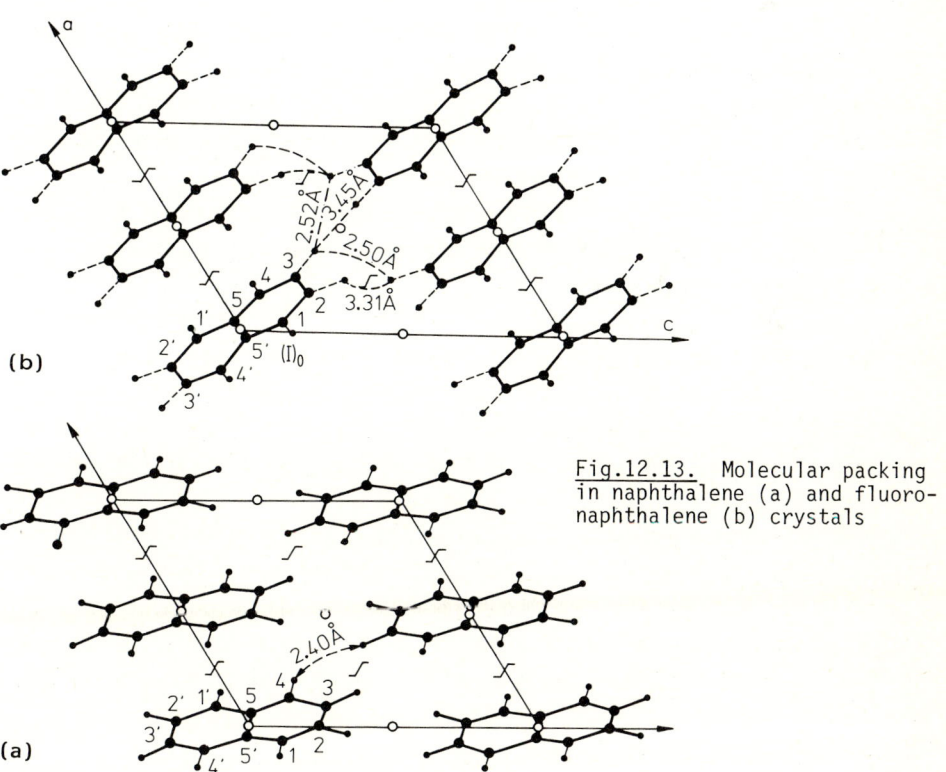

Fig.12.13. Molecular packing in naphthalene (a) and fluoronaphthalene (b) crystals

245

Table 12.3. Solubility data for naphthalene and its derivatives

Matrix \ Guest	H	OH	SH	CH_3	NH_2	F	Cl	Br
H		100%	50%	25%	10%	50%	50%	50%
OH	100%	*				35%		
SH	10%		*					
CH_3	10%			*				
NH_2	90%				*			
F		60%				*		40%
Cl							*	100%
Br	5%					45%	100%	*

leads to the fluoronaphthalene packing (Fig.12.13b), the second to the naphthalene type packing (Fig.12.13a). The geometrical analysis of methylnaphthalene and thionaphthalene cells, which are similar but also rather different, does not show for sure to which structural type they should be assigned. At this point we shall not attempt to predict, on the basis of the cell geometry, the kind of solubility, but instead we shall use phase diagram data to decide whether some particular derivatives should be considered to be of the same or of different structural types, in the sense discussed in Sect.10.1.1.

Data on the reciprocal solubility are summarized in Table 12.3, which contains information on melt-formed crystals, i.e., on their high-temperature form. The vertical columns denote the host crystal. From this table we see, first of all, that it would be desirable to complete the diagram. The available data are self-consistent. A discontinuity of the solubility in the fluoronaphthalene-bromonaphthalene system (about 10% on the eutectic line) spoils the picture a bit. The discontinuity is very short. Perhaps in growing solid-solution single crystals, i.e., in approaching the thermodynamic equilibrium conditions, this gap in solubility disappears. Conclusions on the solubility of the series' extreme members based on the continuous solubility of the bromonaphthalene-chloronaphthalene system and the chloronaphthalene-fluoronaphthalene system are unnecessary. Even though the energy surfaces are similar (the energy minima for all the structures lie in one range of the structural parameters), the differences in the molecules' sizes prevent the formation of a continuous series of solutions. Unfortunately, Table 12.3 does not help us to arrive at any definite conclusion on the crystal structure of methylnaphthalene and thionaphthalene. We have too little data on solubility and there is no information on the orientation of the molecules in the crystals.

Chanh et al. have shown the extent of isomorphism between chloronaphthalene and bromonaphthalene. This unique phase diagram (Fig.12.14) shows that in this system two series of solid solutions create both the high-temperature form I and the lower-temperature form II.

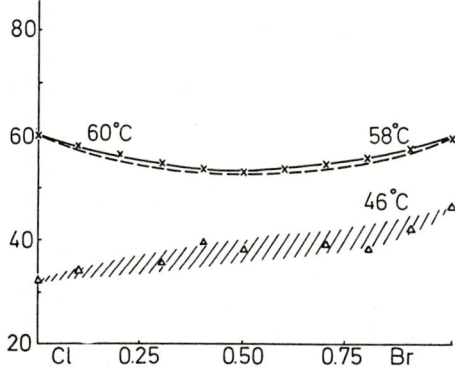

Fig.12.14. Chloronaphthalene-bromonaphthalene phase diagram

The polymorphism of the components results in the decomposition of solid solutions for a variety of cases. When the temperature is lowered, the solid solutions possessing orientational disorder turn into crystals wherein this disorder is obviously absent. In the case of chloronaphthalene the authors observed a transformation which did not destroy the single crystals. This circumstance as well as calorimetric measurements led them to the conclusion that the phase transition proceeds as a result of reorientation of the molecules.

Such a conclusion should be reexamined. The reorientation of the chloronaphthalene molecules is absolutely impossible in a regular lattice. The rotations which would lead from a disordered arrangement to an ordered one require a fantastically large amount of energy. From NMR measurements in solids we know that the reorientation of the naphthalene molecules does not occur even at temperatures approaching the melting temperature. On the other hand, the fact of preserving a single crystal does not in any way exclude the possibility of nucleus growth which can begin at a crystal boundary or from cracks. As to the dependence of the heat capacity on temperature, it is known that such curves are often misleading.

12.7 Solid Solutions of Optically Active Substances or Enantiomers

The types of phase diagrams that form between optical enantiomers were developed at the end of the last century by Roozeboom. We discussed them in Sect. 2.3.7. Among them are cases where the components form solid solutions. It is possible to obtain solid solutions which exhibit a maximum or minimum at a 1:1 composition.

In Chap.2 we indicated that when the liquidus has a maximum or minimum at the place when the liquidus and solidus lines merge, one has no reason to conclude that a new and specific (and not a disordered) solid solution is forming at this extremum. The converse, however, is also equally likely. One may easily assume that for a simple stoichiometric composition this extremum will correspond to an ordered solid solution, similar to the ordered crystal of a binary complex (molecular compound).

In the phase diagram of the laevo and dextro enantiomers of an optically active substance, the extremum corresponds to the 1:1 composition. We may assume that the structure is disordered — the dextro and laevo molecules imitate the center of symmetry or symmetry plane exactly in the same way as it occurs in a one-component solid solution (see Sect.10.2). However, one can suppose that this extremum corresponds to the crystal possessing a well-ordered structure, in which all the cells are identical, the dextro and laevo molecules in each of them forming a pair likely by the inversion center.

In the literature, it is generally accepted to call a crystal *pseudoracemate* if the dextro and laevo molecules have a disordered distribution, and a *true racemate* if the crystal is ordered. Of course, only an X-ray study can determine whether we have a true racemate or a pseudoracemate.

Special conditions for a continuous solubility existing between the *racemate* and the enantiomers were considered in Sect.10.2.2. They are similar for pseudoracemate and true racemate cases or better racemate with orientational disorder and a racemate with ordered structure. In both cases the peculiarity consists in that there is a "continuous" loss of symmetry. In [12.18] we showed that if the geometrical similarity permitted the laevo molecule to replace the dextro molecule without any essential disturbance in the crystal packing, there must necessarily occur a continuous loss of symmetry.

Since we encounter a very small number of space groups in the world of organic crystals, the number of such possibilities is restricted. Thus, for example, the pseudoracemate of symmetry $P2_1/c$ with two molecules per cell would continuously lose its inversion center upon substitution of the laevo molecules for the dextro ones. Finally there would arise for the enantiomer a possibility of a $P2_1$ symmetry crystal accommodating 2 molecules in a cell. Similarly, a transition from a $P\bar{1}$ symmetry pseudoracemate possessing one molecule in a cell to a P1 symmetry crystal would take place.

If we start from a true racemate of symmetry $P2_1/c$ with four molecules per cell, the substitution of the laevo molecules for the dextro ones will lead to the space group $P2_1$ with two pairs of independent molecules; the volume and form of the cell then remain unchanged. But a gradual disappearance of the glide plane with a simultaneous two-fold decrease of the cell size is also possible: in this case the symmetry 2_1 axis may remain, or it too may disappear. The latter case is a "continuous" transition from a monoclinic *syngony* to a triclinic one.

The racemates may also have a syngony without an inversion center. But in this case there must necessarily be two independent crystallographic positions, one occupied by dextro molecules, the other by laevo molecules (4 molecules in the $P2_1$ group or 8 molecules in the $P2_12_12_1$ group). In such cases the formation of solid solutions need not be accompanied by a change in the syngony. We think that a racemic system may exist where the sizes of the cells decrease. There may also be a loss of the symmetry screw axes leading to the transformation of an orthorhombic crystal into a triclinic one with P1 symmetry.

For twenty years such general speculations were not put to an experimental test. The first and the only experiments carried out so far with great accuracy using a perfect X-ray diffraction technique were made by French researchers, Baert and Fouret and coworkers.

In his doctoral thesis BAERT [12.32] described four structural studies which were extended to a fifth system and published recently by FOULTON, BAERT and FOURET [12.33]. Of the five systems investigated by Baert et al. we shall describe in Sect.12.7.1 and 12.7.2 the results obtained in studying the structure of solid solutions of two compounds.

12.7.1 Carvoxime

The carvoxime molecule has an empirical formula $C_{10}H_{15}NO$. The group NOH is added to the hydrocarbon constructed on the basis of a 6-membered ring.

The racemic crystals were derived from a hexane solution containing equimolar amounts of enantiomers. The crystals have a $P2_1/c$ symmetry and the cell contains four molecules with dimensions a = 9.905 Å, b = 11.840 Å, c = 8.521 Å. The monoclinic angle $\beta = 100°40'$, the cell volume equals 982.88 Å^3, and the packing coefficient is 0.73.

By superimposing the left molecule on the right molecule in an optimal manner, we find their geometrical similarity coefficient $\varepsilon = 0.78$ (3.18). At first glance, it may seem that it is not sufficient for the formation of a continuous solid solution. However, structural investigation has shown that the substitution takes place readily.

The coordinates for all the atoms were determined. Although the packing coefficient was high, all the non-bonded distances, with the only exception of one H-H contact (1.97 Å), appeared to be larger than the sums of molecular radii.

The structure consists of dimers bonded around the inversion center by a OH ... N hydrogen bond. The dimer packing results from usual forces, which explains the low melting temperature (92°C). The dense packing is accomplished by a large number of contacts between the molecules.

The X-ray study of the optically active carvoxime has shown that it was impossible to obtain a substance consisting of molecules of one enantiomer. It turned out that the sample used was a crystal made up of 12.5% laevo molecules and 87.5% dextro ones. The crystal belongs to space group $P2_1$, with four molecules per unit cell. Hence, there are two independent crystallographic positions. The cell dimensions are a = 10.247 Å, b = 11.677 Å, c = 8.561 Å. The monoclinic angle is 103°07'; the cell volume equals 998.2 Å^3 and the packing coefficient was slightly lower and equalled 0.713.

The fact that the cell dimensions differ only slightly shows that we are dealing with a case where only a loss of symmetry occurs and one crystallographic position transforms continuously into two. With the aid of an adequate computer program the authors showed that in the crystal under investigation one of the crystallographic positions is occupied by the dextro molecules, while in the second (with weights corresponding to the per cent composition) substitution of the laevo molecules for the dextro ones takes place, thus making the optical active crystal close to a racemate.

Another crystal was studied with 57.5% dextro and 42.5% laevo molecules. The cell dimensions were close to those of the racemate. It was shown that the transition from racemate to pure enantiomer proceeded by substitution of the molecules located in one of the crystallographic positions.

The authors demonstrated the continuity of the loss of symmetry. The extinction of the X-ray reflections associated with the presence of a glide plane was gradually relaxed. The reflections which had been forbidden for the racemate became stronger and stronger with the resemblance between the racemate and the enantiomer crystals becoming more pronounced.

The authors showed that when the substitution of the laevo molecules for the dextro ones took place, the hydrogen bonds between the molecule pairs remained intact. Therefore, the structure pattern did not change and the character of the packing varied only slightly in spite of the essential difference of the coefficient of geometric similarity ε from unity.

12.7.2 Carvoximebenzene [12.34]

The phase diagram of the mixture of laevo and the dextro components resembles that of carvoxime. The racemate melting temperature is again higher than that of the pure enantiomers.

The racemate crystal has a triclinic $P\bar{1}$ symmetry with two molecules per cell, and a packing coefficient of 0.729. In this case the dextro and laevo molecules are more similar in form: the geometrical similarity ε (3.18) equals 0.89. Packing is determined by the usual intermolecular forces; hydrogen bonds are absent.

The active compound has a cell very close to that of the racemate. The molecular packing coefficient is less; it equals 0.717. The enantiomers and racemate structures were determined [12.35].

A detailed X-ray study has proved that the transition of the racemate to the dextro modification consists in the replacement of the laevo molecule by the dextro one. A continuous loss of the inversion center takes place. We pass from a crystal wherein the molecules occupy one crystallographic position, to a crystal wherein the molecules occupy two crystallographic positions. In this case, as in the previous one, the substitution takes place without the creation of orientational disorder. It would be interesting to complement the work by studying the close-range positional order by X-rays or neutron diffuse scattering studies.

12.8 Durene-Para-Dibromobenzene [12.36]

12.8.1 X-Ray Diffraction Analysis

The pure durene phase of this system is of interest for studying the orientation of guest molecules with respect to the inequivalent positions.

The durene (1,2,4,5-tetramethylbenzene) and para-dibromobenzene molecules differ in volume by 16% (the volumes are 151.0 and 126.4 $Å^3$, respectively). Because of this large difference, one does not expect a large mutual solubility. However, the molecules are similar enough in shape, and solid solutions will form with the smaller $p\text{-}Br_2C_6H_4$ molecules in the durene matrix.

The durene-para-dibromobenzene system forms an eutectic phase diagram with a slight mutual solubility of both components in the solid phase. The solid-solution crystals of the durene phase contain from 0 to 7% $p\text{-}C_6H_4Br_2$ molecules, while crystals of the hypereutectic (p-dibromobenzene) phase contain from 0 to 4% durene molecules. Crystals of both phases were grown simultaneously in an eutectic melt and could be easily distinguished from each other by their form. Within the durene structure, the para-dibromobenzene guest molecule can be oriented in two inequivalent positions in which the methyl groups of durene are replaced by Br atoms.

To establish the type of solubility in this phase, X-ray measurements were made of the unit-cell parameters for both pure durene and maximally doped durene. The results are given in Table 12.4. In the durene phase only an insignificant shift occurs in the unit-cell parameters. Replacing large molecules by smaller ones results in the formation of a small number of voids (in six sites out of a hundred) equal to the difference in the volume increments of the groups $CH_3(\Delta v = 22.5 \text{ Å}^3)$ and $H(\Delta v = 2.0 \text{ Å}^3)$, the volume occupied by a 100 molecules being equal to 21475 Å^3. Neglecting the redistribution of the stresses during the formation of a solid solution, one can evaluate the distribution of the "new" voids $[6 \cdot (22.5 - 2.0)]/21475 \approx 0.5\%$. The molecular packing coefficient in the durene structure equals 0.7, i.e., voids between the molecular bodies constitute 30% of the crystal bulk. A 0.5% increase undoubtedly allows this solid-solution crystal to remain a densely packed structure.

Table 12.4. The unit-cell parameters for durene and a solid-solution crystal containing 6% of p-dibromobenzene

Parameters	100% durene [Å]	94% durene [Å]
a	7.024 ± 0.004	7.015 ± 0.005
b	11.570 ± 0.006	11.571 ± 0.007
c	5.734 ± 0.0006	5.742 ± 0.0006
γ	112.863° ± 0.002 ($P2_1/b$ Z = 2)	112.845° ± 0.002

As has been noted the preservation of some small voids in the packing arising as a result of substitution does not require a great expenditure of energy. As we have repeatedly emphasized, this conclusion depends on the asymmetrical form of the potential energy curve for molecular interaction.

Hence the small observed variation in the unit-cell parameters in the durene-phase solid solution (especially if they are compared with the absence of distortions in the hypereutectic phase with a para-dibromobenzene structure) may be due to a true substitution. This conclusion is supported by direct X-ray diffraction analysis.

The next task was to determine the orientational disorder in the crystals. In the durene-phase solid solution, para-dibromobenzene admixture molecules can simulate the orientation of the durene matrix molecules in two crystallographically inequivalent ways. These are illustrated in Figs. 12.15,16.

With a small concentration of guest molecules it is evidently impossible to "see" the guest molecule (even though it has two heavy halogen atoms) on an electron-density map. Analysis of the changes in the heights of two independent electron-density maxima requires extremely accurate work, especially in the case of a uniform distribution of the guest molecules over both positions being substituted.

A careful X-ray study of the durene-phase solid-solution crystal containing 6% impurity was carried out. Durene-phase solid-solution single crystals used in the experiment were grown from a melt (at minimal supercooling $\Delta t = 0.15°C$, $t_{cr} = 50.80 + 0.05°C$), whose composition was close to eutectic: 57 mol% durene + 43 mol% para-dibromobenzene. The natural facets of these crystals were pre-

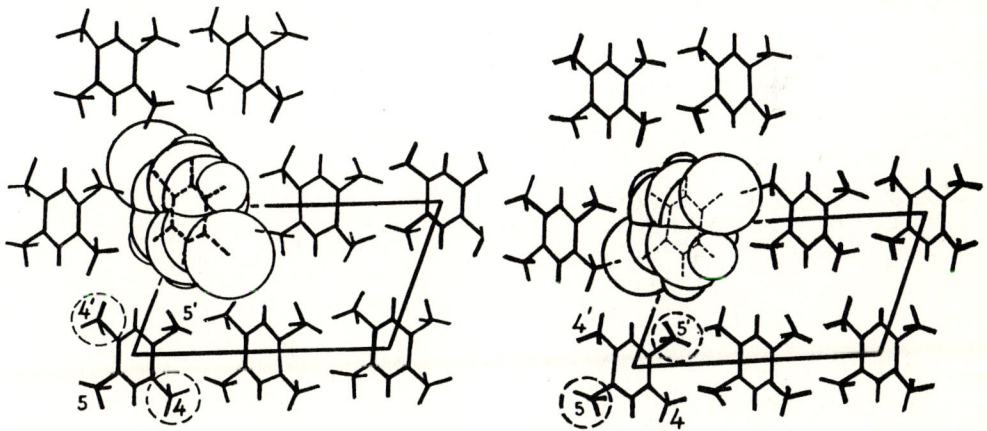

Fig.12.15. First mode for substituting a durene molecule by a para-dibromobenzene molecule

Fig.12.16. Second mode for substituting durene molecule by a para-dibromobenzene molecule

served; to protect them from sublimation they were coated with a thin layer of shellac before the X-ray experiment.

The reflected intensities measured in two independent octants of the reciprocal lattice were determined; 479 nonzero structure factors were calculated. The signs of the corresponding pure durene F_{calc} amplitudes were assigned to the set of experimental F_{ob} values for the solid solution. The mean atomic positions were refined by conventional methods, after which the electron-density map was calculated. The purity of the resulting electron-density maps was appreciable; the peak heights were: 602, 588, 596 (relative units) for the benzene-ring atoms $C(1)$, $C(2)$, $C(3)$, and 452 and 458 for the carbon methyl atoms $C(4)$ and $C(5)$, respectively.

By comparing these results with those for pure durene, one could notice a marked change in the peaks for the methyl atoms $C(4)$ and $C(5)$, which are considerably lower (433 and 437, respectively) in pure durene. That this difference is not accidental is seen from the fact that the maxima for the benzene ring atoms do not change. Thus the electron-density map shows the formation of a true solid solution. In order to determine the guest-molecule orientation, a comparison was made with the experiment on structure amplitudes calculated for three models of the impurity molecules' positioning.

The first model assumes that the substitution occurs only at one of the two distinct positions in the durene crystal matrix, viz. all 6% of the para-bromobenzene molecules are located so that all the coordinates of the benzene-ring atoms of both molecules coincide, while the bromine atoms replace the methyl group only at position (4-4'). In this case the bromine atoms were found at a distance of 1.85 Å in the C_{ar}-C bond direction. Figure 12.17 illustrates schematically the "average molecule" corresponding to the case being discussed.

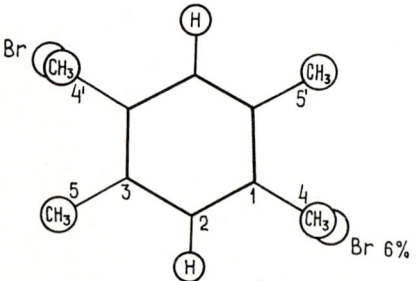

Fig.12.17. First substitution model used for calculating X-ray intensities of durene-para-dibromobenzene crystals

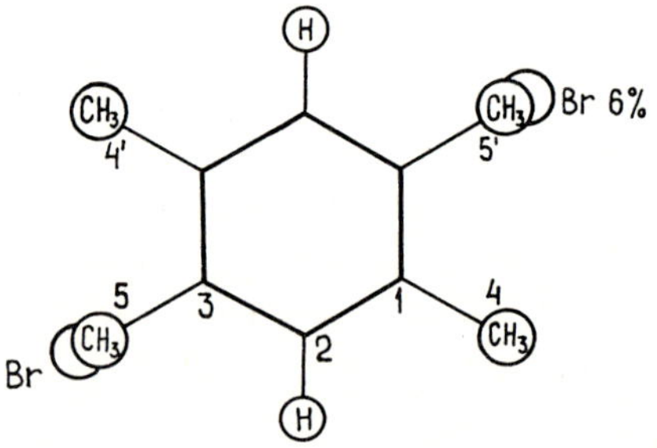

Fig.12.18. Second substitution model used for calculating X-ray intensities of durene-para-dibromobenzene crystals

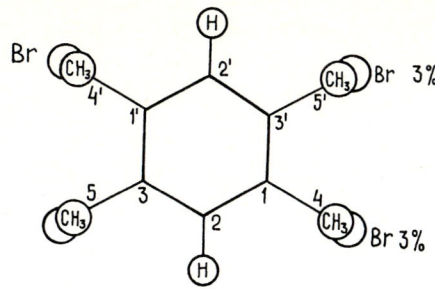

Fig.12.19. Third substitution model used for calculating X-ray intensities of durene-para-dibromobenzene crystals

In the second model, the bromine atom was positioned over the other methyl group (5-5') as shown in Fig.12.18. Finally, the third model presupposed that the guest molecules were distributed with equal probability among the two crystallographically nonequivalent sites (4-4') and (5-5'). This case is illustrated in Fig.12.19.

Of the three models, the closest to being a practically realizable method of substitution is that of Fig.12.19.

A definite increase in the "thermal vibrations" of all the atoms in the crystal is observed. An increase in the vibrational amplitude is also noticeable in a direction in the molecular plane perpendicular to the C-Br bond. Also noticeable is the relatively large increase in the "thermal vibration" amplitudes of the atoms in the direction of the monoclinic c-axis. One may suppose that these changes characterize the static distortion of the lattice which takes place predominantly in this direction. This is evidenced also by a slight elongation of c (Table 12.4).

12.8.2 Substitution Energy Computation

Even though careful X-ray diffraction analysis gave unambiguous results of true solid-solution and orientational disorder over two possible orientations for the durene-p-dibromobenzene system, nevertheless, it is desirable to confirm computationally the results of this subtle experiment.

All the energy estimates presented in this section were made on the assumption that the host molecules occupy the same positions around the guest molecules as in the pure matrix crystal, i.e., changes in the orientation of the host molecules surrounding the guest molecule are ignored.

The interaction energy calculations between the molecules were made using the atom-atom potentials in the form

$$V = -Ar^{-6} + B \exp(-\alpha r)$$

with parameters A, B for H-H, C-H, C-C; and Br-H, Br-Br, and Br-C interactions taken from [12.37,38]. The summation radius of the atom-atom interactions was equal to 15 Å. With such a limitation of the summation range the error in the energy values did not exceed 1%.

The calculations were performed according to the following scheme. To determine the difference in the "behavior" of the guest molecule in each of the two symmetrically independent positions, the energy values were calculated for several rotation angles of the para-dibromobenzene molecule as a unit in the plane coinciding with the aromatic plane of the durene molecule, in the range of 6°(Fig.12.20). Studying these data it was found that the solid-solution lattice energy is minimized when the impurity molecule rotates from its optimally substituted position so that its bromine atoms can avoid the densely packed (100) layer of the durene matrix structure.

These interaction calculations yield V^{durene} = -20.8 kcal/mole for the pure durene lattice energy. If the bromine atoms substitute for the methyl groups in position (4-4'), the interaction energy of the para-dibromobenzene molecules equals -20.2 kcal/mole, and -20.4 kcal/mole if they are in position (5-5'). In both cases the addition of the para-dibromobenzene molecules into the matrix lattice is associated with a small increase in the lattice energy: Δv = 0.4 - 0.6 kcal/mole.

Substitution energy calculations were also performed for the para-dibromobenzene-phase solid solution. Experience shows that when the guest molecule (durene, v_{mol} = 151.0 Å^3) is larger than the host molecule (para-dibromobenzene v_{mol} = 127.0 Å^3), the formation of a true solid solution proceeds with a noticeable increase in the unit-cell volume, even with a low concentration of the guest molecules. Indeed, when the smaller para-dibromobenzene molecules are truly replaced by the larger durene molecules, one pair of the methyl group

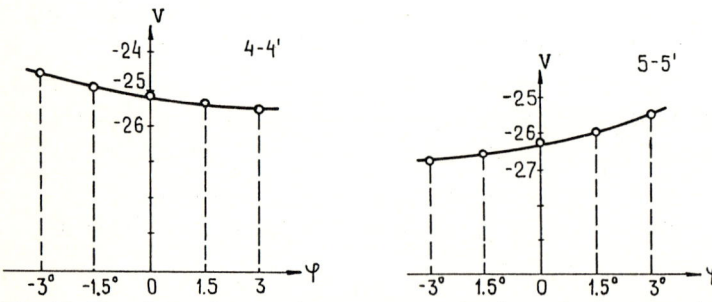

Fig.12.20. Dependence of energy on rotation of the para-dibromobenzene molecule in the plane of the benzene ring of the durene molecule

(Δv_{CH_3} = 22.5 Å^3) can occupy the sites of the bromine atoms (Δv_{Br} = 26.5 Å^3) apparently without introduction of any significant distortions; the other methyl group will occupy hydrogen atoms' positions (Δv_H = 2.0 Å^3). A considerable distinction in the sizes and forms of the substituent obviously leads to the appearance of high stresses in the para-dibromobenzene matrix, which can significantly influence the course of solubility, even preventing the possibility of true substitution in this phase. As was indicated in Sect.12.8.1, no variations in the solid-solution parameters of this phase were observed, but since the solubility is so small one cannot state with assurance whether the impurity molecules replace the matrix molecules at the mosaic block boundaries or at defect sites.

Energy calculations offered an answer to this question. The interaction energy was calculated for two different substitution modes differing from each other in that the substituent, i.e., the methyl group replacing the aromatic hydrogen atom, occupies positions either to the right or to the left of the basic substituted position (Br-Br).

For one mode of substitution the total value of the interaction energy for the durene host molecule with its surrounding was -2.4 kcal/mole, for the other 180 kcal/mole. In both cases this value appreciably exceeds the interaction energy of the host molecule with its surrounding, amounting to -18.3 kcal/mole.

Since the value of $V_n - V_m$ in a solid organic substitution solution usually does not exceed 2 - 3 kcal/mole, one can state that only an interblock solubility is possible in the para-dibromobenzene phase.

12.8.3 Probability Calculations of Different Guest Molecule Orientations

A solid-solution durene-phase lattice site can be occupied either by the host molecule or by the guest molecule in two different orientations. Each lattice site can be, so to say, in three different states. Generalizing (4.47), one can write the configurational free energy of the durene phase as follows:

$$F = \Theta \sum_r \sum_{\alpha=1}^{3} n_\alpha(\underline{r}) \ln n_\alpha(\underline{r}) + \frac{1}{2} \sum_{r,r'} \sum_{\alpha,\alpha'=1}^{3} V_{\alpha,\alpha'}(\underline{r}',\underline{r}) n_\alpha(\underline{r}) n_{\alpha'}(\underline{r}') \quad , \quad (12.2)$$

where $n_1(\underline{r})$ denotes the single-particle probability (the mean number of occupations) for finding the matrix molecule in the crystal lattice site $\underline{r}$; $n_2(\underline{r})$ and $n_3(\underline{r})$ are the single-particle probabilities for finding the guest molecule at lattice site $\underline{r}$ in the first and the second orientation, respectively;

$V_{\alpha,\alpha'}(\underline{r}',\underline{r})$ are the pairwise energies of the molecular interaction in sites $\underline{r}$ and $\underline{r}'$ and in the states α and α', respectively, $\Theta = kT$.

The single-particle probabilities $n(\underline{r})$ obey two additional conditions. The first condition is

$$\sum_\alpha n_\alpha(\underline{r}) = 1 \quad , \tag{12.3}$$

where the $\underline{r}$ site may be occupied either by a host molecule or by a guest molecule with two different orientations. The second condition, which retains the complete number of the host molecules, is

$$\sum_r n_1(\underline{r}) = N(1 - c) \quad , \tag{12.4}$$

where N is the full number of the crystal lattice sites, and c is the guest molecule concentrations.

The single-particle probabilities $n_\alpha(\underline{r})$ describe the positioning of particles of different types and orientations. Since it is assumed that para-dibromobenzene molecules are randomly distributed throughout the host lattice sites, we obtain the following relations:

$$n_1(\underline{r}) = 1 - c \quad , \qquad n_2(\underline{r}) + n_3(\underline{r}) = c \quad . \tag{12.5}$$

Substituting (12.5) into (12.2) we obtain

$$F = F_0 + \sum_r \varphi n(\underline{r}) + \Theta \sum_r n(\underline{r}) \ln n(\underline{r}) + \Theta \sum_r [c - n(\underline{r})] \ln [c - n(\underline{r}')]$$

$$+ \frac{1}{2} \sum_{r,r'} w(\underline{r},\underline{r}')n(\underline{r})n(\underline{r}') \quad , \tag{12.6}$$

where F_0 is a constant, independent of the molecular arrangement configuration,

$$\varphi = (1 - c)[\Phi_{12}(0) - \Phi_{13}(0)] + c[\Phi_{23}(0) - \Phi_{33}(0)] \quad . \tag{12.7}$$

Here

$$\Phi_{\alpha\alpha'}(0) = \sum_r W_{\alpha\alpha'}(\underline{r}) \qquad (\alpha, \alpha' = 1, 2, 3) \quad .$$

In (12.6)

$$w(\underline{r},\underline{r}') = V_{22}(\underline{r},\underline{r}') - V_{23}(\underline{r},\underline{r}') - V_{32}(\underline{r},\underline{r}') + V_{33}(\underline{r},\underline{r}')$$

represents a mixing energy analog and $n(\underline{r})$ is identically equal to $n_2(\underline{r})$.

We can obtain the single-particle probability equation if we set the first variation with respect to $n(\underline{r})$ of the free energy determined by (12.6) equal to zero. Then we obtain for the single-particle probability $n(\underline{r})$ the following equation, which is similar to (4.37):

$$n(\underline{r}) = \frac{c}{1 + \exp\left\{1/\Theta\left[\varphi + \sum_{\underline{r}'} w(\underline{r},\underline{r}')n(\underline{r}')\right]\right\}} \quad . \tag{12.8}$$

The radius vector of the molecule center is

$$\underline{r} = \underline{R} + \underline{h}_p \quad , \tag{12.9}$$

where $\underline{R} = \underline{a}x + \underline{b}y + \underline{c}z$, and x, y, z are the unit-cell center coordinates; $\underline{a}, \underline{b}, \underline{c}$ are the basic lattice translations; $\underline{h}_p$ has two values ($Z = 2$).

Since the pairwise potential of interaction of two molecules is independent of the coordinate system, it can be presented in the following form:

$$w(\underline{r},\underline{r}') = w^{pp'}(\underline{R} - \underline{R}') \quad , \tag{12.10}$$

where $\underline{r}' = \underline{R}' + \underline{h}_{p'}$, $(p = 1, 2)$.

Using (12.9,10), we can rewrite the self-consistent field equation (12.8) as

$$n^p(\underline{R}) = \frac{c}{1 + \exp\left\{1/\Theta\left[\varphi + \sum_{\underline{R}'}\sum_{p'} w^{pp'}(\underline{R} - \underline{R}')\, n^{p'}(\underline{R}')\right]\right\}} \quad . \tag{12.11}$$

The equation is a nonlinear finite difference equation of the integral type relative to $n^p(\underline{R})$, the probability of finding the guest molecule in the first orientation at the $(p; \underline{R})$ "site". This equation at high temperatures has the solution $n^p(\underline{R}) = \gamma$. The solution is independent of the crystal lattice site coordinates and indicates that the guest molecules can be arranged along two different orientations with probabilities γ and $c - \gamma$.

The dependence of the probability γ on T and c is

$$\gamma = \frac{c}{1 + \exp\{\varphi/\Theta + (\gamma/\Theta)[\Phi^{11}(0) + \Phi^{12}(0)]\}} \quad . \tag{12.12}$$

The values $\phi^{pp'}(0) = \sum_R \phi^{pp'}(\underline{R})$ are Fourier components of the elements of the mixing energy matrix.

To calculate the values of γ and $c - \gamma$, the substitution probabilities at positions (5-5') and (4-4'), respectively, the interaction energy values, $V_{\alpha\alpha'}^{pp'}(\underline{R} - \underline{R}')$, were determined for the pairs of matrix and admixture molecules located in the $\underline{R}$, $\underline{R}'$ sites in the α, α' states, respectively ($\alpha, \alpha' = 1, 2, 3$). The $V_{\alpha\alpha'}^{pp'}(\underline{R} - \underline{R}')$ calculations were carried out by computer.

Using the pairwise interaction energy values, $V_{\alpha\alpha'}^{pp'}(\underline{R} - \underline{R}')$, we find $\phi^{11}(0)$, $\phi^{12}(0)$, φ, and with the aid of (12.7) we obtain the following numerical values: $\phi^{11}(0) = +1.039$ kcal/mole, $\phi^{12}(0) = -3.944$ kcal/mole, $\varphi = -0.5$ kcal/mole. Substituting these values into (12.12), we obtain $\gamma = 0.041$ and $c - \varphi = 0.019$ for $T = 300$ K (the experiment was carried out at room temperature). The numerical results show that the first orientation of the impurity molecules [substitution of position (5-5')] is twice as probable as that of the second orientation. We can consider the agreement between experiment and the substitution energy estimates to be fairly satisfactory.

12.8.4 Calculation for Oriented Short-Range Order Correlation

The interaction energies of the host and guest molecules in different orientations are, in general, different, and therefore each host molecule will strive to surround itself either with guest molecules with different orientations or with host molecules. The difference in the interaction energy between the host and guest molecules with different orientations may be enough for each host molecule to prefer guest molecules of a particular orientation.

This short-range order is characterized by a distribution of the molecules over the solution's positions and orientations. For the positional and orientational characteristics of the short-range order in organic solid solutions, one may use the following correlation parameters:

$$\varepsilon_{\alpha\alpha'}(\underline{r},\underline{r}') = \overline{n_\alpha(\underline{r})n_{\alpha'}(\underline{r}')} - n_\alpha(\underline{r})n_{\alpha'}(\underline{r}') \quad , \quad \alpha\alpha' = 1, 2, 3 \quad (12.13)$$

where $\overline{n_\alpha(\underline{r})n_{\alpha'}(\underline{r}')}$ are the probabilities of the simultaneous residence of different pairs of host and guest molecules in the $\underline{r}$, $\underline{r}'$ sites and in the α, α' states. It can be shown that the number of independent parameters in the durene-para-dibromobenzene system is reduced to 12.

Coordination spheres which envelop 12 molecules surrounding the central one were taken into consideration in the calculations of the orientational short-range order parameters. The numerical results are the following:

$$\varepsilon_{\alpha\alpha'}^{22}(1) = \varepsilon_{\alpha\alpha'}^{11}(1) = \begin{array}{l} +0.0011, \; -0.0007, \; -0.0004 \\ -0.0007, \; +0.0004, \; +0.0003 \\ -0.0004, \; +0.0003, \; +0.0001 \end{array}$$

$$\varepsilon_{\alpha\alpha'}^{12}(2) = \begin{array}{l} +0.0010, \; -0.0005, \; -0.0005 \\ -0.0005, \; +0.0004, \; +0.0001 \\ -0.0005, \; +0.0001, \; +0.0005 \end{array}$$

$$\varepsilon_{\alpha\alpha'}^{21}(2) = \widetilde{\varepsilon}_{\alpha\alpha'}^{12}(2)$$

$$\varepsilon_{\alpha\alpha'}^{22}(3) = \varepsilon_{\alpha\alpha'}^{11}(3) = \begin{array}{l} +0.0003, \; -0.0000, \; -0.0003 \\ -0.0000, \; -0.0003, \; +0.0003 \\ -0.0003, \; -0.0003, \; -0.0000 \end{array}$$

$$\varepsilon_{\alpha\alpha'}^{12}(4) = \begin{array}{l} -0.0004, \; +0.0004, \; -0.0000 \\ +0.0005, \; -0.0004, \; -0.0001 \\ -0.0001, \; +0.0000, \; +0.0001 \end{array}$$

$$\widetilde{\varepsilon}_{\alpha\alpha'}^{21}(4) = \widetilde{\varepsilon}_{\alpha\alpha'}^{12}(4) \quad ,$$

where the Figures 1, 2, 3, 4 in the correlation parameter arguments indicate the number of the corresponding coordination sphere $\widetilde{\varepsilon}_{\alpha\alpha'}^{pp'}$ matrix transposed in respect of $\varepsilon_{\alpha\alpha'}^{pp'}$.

Using only the correlation parameter signs, it is possible to construct a schematic picture of the most probable distribution of different pairs of matrix and admixture molecules.

Figure 12.21 emphasizes the major results of the calculation. There occurs a planar *segregation* of the host molecules. The guest molecules also

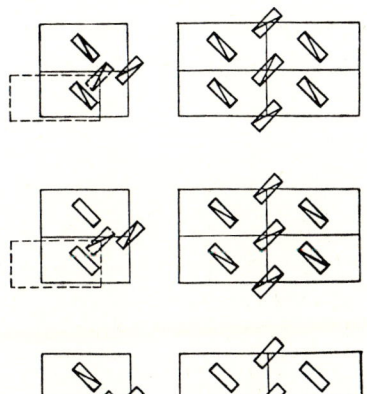

Fig.12.21. Most probable structure of the first coordination sphere of the host and guest molecules. Shown on the right is the mutual arrangement of molecules in the ab plane. Shown on the left are the three molecules in the nearest ab plane (the initial molecule is in the vertex of the dotted rectangle)

strive to form *islands* out of molecules of one kind. Here the difference in the orientation does not play a particular role (Fig.12.21, second and third from the top). Consequently, it can be concluded that the solid solution must decompose with decreasing temperature.

12.9 Tolane Diphenylmercury [12.39]

This system seems to be the most thoroughly studied. The presence of a heavy atom in one of its components substantially facilitates X-ray diffraction analysis of solid solutions and allows a deeper study of the orientational disorder in the tolane ($C_6H_5 - C \equiv C - C_6H_5$) phase. The reasons for such a disorder differ from those of the para-dibromobenzene-durene system. In this system we see only one way in which the replacement of the tolane molecule by that of diphenylmercury takes place. But we find that in the tolane lattice where the molecules occupy two independent crystallographic positions, the substitution can take place in two ways. Therefore, it is of interest, first of all, to study the manner in which admixture molecules are distributed among the two positions in the lattice.

Tolane crystals [12.40] belong to the $P2_1/a$ space group with the following unit-cell parameters: a = 12.71, b = 5.77, c = 15.68 Å, β = 114°38'. The unit cell contains 4 molecules; they occupy two independent families of inversion centers thus forming a double layer packing.

Diphenylmercury crystals [12.41] belong to the $P2_1/a$ space group with the unit-cell parameters: a = 11.56, b = 8.30, c = 5.59 Å, β = 112°20'. The unit cell now has 2 molecules, i.e., one molecular stack located on the inversion centers.

Tolane and diphenylmercury molecules (Fig.12.22) differ in length by not more than 0.2 Å; this value does not exceed the precision of the values de-

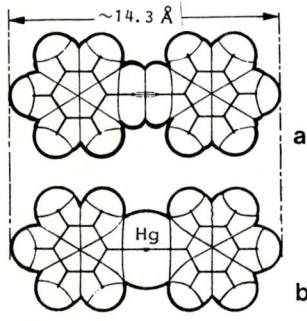

Fig.12.22. Tolane (a) and diphenylmercury (b) molecules

termined for intermolecular radii. The width and thickness of the molecules are also the same, since they are only determined by the width and thickness of the phenyl rings. The volumes per molecule in the crystal differ by 0.6%.

Despite the fact that the molecules' sizes are practically the same, one may expect only a limited reciprocal solubility of the components in solid state, since the differences in their crystal structures would interfere with the continuous solubility.

The phase diagram of the tolane-diphenylmercury system has been determined by growing crystals. The phase diagram is shown in Fig.12.23; the coordinates of the eutectic point are $T = 56.0°C$, diphenylmercury content 16.8%.

Data for the solidus line were determined both by microanalysis of the percentage of mercury in the single crystals grown and by visually determining the end of their melting interval. The maximum tolane content in crystals with a diphenylmercury structure amounted to 8.2%, while that of diphenylmercury in a tolane phase equaled 14%.

At this point it is worthwhile to say a few words about the guest molecule's influence on crystal growth. The seeding needle dipped into a $0.1 - 0.2°C$ supercooled pure tolane melt induced immediate formation of several nuclei which grew simultaneously at the same rate and then coalesced. As a result, at the tip of the needle there appeared a single crystal, but without a faceted droplet. Only by means of substantial supercooling of the seeding needle (by $0.3 - 0.4°C$), followed by its cooling to room temperature before dipping it into the melt, were one or two perfectly faceted little crystals obtained whose maximum sizes did not exceed 1 mm along the b axis. An increase in the amount of diphenylmercury in the melt significantly facilitated the production of such well-faceted single crystals. As a result it was rather easy to grow fine-faceted tolane-phase crystals of rather large sizes (3×1, 5×1 mm^2) in

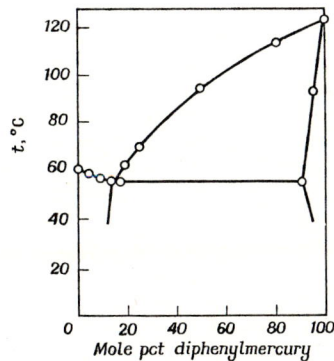

Fig.12.23. Tolane-diphenylmercury diagram of state

the immediate vicinity of the eutectic point (in a melt containing 16% diphenylmercury).

In turn, the tolane solute also affected the growth of crystals with a diphenylmercury structure. If a cluster of extra fine long needlelike crystals was grown rapidly in the melt of a pure substance on the needle tip, an admixture of tolane retarded the growth rate and caused the crystals to grow both in length and in thickness.

Thus, in this system, an increase in the solute concentration in the melt facilitates the production of fine-faceted solid-phase single crystals, whose sizes are quite suitable for X-ray study.

12.9.1 Diphenylmercury Phase

Figure 12.24 illustrates the variation of the molecular volume upon changes in the tolane concentration. The dash-and-dot line connects the molecular volumes of the components. We encounter a rare case of where a solid solution becomes more dense. Can this result be interpreted on the basis of simple geometrical considerations? The tolane molecules are slightly smaller (shorter by 0.2 Å) than the diphenylmercury molecules, but the volume per tolane molecule in the crystal (261.5 Å) is greater than the molecular volume of diphenyl mercury (248.1 Å). This is due to the fact that the diphenylmercury molecules are considerably more densely packed in the crystal. Thus the slope of the dash-and-dot line is opposite to that of a line which would have connected the intrinsic volumes of these two particular molecules so similar in form. When tolane molecules enter into the diphenylmercury crystal and are positioned

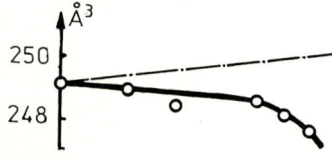

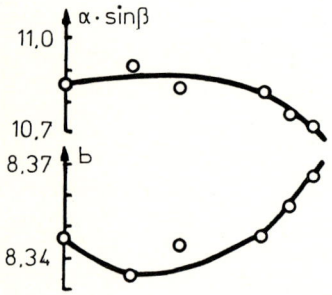

Fig.12.24. Volume per average molecule as a function of tolane concentration in a tolane-diphenylmercury solid solution

in the same way as the host molecules are arranged, the guest tolane molecules will be packed more densely compared to their own phase. Furthermore, the mutual arrangement of the diphenylmercury and tolane molecules is such that the free space located around the mercury atoms and the bridge $C \equiv C$ groups can be used more efficiently than in the diphenylmercury phase. Thus the reason for the increase in density quite clear from a geometrical viewpoint results from utilizing more efficiently the central "hollows" of the molecular components.

The information that could be derived from the diffraction pattern of the X-ray study was meagre. No orientational disorder was observed in the phase. It is practically impossible to study short-range order when weakly scattering molecules are mixed in a matrix crystal constructed from strongly scattering molecules.

12.9.2 Tolane Phase

As distinct from the diphenylmercury phase, addition of the admixture molecules into the tolane phase can be traced with extremely high precision.

We conducted an X-ray diffraction analysis of the tolane phase and the results obtained are presented not only in this book, but also in [12.19]. Therefore, we shall make a cursory survey of the investigations, now more than ten years old, and then deal in more detail with recent studies of this interesting solid solution.

The specific volume versus concentration curve has a maximum, and the volume per one "average" molecule begins to grow at first. We can ascribe this to a guest molecule possessing a linear dimension 0.2 Å greater than that of the tolane molecule.

With an increase in the amount of the guest molecules in a crystal, the effect of short-range order begins to be felt more strongly. As will be seen in the next section, the data obtained by the X-ray diffuse-scattering technique indicate that guest molecules tend to aggregate. Because diphenylmercury molecules are more densely packed than tolane molecules, it is quite natural that at the moment when diphenylmercury molecule "islands" form, there appears a decrease in the molecular volume. This seems to be a reasonable explanation of such a strange behavior in the plot of cell volume versus concentration.

It is not so simple to interpret the behavior of the Debye-Waller temperature factor B when one considers the exponential function responsible both for the heat variation and stable distortion of the lattice. This function

(Chap.6) is determined by the decrease in the X-ray scattered intensities with an increase in the scattering angle.

The B factor is related by an equation to the mean square deviation of the molecule's position in the direction perpendicular to the reflecting plane. (This is a very rough approximation; actually the intensity of each ray is determined by a sum the terms of which are related to the atoms of the molecule, each term of the sum having its own temperature factor). This deviation may be due to two reasons: static distortions and thermal motion. Experiments have shown that the curve $B(c)$ has a pronounced minimum at room temperature which vanishes at liquid nitrogen temperature. The static distortions are sure to depend on concentration and temperature (via the short-range order). Consequently, the dynamic effects depend both on concentration and on temperature, and it is impossible to separate these effects.

In contrast to the scanty results of these experiments, an abundance of information is offered by an electron-density map obtainable from the observed X-ray scattering intensities. X-ray diffraction analysis has made it possible to establish (for the first time for this system) the nature of the orientational disorder. The task was to find out which of the two independent positions the tolane molecules occupy as they replace diphenylmercury molecules.

The electron density map (Fig.12.25) is exceptionally expressive in this case. This pattern, in our opinion, is to become a classic. The result of the experiment is quite obvious: the diphenylmercury molecules replace the tolane molecules in one of the two independent host crystal layers (at bottom left in Fig.12.25 only a). In one of the layers we can also see, as in the pure

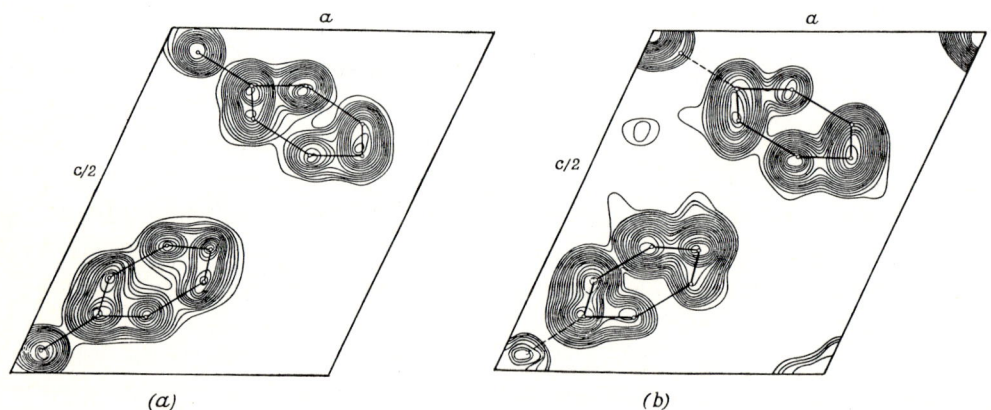

Fig.12.25. Electron density of (a) pure tolane and (b) a tolane solid-solution crystal containing the maximum amount of diphenylmercury

tolane map, the maximum corresponding to the hydrogen bridge atom; in the other layer we see a mercury atom located at the lattice symmetry center.

The calculations of the substitution energy by the atom-atom potential method proved to be in excellent agreement with the experiment. The substitution of a diphenylmercury molecule for one tolane molecule in layer I changes the lattice energy by -4.88 kcal/mole; the substitution in layer II changes it by -2.52 kcal/mole, the difference between them being $\Delta = 2.36$ kcal/mole. By using the Boltzmann formula, we can calculate the relative probabilities of the admixture molecule distribution between the nonequivalent layers:

$$\frac{W_I}{W_{II}} = e^{-\Delta/RT} \approx 51 \quad .$$

Thus the probability of substitution in one layer is 50 times as great as that in the other. We may assume that all the guest molecules are disposed in one layer, and this is clearly shown by the X-ray diffraction analysis.

There is still another way of verifying the validity of the substitution energy calculations by the atom-atom potential method. The theory explained in Sect.4.3 and successfully applied to the diphenyl-dipyridyl system can also be applied here. It is necessary, however, to derive new formulas.

12.9.3 Diffuse X-Ray Scattering Studies

As was mentioned in Chap.6, the X-ray scattering technique can obtain a great deal of information on the structure of solid solutions and is also generally used for studying short-range order. It has been shown in recent papers that this technique can also be employed successfully to determine the parameters of atomic (molecular) interaction, the chemical activity of the components, the free energy; to construct equilibrium diagrams; to study elementary diffusion; and to find macroscopic diffusion characteristics (activation energy of diffusion, coefficients of diffusion and self-diffusion).

These parameters can be determined from quantitative X-ray diffuse-scattering intensity measurements for single crystals of disordered alloys, the scattering resulting from the existence of short-range order. This has been done so far only for the metallic alloys Fe-Al and Fe-Si [12.42,43].

The first study of short-range order and energies of intermolecular interaction in solid solutions of molecular crystals coupled with theoretical atom-atom potential calculations was made for binary solid substitutional solutions of tolane-diphenylmercury [12.44,45]. The quantitative determination by the

X-ray diffuse-scattering technique of the short-range order parameters $\alpha(\underline{r})$ and the mixing energies $w(\underline{r})$ is based on (6.30,31). It can be seen from the formulas that in the case of an ideal solid solution, in which A and B molecules are distributed randomly over the lattice sites [when the mixing energies $w(\underline{r})$ are zero for any $\underline{r} \neq 0$], the short-range order also vanishes $[\alpha(\underline{r}) \equiv 0$ for any $\underline{r} \neq 0]$. In this case the expression for the intensity takes the form

$$|F_A - F_B|^2 \, c(1 - c) \quad .$$

Thus the experimental distribution of the diffuse scattering intensity at various points of the reciprocal space, treated with the aid of the formulas in Sect.6.5.1, reveals the character of the short-range order, i.e., the arrangement of A and B molecules in the lattice sites. This distribution also yields the short-range order parameters $\alpha(\underline{r})$ and mixing energies $w(\underline{r})$ for different values of $\underline{r}$. If the $\alpha(\underline{r})$ are known, the average number of molecules of any kind at a given distance from a given molecule can be calculated.

The short-range order and molecular interaction were investigated [12.44, 45] in single crystals of disordered solid solutions of tolane-diphenylmercury, containing 7.8, 9.5, and 13 mol. % of diphenylmercury, respectively. Diffuse scattering from single crystals of pure tolane was also studied. The single crystals were grown by Bridgman's method and also from molten solutions in a thermostat, the latter technique providing more precise equilibrium crystallization conditions. The solid solution composition was determined to within 0.5 mol. % . This was accomplished by sealing the crystal in a capillary tube and measuring the end melting temperature of the crystal, and comparing it to the known data on the equilibrium diagram. Samples in the form of thin plates 0.2 - 1 mm thick and with an area of 10 - 50 mm^2 were prepared from the single crystals. The X-ray scattering intensity from a CuK$_\alpha$ source was measured by a scintillation counter in a small-angle scattering chamber, continuously evacuated to a pressure not higher than 10^{-1} mm Hg.

The scattering chamber permitted intensity measurements in the range of small angles corresponding to the first Brillouin zone centered at the origin of the reciprocal lattice ($2\pi \underline{H} = 0$). The angular range 2Θ was from 30 - 35' to approximately 8°, just beyond the maximum size of the first Brillouin zone. This experimental method differs from the conventional technique for studying short-range order in inorganic materials which makes measurements in the range of large angles. On the one hand, there are complications with this method due to the large air scattering at small angles and large parasitic scattering at the edges of the diaphragm. However, these complications can be eliminated by continuous evacuation of the chamber and plotting a reference curve for parasitic scattering. The total parasitic scattering, which amounted to a considerable portion of the total scattering measured in the angle range $2\theta \leq 1°$, was determined by placing a small empty capron bag in the X-ray beam. On the other hand, there are significant advantages with this method as compared to the conventional technique: almost complete absence of the Compton scattering ($I_{Comp} = 0$ for $2\theta = 0$) and very small scattering caused by static and thermal distortions of the lattice. Moreover, at small angles the scattered intensity $I_{sh-range}(\underline{s})$ due to the short-range order is higher than at large angles.

The intensity was reduced to absolute scale differently than in small-angle experiments or in short-range order studies at large angles. This reduction to absolute scale is based on the summation of the intensity $I_{sh-range}(\underline{s})$ throughout the entire first Brillouin zone. We obtain from (6.29)

$$\frac{1}{N} \sum_s \frac{I_D(\underline{s})}{|F_A - F_B|^2} = c(1 - c) \quad , \tag{12.14}$$

where N is the number of allowed values of the wave vectors $\underline{k}$ within the first Brillouin zone.

Since the composition "c" of the solid solution is known, (12.14) allows the conversion of the intensities to electron units.

The study of the short-range order in solid solutions of tolane-diphenylmercury (from the tolane side) is simplified because disordered solid solutions are present from the tolane side in a wide range of concentrations. The large difference in the scattering amplitudes for tolane and diphenylmercury molecules ($F_T - F_D$) leads to a large value of $I_{sh-range}(\underline{s})$, but little scattering caused by static distortions of the lattice (especially in the range of small angles). Figure 12.26 presents the intensity curves (in arbitrary units) for diffuse scattering along the [010]* direction, which coincides with the symmetry axis 2_1 (along b), for solid solutions of various compositions and for pure tolane.

It can be seen from the figure that the diffuse scattering intensity for solid solutions is an order of magnitude higher than for pure tolane, the intensity increasing with an increase in the content of diphenylmercury in the alloy. This can be related to the short-range order in solid solutions. In fact, the scattering by pure tolane includes only the thermal I_{therm} and Compton I_{Comp} components, whose values are very low in the range of small angles.

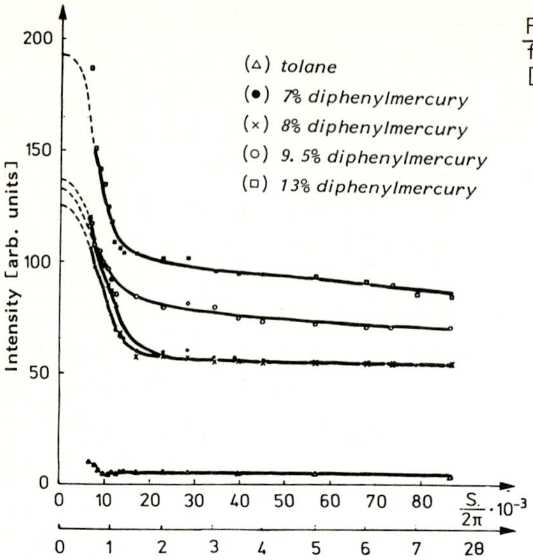

Fig.12.26. The intensity curves for diffuse scattering along the [010]* direction

(△) tolane
(●) 7% diphenylmercury
(×) 8% diphenylmercury
(○) 9.5% diphenylmercury
(□) 13% diphenylmercury

The scattering by a solid solution includes two more components, scattering due to lattice distortions I_{dist} and scattering caused by the short-range order $I_{sh-range}$:

$$I_{solid\ sol} = (I_{therm} + I_{Comp})_{solid\ sol} + I_{dist} + I_{sh-range} \quad .$$

By calculation, the sum $I_{Comp} + I_{therm}$ for pure tolane at small angles does not exceed 15 - 25 electron units, while calculations of $I_{sh-range}$ from the formula valid for a solid solution with random distribution of molecules yield values which are 10 - 20 times higher. Since addition of diphenylmercury should not lead to a considerable change in I_{therm} and I_{Comp} (for an angle of ~8° the value of I_{Comp} varies from 6.6 e.u. for pure tolane to 7.3 e.u. for a solid solution with 10 mol. % diphenylmercury), and I_{dist} is of the same order as I_{therm}, therefore it is obvious that the difference in the scattering by tolane and solid solutions is due mainly to $I_{sh-range}$. Also, scattering due to short-range order should increase with the diphenylmercury concentration, as observed by experiment.

In solid solutions the scattering intensity in the direction [010]* of the reciprocal lattice gradually increases with decreasing angle. At large angles (> 2°) the intensity barely changes and is close to the value obtained for a random distribution of molecules.

The change in the scattering intensity as a function of reciprocal space direction has been studied most thoroughly for an alloy containing 8 mol. %

diphenylmercury. For this alloy the measurements were performed in the directions [100]*, [010]*, [001]*, [101]*, [110]*, [011]*, and [111]*. The results reveal two characteristic features. First, a diffuse maximum forms at the center of the Brillouin zone. This corresponds to a short-range order of the *segregation* type formed in a solid solution when molecules of a given kind have a tendency to surround themselves with similar molecules. As the temperature is lowered, such a solid solution decomposes into two solid solutions of different compositions.

The second feature is that for all the above-mentioned directions (except [001]*), the diffuse scattering intensity does not depend on the wave vector $\underline{k}$, i.e., on the direction in the reciprocal space: the intensity distribution except for the [001]* shows an anomalous increase in the intensity in the form of a narrow diffuse *necking*.

Since the intensity of the [001]* necking is negligibly small as compared to the total diffuse scattering, caused by the short-range order throughout the entire first Brillouin zone, the [001]* intensity can be neglected without loss of accuracy in the analysis of short-range order. The remaining spherically symmetric diffuse scattering was used to determine the short-range order parameters $\alpha(\underline{r})$ and mixing energies $w(\underline{r})$ by the technique of Chap.6. In general such a calculation would be formidable. In our example a simplification is possible.

The substitution of a diphenylmercury molecule for a tolane molecule can be considered as the substitution of mercury atoms for the central carbon acetylene bridges; also a pair of carbon atoms can be replaced by one "doubled atom" at the center of the bridge. The solid solution can, therefore, be described by a face-centered monoclinic lattice, whose sites are occupied either by mercury atoms or by a pair of carbon atoms. With this approximation the short-range order parameters and mixing energies were readily calculated. The region of the reciprocal lattice, sufficient for the summation, was divided into 32000 points.

The master formulas are, using the shorthand I_{sr} for $I_{sh-range}$,

$$\alpha(\underline{r}) = \frac{2}{32000} \sum_\xi \sum_\varphi \sum_\eta \left[\frac{I_{sr}(|\underline{k}|)}{\frac{2}{32000} \sum_\xi \sum_\varphi \sum_\eta I_{sr}(|\underline{k}|)} - 1 \right] \times \cos 2\pi(\xi n_1 + \varphi n_2 + \eta n_3) ,$$
(12.15)

$$w(\underline{r}) = \frac{2}{32000} \frac{T}{n(1-n)} \sum_\xi \sum_\varphi \sum_\eta \left[\frac{\frac{2}{32000} \sum_\xi \sum_\varphi \sum_\eta I_{sr}(|\underline{k}|)}{I_{sr}(|\underline{k}|)} - 1 \right]$$
$$\times \cos 2\pi(\xi n_1 + \varphi n_2 + \eta n_3) ,$$
(12.16)

where $k = 2\pi(\xi \underline{H}_{200} + \varphi \underline{H}_{110} + \eta \underline{H}_{001})$; ξ, φ, η are the coordinates of the vector $\underline{k}$ varying in the interval:

$$0 \leq \xi \leq 1/2 \quad ; \quad -1/2 \leq \varphi \quad , \quad \eta \leq 1/2 \quad ;$$

$$k = 2\pi \sqrt{\xi^2 \underline{H}_{200}^2 + \varphi^2 \underline{H}_{110}^2 + \eta^2 \underline{H}_{001}^2 + 2\xi\varphi \underline{H}_{200}\underline{H}_{110} \cos \gamma^* + 2\xi\eta \underline{H}_{200}\underline{H}_{001} \cos \beta^*}$$
$$+ 2\varphi\eta \underline{H}_{110}\underline{H}_{001} \cos \alpha^* \quad ;$$

α^*, β^*, γ^* are the angle between the axes of the parallelepiped constructed from the vectors $\underline{H}_{200}$, $\underline{H}_{110}$ and $\underline{H}_{001}$. In (12.15,16) the quantities n_1, n_2 and n_3 are integers determining the vector $\underline{r} = n_1\underline{r}(1/2, 1/2, 0) + n_2\underline{r}(010) + n_3\underline{r}(001)$.

The results of calculation of $w(\underline{r})$ from the experimental diffusion scattering intensities for the alloy containing 8 mol. % diphenylmercury are presented in Table 12.5 and in Fig.12.27. The experimental data are compared with calculations performed by the atom-atom potential method. The calculated results are shown by the solid line, the X-ray data by the dashed line. The ordinate axis reads the negative values of the mixing energy in K(w/k). The general agreement of the data is rather satisfactory. Both methods lead to a segregation type of short-range order.

Table 12.5. Calculated parameters of short-range order for tolane-diphenyl-mercury solid-solution crystal with 8% tolane

lattice direction	r_{min} [Å]	from atom-atom potential calculations	from diffuse scattering			
			$\alpha(r_{min})$	$\alpha(2r_{min})$	$\alpha(3r_{min})$	$\alpha(4r_{min})$
010	5,77	0,078	0,02948	0,01312	0,00554	0,00267
110	6,98	~0,029	0,02658	0,00962	0,00376	0,00174
130	10,74	~0,006	0,01512	0,00326	0,00128	0,00064
100	12,71		0,01261	0,00303	0,00161	0,00056
112	14,54		0,00942	0,00096	0,00060	-0,00009
001	15,68		0,00970	0,00090	0,00047	0,00001
111	16,57		0,00724	0,00119	0,00029	-0,00008

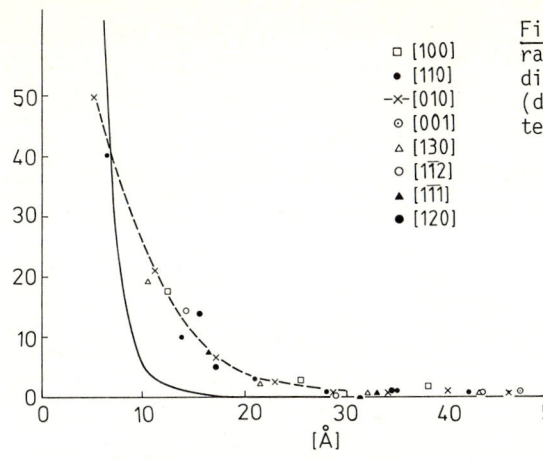

Fig.12.27. Calculation of short-range parameters from experimental diffusion scattering intensities (dashed line) and the atom-atom potential method (solid line)

The possibilities for experimentally determining the mixing energy as a function of distance appear to be very significant. For organic solid solutions such a possibility was demonstrated [12.44,45] for the first time. The experimental values can be used, if necessary, to improve the parameters of the atom-atom potentials, and to reach complete agreement between the calculated and experimental data.

In disordered solid solutions of organic substances, as in metallic solid solutions, the short-range order is a stable equilibrium state of a solid solution. The physical reason which causes the short-range order is the interaction energy, since in the absence of the interaction the molecules are distributed randomly over the lattice sites. In tolane based solid solutions of tolane-diphenylmercury, the intermolecular interaction explains why tolane molecules tend to surround themselves with tolane molecules, and diphenylmercury molecules with diphenylmercury molecules. This type of the short-range order is established in these solutions at high temperatures. According to existing theories, such solid solutions decompose at low temperatures into two segregated disordered solid solutions, α and α', of different compositions (with the same lattice). This corresponds to a two-phase region ($\alpha + \alpha'$) on the phase diagram.

KHACHATURYAN showed [12.46] that the mixing energy as a function of distance is the parameter through which the free energy of a solid solution can be expressed in terms of the concentration and temperature, and hence the phase diagram can be constructed.

The calculation performed for the diphenylmercury-tolane system showed that diphenylmercury solid solutions can decompose in tolane. The bell-shaped region of $\alpha + \alpha'$ in the phase diagram lies entirely inside the two-phase region $\alpha + \beta$

(α is a solution of diphenylmercury in tolane; β is a solution of tolane in diphenylmercury).

Comparison of the experimental results for X-ray diffuse scattering with theoretical data on the phase diagram leads to the conclusion that the solid solution investigated (8% diphenylmercury) at room temperature is on the left of the $\alpha + \alpha'$ boundary and, evidently, near the boundary of the two-phase region $\alpha + \beta$. The latter conclusion follows from the presence of the diffusion necking in the scattering pattern. Such an intensity distribution is known to indicate the presence of platelike segregations, and the necking dimensions tell us that the thickness of these segregations does not exceed two or three intermolecular spacings. This, in turn, means that the solid solution is in a state close to decomposing the thinnest plates of the β-phase. It was shown that the platelike segregations are arranged so as to minimize the elastic energy which necessarily arises under coherent relationships between the host lattice and that of the segregated phase.

13. Complexes

In this chapter we consider *ordered binary* molecular crystals. The main emphasis is on organic systems simply because molecules which occur in all condensed states as definite combinations of atoms are, mainly, organic molecules. However, several pages are devoted to gas hydrates. We will also dwell upon the specific features of complexes consisting of organic and iodine molecules.

Note that here we use the term "ordered" not in a strict sense of the word. The complexes or molecular compounds can be crystals with some elements of disorder. A stoichiometric composition is not a necessary feature of the complex either.

The majority of complexes may be divided into two classes: complexes in which an *additional bond* between the molecules of the components is formed, and *packing complexes*.

A separate section is devoted to layer complexes such as graphite compounds.

13.1 Quasi-Valence Bonds Between Molecules of the Components

There is a rather large group of complexes formed by organic molecules with small inorganic molecules such as iodine, iodoform, or antimony trichloride. The main feature of these complexes is that in a crystal the molecules of the components form a quasi-valence bond. The distances between the atoms belonging to different components are somewhat greater than a valence bond distance, but are considerably smaller than the sum of intermolecular radii. Upon dissolution, complexes disintegrate, and consequently, thermodynamically, we are dealing with a binary system.

No phase diagram data are available for such binary systems. However, from general structural considerations, we may suppose that the complexes described in this section do not form solid solutions when their composition is non-stoichiometric.

Since we cannot review here all the material concerning these interesting compounds, we shall restrict ourselves to several examples. In the first of their papers CHAO and MC CULLOUGH [13.1-3] studied the structure of the complex formed by a *1,4-dithiane molecule* with two *iodine* molecules. As is shown in Fig.13.1, the complex is very prominent in the crystal. The unit cell (space group $P2_1/c$) contains two molecules of the complex. This means that each 1,4-dithiane molecule is located at an inversion center and flanked by two iodine molecules. A bond is formed between the iodine and sulphur atoms. The distance between them (2.87 Å) points to the formation of a quasi-valence bond. The I-I distance in the (1,4-dithiane) I_2 crystal is 0.1 Å longer than the I-I distance in pure I_2.

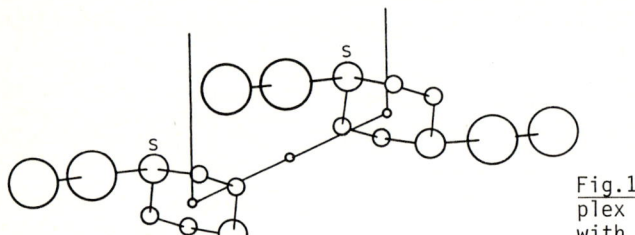

Fig.13.1. Structure of a complex formed by 1,4-dithiane with two iodine molecules

The molecules of the complex form an ordinary close-packed structure typical of monocomponent crystals. All the distances between the organic molecules of the complex shown in Fig.13.1 are normal van der Waals separations. The structure is characterized by a packing coefficient usual for organic molecules and by a positive increase in density Δ. This means that the formation of the complex does not lead to a loss in the lattice energy as compared to the sum of the lattice energies of the components.

A similar situation takes place when a complex is formed consisting of *1,4-diselenane and two iodine* molecules. As in the previous case, one of the iodine atoms is bonded to a selenium atom; the distance between them is 2.83 Å. In this case the I-I distance is 0.2 Å larger than the I-I distance in an iodine crystal.

The authors indicate that the geometry of these two complexes differ considerably and believe that one deals here not with a real isomorphism but with pseudoisomorphism. But we do not see any grounds for such a conclusion. It would be interesting to try to obtain solid solutions of these two complexes.

In the case of the diselenane complex, the packing coefficient is as usual but Δ is several hundredths less than zero. We do not believe that this circumstance deserves serious attention. The gain in energy caused by the formation of a quasi-valence bond more than compensates, undoubtedly, the insignificant loss in the lattice energy.

Among the complexes of the above type is *tetrahydroselenophenone* which forms a compound with an *iodine* molecule. In this case as well, the formation of a bond between one of the iodine atoms and a selenium atom is beyond doubt. The Se-I distance 2.76 Å is only by 0.26 Å greater than the sum of the covalent radii. The authors reported a definite decrease in the intermolecular distance between the end iodine atom and the neighboring molecule of the complex. It equals 3.64 Å, whereas the sum of the intermolecular radii is 4.15 Å. It should be noted that in this crystal all the other intermolecular distances are also slightly less than normal, which results in an increase in the packing coefficient. Nevertheless, Δ is somewhat less than unity, which is due to very short distances between the neighboring iodine atoms in the crystal of iodine itself.

The molecule is asymmetrical and the straight line, passing through 2 iodine atoms and a selenium atom, forms an obtuse angle with the plane of the five-membered ring.

HOLMESLAND and ROMMING [13.4] studied the complexes of *diiodoacetylene* with *1,4-dithiane* and *1,4-diselenane*. The structures of the two compounds are very similar. But the complex with dithiane crystallizes in the $P2_1/c$ space group with four molecules per unit cell, whereas that with diselenane there are only two molecules per unit cell. The cells are also very similar in shape. In the diselenane complex the axis is half that of the dithiane complex. Thus the selenium atom gives a centrosymmetric complex and the sulphur atom gives rise to an asymmetrical molecule. The distances between the sulphur and iodine atoms and between the selenium and iodine atoms belonging to the neighboring atoms of the molecular components are 3.27 and 3.34 Å, respectively.

As compared to the complexes formed by the diselenane and dithiane molecules with iodine, these bonds are weaker. The authors believe it more reasonable to consider in this case not the valence bonds between the molecules of the components, but a charge-transfer complex as the diiodoacetylene molecules are electron acceptors. However that may be, the existence of electron transfer between the molecules of the components is beyond doubt. An increase in the distances between iodine atoms and selenium and sulphur atoms may be due to the fact that we deal here with chain structures. In contrast to the cases just considered, the diiodoacetylene molecule exchanges its electrons not with one but with two molecules of other components.

The distances between these chains do not differ from ordinary van der Waals separations. The chain structure is shown in Fig.13.2.

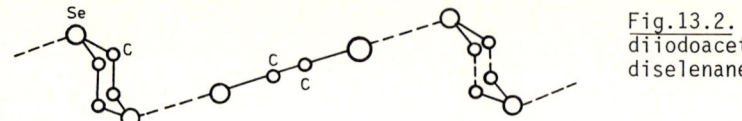

Fig.13.2. Chain formed by diiodoacetylene and 1,4-diselenane molecules

Dithiane molecules form a 1:1 complex with *iodoform*. BJORVATTEN and HASSEL [13.5] studied the structure of the complex which crystallizes with two molecules per unit cell in the $P2_1/m$ space group. However, we believe that this conclusion is wrong and the true space group is $P2_1$. Nevertheless, the main result of the research is probably correct. The iodoform and dithiane molecules form a chain very similar to that shown in Fig.13.2. It is natural that in this case one of the iodine atoms does not participate in an electron-exchange bond. The authors reported that the distance between the valence-bonded atoms of iodine and sulphur is 3.32 Å.

BJORVATTEN and HASSEL [13.6] also determined the structure of the complex formed by one molecule of *quinoline* and three molecules of *iodoform*. The complex forms crystals of high symmetry: the hexagonal cell dimensions are a = b = 22.4 Å normal to the threefold axis and c = 4.59 Å parallel to it. In this case as well, the complex is considered a Mulliken donor-acceptor complex. Besides quinoline, the role of donors in this type of complex is also played by sulphur, 1,4-dioxane, and 1,4-dithiane. In all such cases the halogen atom of the acceptor molecule bond to the oxygen, sulphur, or nitrogen atom of the donor molecule and the three atoms lie in a straight line. Electron exchange in this structure reduces the iodine-nitrogen distance by 0.6 Å as compared to the sum of van der Waals separations. Neither symmetry centers nor symmetry planes are present in the crystal. The iodoform molecules arè arranged on the three-fold symmetry axis. The quinoline molecules are in a general position. Though the authors did not consider the packing geometry, it seems that here, too, the chain bond is arranged along the three-fold axis.

BJORVATTEN [13.7] also determined the structure of the 2:1 complex formed by *antimony triiodide with 1,4-dithiane*. The space group of crystals of the complex is C2/c. The dithiane molecules occupy the symmetry center and the molecules of the second component are in a general position. The antimony atom exchanges electrons with the two neighboring sulphur atoms. Hence, in this case as well, infinite chains are formed. Two distances between the antimony and sulphur atoms are 3.27 and 3.34 Å. The molecules' arrangement pattern can be seen in Fig.13.3. The distance between the sulphur and antimony atoms (black circles) is 0.7 Å smaller than the intermolecular distances. Electron

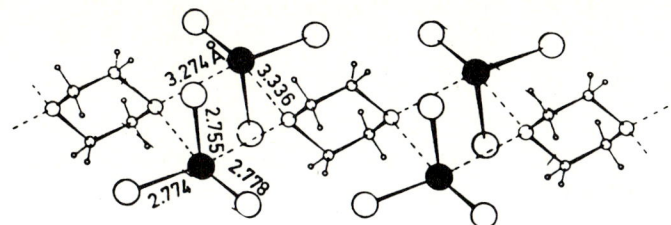

Fig.13.3. Arrangement of molecules in the 2:1 complex of antimony triiodide with 1,4-dithiane

transfer takes place between these atoms; the iodine atoms participate, evidently, only in intermolecular interactions. The distances between the iodine atoms in the neighboring molecules (about 4 Å) are normal contact distances. A somewhat reduced distance between the sulphur and iodine atoms (3.73 Å) may be a result of packing. In comparison a molecule of antimony triiodide entering into the complex has somewhat greater Sb-I distances than those found in gas molecules. The difference, however, does not exceed 0.03 - 0.04 Å.

BRUN and BRANDEN [13.8] determined the structure of the 1:1 complex formed by *antimony pentachloride* and *dimethylformamide*. This complex was known as early as the beginning of the century. In the structure a quasi-valence bond is formed between the two components. The antimony atom acquires octahedral coordination. Chlorine atoms occupy five vertices at ordinary distances (2.34 Å) from the antimony atom; at the sixth vertex of the octahedron is a carbonyl group of dimethylformamide. The antimony-oxygen distance is 2.05 Å. The unit-cell parameters are a = 9.71, b = 13.52, c = 8.76 Å; β = 91°60'. The space group is $P2_1/n$, and there are four molecules per unit cell.

The general conclusions which can be made when considering the structures of such complexes are as follows. The components in solid state can form ordinary molecules or chains of molecules which are packed in a crystal in accordance with the same rules which govern the packing of molecules in single-component systems. Different types of bonds arise between the molecules forming the complex. The electron transfer can be as pronounced as in any valence compound, but sometimes it can be weaker than an ordinary valence bond. The distances between the neighboring molecules of the components which exchange their electrons are a sufficiently good characteristic of the nature of the intermolecular bond. In all cases without exception such molecular compounds are formed due to the fact that the generation of an electron bond is favorable and at the same time it does not violate the packing density of the molecules.

13.2 Complexes of Aromatic and Nitro Compounds

A large number of complexes are known (mostly of 1:1 stoichiometry) which are formed by molecules of *trinitrobenzene, trinitrophenol,* and *trinitrotoluene* with aromatic molecules of approximately the same size. *Naphthalene, anthracene, fluorene,* and *perylene* are suitable components for complexing. New forces arise between the molecules of the components which are absent in the crystals of the components. But these forces are not quasi-valence. A specific interaction arises because of an induction effect, namely, a dipole moment induced in an easily polarizable hydrocarbon molecule by the rigid dipole moment of a nitrogroup of the neighboring molecule.

An approximate calculation shows that the attraction energy of a constant dipole moment of 4 Debyes (nitrogroup) by an isotropic body with polarizability $20 \cdot 10^4$ (aromatic hydrocarbon) is about 2 kcal at a distance of about 3 Å. Experiments indicate that the force binding the molecules of the complex is proportional to the polarizability of the hydrocarbons and that the complexes have an electrical conductivity that differs only slightly from that of the crystals of the components. Possibly, an electron exchange takes place between the molecules of the components. Nevertheless, the main energy gain during the formation of such complexes is due to the induction effect.

The formation of complexes between nitro compounds and aromatic hydrocarbons is accompanied with neither a gain nor a loss in the packing density. To be more exact, because of the different possibilities of the molecules to form dense-packed structures in such a way that projections of one molecule coincide with hollows of another, we have in some cases positive Δ factors and in others negative ones.

Thus, for instance, the geometry of a naphthalene molecule proves to be suitable for the formation of complexes with an enhanced density. *Naphthalene* molecules form crystals with *trinitrobenzene* for which $\Delta = 1.23$. The density of the complex with γ-*trinitrotoluene* is 30% higher and that of the complex with α-*trinitrotoluene* is 14% higher. In contrast the geometry of *acetonaphthene* causes a decrease in density in the complexes with all the isomers of trinitrotoluene, the density losses being not more than 10%. In many cases Δ is practically zero. The same conclusion can be drawn from calculating the packing coefficients of the complexes and comparing them with the packing coefficients of the components. We performed such calculations more than 30 years ago [13.9]. It turned out that packing coefficients vary from 0.63 to 0.73 both in the crystals of the components and the crystals of the complexes.

How does one observe an additional bond arising between the molecules of the complex and how do crystals retain the normal packing density of the molecules? Numerous structural investigations conducted for the last decades answer these questions. Let us consider some examples.

The complex of *anthracene* with *trinitrobenzene* forms monoclinic crystals with space group C2/c. The cell parameters are a = 11.7, b = 16.2, c = 13.2 Å, the monoclinic angle being 133°. The molecules of the complex, arranged so that their planes are almost parallel, form one-dimensional sequences with alternating molecules of the components. Figure 13.4 illustrates molecular arrangement and Fig.13.5 shows contact distances.

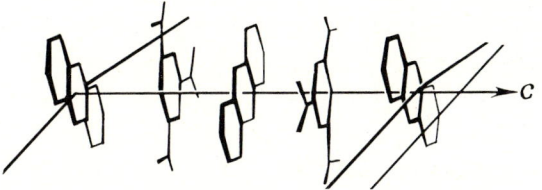

Fig.13.4. Arrangement of molecules in the anthrancene-trinitrobenzene complex

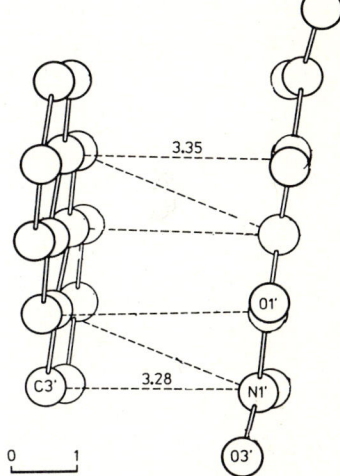

Fig.13.5. Contact distances in the anthracene-trinitrobenzene crystal

We see that an induction effect does not cause any reduction in distance between the carbon atoms as compared to those between carbon atoms in the adjacent layers of the graphite crystal. Intermolecular contacts between the atoms of different kinds also remain the same. The packing of the molecules inside the column and between the columns is optimal so that the packing is dense. Thus we can judge the existence of the additional bond not by the intermolecular distances, but by the specific mutual arrangement of the components which arises due to the interactions. This remarkable investigation was performed by BROWN et al. [13.10].

Next, let us consider the 1:1 complex that *sym-trinitrobenzene* forms with either *skatole* or *indole* molecules that are similar in geometry [13.11]. Both complexes form crystals belonging to space group $P2_1/a$ with four molecules in a cell. Cell parameters of the complex with skatole at -144°C are a = 16.76,

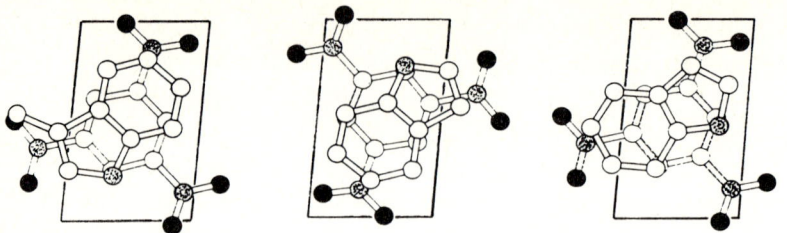

Fig.13.6. Superposition of skatole (left) and two modes of superposition of indole molecules on the trinitrobenzene molecule

b = 6.61, c = 13.45 Å, and the monoclinic angle is 96°; for the complex with indole, a = 15.87, b = 6.58, c = 13.47 Å, monoclinic angle 95°. Figure 13.6 shows that the molecules are superimposed with the maximum density permissible by their respective geometries. It is of interest that skatole (left) can be packed with less difficulty than indole. The dense crystal is formed due to orientational disorder in the arrangement of the indole molecules which can have two different orientations. A columnlike arrangement of the molecules occurs in this case as well. The columns lie along the b axis so that the distance between the mean planes of the molecules forming the complex is half the cell period, i.e., 3.30 Å. Thus we arrive at the same conclusion: the presence of the additional bond manifests itself only as a specific mutual arrangement of the molecules, i.e., as the formation of columns with alternating molecules of the components. The reduction of distances between the atoms of the neighboring molecules is insignificant.

The complex of *sym-trinitrobenzene* with *azulene* has a similar structure [13.12]. We have here the same space group, the same number of molecules in the cell, and again the columns of the molecules arranged along the b axis at a distance of 3.33 Å. The stacking overlap of the neighboring molecules is shown in Fig.13.7. This elegant figure shows that nature has found a way to pack in the densest manner even molecules of such different geometry. If the reader wants to calculate the packing coefficients, we give here the cell parameters: a = 16.4, b = 6.66, c = 13.8 Å; monoclinic angle is 96°. This structure is also characterized by orientational disorder. About 7% of the azulene molecules are arranged so that the five-membered rings occupy the positions of the seven-membered ones (as in the crystals of azulene itself). It would be very interesting to calculate the lattice energy in order to check whether this disorder results from a gain in entropy or from a gain in the lattice energy. Probably in 1:1 complexes the molecules are always arranged in a columnlike manner.

In some cases complexes of other composition are also formed. DAMIANI et al.

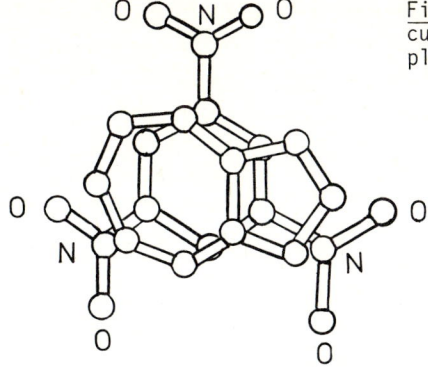

Fig.13.7. Superposition of neighboring molecules in the sim-trinitrobenzene-azulene complex

[13.13] showed that the 2:1 complexes formed by 1,3,7,9-*tetramethyluric acid* with *aromatic hydrocarbons* are sandwiches in which aromatic molecules are pressed between two molecules of the second component. The sandwiches are densely packed, following the scheme typical of all organic molecules.

13.3 Packing Complexes

Earlier we considered complexes whose formation is conditioned by the fact that the creation of specific bonds between the molecular components provides a gain in energy. Two examples were considered. In Sect.13.1 we described complexes in which a considerable electron transfer (exchange) takes place between the molecules of the components; we can even speak there of a chemical bond. In Sect.13.2 we dealt with the compounds in which interaction of the induced dipoles gives rise to a specific, mainly columnlike arrangement of the molecules, resulting in a small charge exchange between the molecules. The complexes crystallize readily and form stable, closely packed structures. In the course of complexing, the packing density remains unchanged; Δ is close to zero and in some cases it can even take on small negative values.

Now we shall consider a rather vast class of complexes that chemists call *"inclusion compounds"*. As we have already mentioned, these compounds are of interest for applied chemistry since they can be used for separation of compounds and for chemical reactions differing from those occurring in a liquid phase. In a number of cases, inclusion compounds make it possible "to store" the guest molecules. These complexes turned out to have other numerous applications so that many researchers became engaged in their synthesis. Many monographs and reviews are devoted to inclusion compounds, but they have all been written by chemists. The crystallographic aspect of the problem has not been considered in any detail, and in many cases the ideas about the formation of the complexes are erroneous. The reader interested in inclusion compounds already synthesized can obtain the information in the review by MAC NICOL et al. [13.14], where all the relevant references are given.

To make clearer the standpoint from which we shall consider inclusion compounds, we shall use the term *"packing complexes"* instead of "inclusion compounds". This term is preferable since it reveals the reason for the complex formation: a loose structure of the crystal of one or both components and increases in packing density (a positive value of Δ) which occur when a packing complex is formed. In all cases, no additional bonds appear between the molecules of the components other than those in single-component crystals. A complex is formed due to the tendency of the particles to pack as densely as possible in the crystal. In terms of energy it means that the formation of a complex is accompanied by a considerable gain in the short-range part of the energy of interaction between the molecules.

This does not mean that the term "inclusion compound" should be abandoned. In all packing complexes the *host* molecules form a framework with cavities of different shapes. These cavities are occupied by *guests* (the molecules of the second component). However, guest molecules are not necessarily smaller than the host molecules. Thus the molecules located in the cavities of clathrate hydrates are almost always larger than water molecules. When host molecules form frameworks with channels, even polymer molecules can be guests.

We classify a complex as a packing complex when it is quite obvious that the main reason for its formation is an agreement between the shape and size of the cavities in the structure formed by the host molecules and those of the guest molecules. The interaction between the molecules of the components can, in the majority of cases, be successfully described by a *universal short-range potential*. Sometimes electrostatic forces act between the guest and host molecules or hydrogen bonds are formed. But they are of secondary importance and depend on geometric correspondence of the cavities and dimensions of incorporated molecules.

Of course, the picture depicted above justifies the term "inclusion". Erroneous associations which may arise when this term is used are as follows. Except for very rare cases there is no such process as the incorporation of guest molecules into the framework of host molecules. Molecules of the components crystallize simultaneously. Moreover, the frameworks formed by the host molecules are so unstable that, as a rule, they do not form if there are no molecules capable of filling the vacant sites. The host component either does not form crystals at all or forms crystals of other structures.

Cases are known where complexes are formed as a result of interacting small gaseous molecules with a crystalline host. But in these cases the crystals of the host decompose and new crystals grow which are energetically more advantageous.

Unfortunately, there are many cases where the host's crystalline structure is not known. But when it is known, the crystals form structures with a very low packing coefficient. However, if the host molecules do form crystals, why do we still consider the void frameworks formed from the same molecules to be unstable? We believe the reason for this to be as follows. Due to a specific shape of the molecule or because of the necessity to find a compromise between the tendency to form a dense-packed structure and not to saturate hydrogen bonds, a crystal with a low packing coefficient is formed. In such a crystal, we find a great number of small cavities which cannot be filled even by the smallest molecules of the second component.

The formation of a complex is accompanied by a complete rebuilding of the structure. In this framework, a small number of cavities of large size are formed. Such a rebuilding of the structure favors the formation of a maximum number of normal contacts between all the other atoms of the host molecules (except the atoms bounding the void) and for the maximum saturation of all the hydrogen bonds between the host molecules, if they exist. A void framework is unstable since there is a great number of atomic pairs located at distances that greatly exceed the equilibrium distances. When the voids are filled in by molecules of the suitable size ("suitable" in the sense that the framework should not necessarily change its geometry), a structure is formed in which all the pairs of the adjacent atoms make an optimum contribution to the lattice energy. In this section the validity of the above-given general principles for constructing packing complexes shall be illustrated by many examples.

An exact classification of the inclusion compounds is difficult and not really necessary. Nevertheless, we may divide them into the following groups. First of all, there is a distinct class of molecules, containing hydroxy and amino groups, whose geometry prevents the combination of a dense-packed structure with the formation of an optimal number of hydrogen bonds between neighboring molecules. Among them are thoroughly studied clathrate compounds of hydroquinone and poorly studied, but most likely similar, complexes of phenol, cresol, and their derivatives of simple shape.

We think it reasonable to use the term *"clathrate"*, proposed by Powell, for those packing complexes in which the host molecules form a framework built of hydrogen bonds and having cavities in the form of closed cages.

Hydrates, as well as other *solvates*, should be classed with packing complexes. If the number of water molecules is small, the crystal is formed with a lower lattice energy and the competition which may arise between the formation of dense-packed structure and a net of hydrogen bonds can be easily resolved.

Of particular interest are clathrate hydrates in which the water molecules form cages of extremely interesting geometry, in which quite different molecules and ions can be located. We devote Sect.13.3.2 almost exclusively to clathrate hydrates, and only mention briefly some other hydrates, whose structure and thermodynamics are much less interesting.

Complexes based on molecules forming a framework of hydrogen bonds with channel-shaped cavities deserve consideration as a separate class. Among them are the inclusion compounds of urea and thiourea.

Organic chemistry puts at the crystallographer's disposal an inexhaustible variety of molecular shapes. Among them are molecules whose parts are constructed so that a large open cavity is formed inside the molecule. Sufficiently rigid molecules are also interesting. If we build a ring molecule from methylene groups, its cavity will not be a physical reality. Such a molecule, due to the freedom of rotation about the ordinary bonds, can always eliminate the void without the aid of other molecules. The situation is completely different in the case of rigid molecules. Their packing in a crystal is either impossible or gives crystals with a very low packing coefficient. Such molecules readily form complexes with guest molecules which enter into their cavities. Finally, there is another class of packing complexes formed by molecules whose shape is inconvenient for dense packing. They will form complexes with molecules entering into intermolecular voids. It should be noted that the thermodynamics of packing complexes have not been sufficiently studied as yet.

A priori we may ascertain that among the inclusions there are *complexes*, i.e., crystals of definite (not necessarily stoichiometric) compositions, and *interstitial solid solutions*. In turn, the interstitial solid solutions can be divided into solutions formed on the basis of a host molecule framework and solid solutions of intermediate composition possessing a structure unlike that of the host molecule's crystals.

Clathrate hydrates having a structure of the 2nd type (cubic lattice with large cell edges) are stable. Undoubtedly, inclusion compounds of the type Bu_4NF 28.7 H_2O (Bu = butyl) are also stable. Among the stable complexes are complexes whose phase diagrams contain a pronounced maximum as well as complexes whose phase diagrams contain a latent maximum. Among the clathrate hydrates we can also find many thermodynamically unstable complexes. There may be inclusion compounds unstable both with respect to the initial components and with respect to other inclusion compounds. For instance, a case is known when a complex based on the cation Bu_4N with 65 H_2O is unstable with respect to a complex having 63 H_2O molecules.

On the basis of the host molecule framework, there probably exist solutions, provided the framework is rigid and the cavities wherein the guest molecules can be included are not too large. Such are zeolites and mixed silicates but we do not touch upon these in this book. It might be (but is not yet proven) that some of the packing complexes of the Dianin type compounds are able to give an interstitial solid solution of this type [13.14].

We believe it to be fairly certain that there exist interstitial solid solutions of argon in a beta hydroquinone framework. Argon clathrates can be kept in an ordinary glass vessel for a long time if the equilibrium pressure equals several atmospheres at room temperature.

As will be emphasized below, packing complexes should not be considered as interstitial solid solutions. They are actually complexes, i.e., compounds of definite, not variable, compositions.

13.3.1 Hydroquinone-Based Clathrates

No data are known to be available on the crystal structure of hydroquinone. There are conflicting reports on the size of the elementary cell of this seemingly very simple substance. CHEKHOVA et al. [13.15] have grown a single crystal of α-*hydroquinone* and determined it to have a rhombohedral structure. The space group was not established unambiguously. The cell parameter was found to be 22.34 Å, angle 119°16'. The cell contains 18 molecules. The volume of one molecule is about 135 Å, which gives a very high packing coefficient. At the same time, the ability of hydroquinone to readily form complexes with molecules of small size points to the low stability of the α-hydroquinone structure. If the cell parameters are correct, this low stability may be due to a geometry unfavorable for the formation of strong hydrogen bonds. It is very desirable to perform a complete structural study of an α-hydroquinone crystal.

The framework structure formed by hydroquinone molecules in a complex with small molecules has been determined rather reliably. It differs from the structure of pure hydroquinone. To distinguish between these two structures, the framework of hydroquinone molecules in the crystals of the complexes is termed β-*hydroquinone*. Attempts to obtain pure β-hydroquinone have been unsuccessful, although its framework does not have very many cages and their sizes are not very large.

In all inclusion compounds the crystals of β-hydroquinone have almost the same hexagonal cell with the parameters a = 16.3 and c = 5.8 Å. The cell contains three formula units of $3C_6H_4(OH)_2M$, where M are different small molecules. The cell volume is 1325 Å^3. The M molecules are located in the cages of the framework. Had all the cages of the framework been empty, the volume of one hydroquinone molecule would have been 148 Å^3. The proper volume of the hydroquinone molecule is about 95 Å^3. This gives a packing coefficient of 0.64, a figure common for molecular crystals which have no tendency to form complexes.

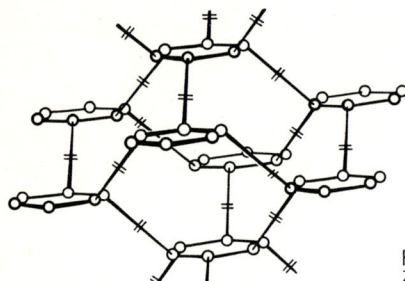

Fig.13.8. Lattice framework formed by hydrogen bonds in hydroquinone-based clathrates

Figure 13.8 is a classic pattern of the nature of hydrogen bonds forming the lattice framework and the shape of the voids into which the molecule of the second component enters. There are only three voids per cell. The cavity center is uniformly surrounded by 24 oxygen and carbon atoms. They are located on an almost spherical surface with a radius of 3.95 Å. If an average intermolecular radius is taken to be 1.7 Å, such a cage will accommodate a molecule 4.5 Å in diameter. This is less than the size of a methane molecule.

Even this example shows that for the structure to be stable not only should the packing coefficient be satisfactory, but the size of the voids is very important too. Had the number of the voids been twice that existing in reality, and their size, correspondingly, half as large, the crystal would have been stable. Because of large voids, the β-hydroquinone framework cannot exist without incorporated molecules. The number of atomic pairs, the distance between which is considerably greater than the equilibrium distance, should be decreased.

POWELL [see his article in 13.16], the first investigator of this class of the compounds, reports the following compositions of the hydroquinone complexes with incorporated molecules M.

M =	Kr	Xe	HCL	HBr	H_2S	SO_2	CO_2	CH_3OH	HCO_2H	CH_3CN
M/3Q	0.74	0.84	0.85	0.36	0.64	0.88	0.74	0.97	0.82	0.97
[Å]	4	4.4	4.1	4.5	4.4	5.5	5.5	5.8	6.0	6.1

The first line of the table gives the ratio M/3Q and the second line, the maximum size of the incorporated molecule in Å units. As was already mentioned, the cell dimensions of all these complexes are almost the same. Only the cell of the crystal which contains the largest molecule, acetonitrile, as a guest component is slightly distorted. It is believed that in the case of acetonitrile all the cavities of the crystal-matrix are occupied. CHEKHOVA and DYADIN [13.17] obtained a complex with methanol where all the cages of hydroquinone are occupied by guest molecules.

Other M/3Q ratios, closer to 1, are reported in literature for H_2S [13.14]. CHEKHOVA et al. [13.15] are of the opinion that the compound with sulphur dioxide is also thermodynamically stable when all the voids are filled.

The complexes described above are prepared by recrystallization of hydroquinone from the second component used as a solvent (for example, methanol or acetonitrile). If this is impossible, then both components are dissolved in a non-complexing solvent. We may assume that in the second case the reasons why not all the voids are filled are of a kinetic nature.

Reliable data has not been found which would allow inclusion compounds of small molecules into the framework of β-hydroquinone to be considered as solid solutions. Evidently, the lattice whose voids are completely occupied by guest molecules is thermodynamically stable. At the same time, the existence of some crystals whose voids are only partially filled cannot be doubted. This points to the "strength" of the framework. The conclusion that the framework of β-hydroquinone has a structure most suitable for the formation of stable hydrogen bonds is supported by the fact that any essential distortion of this framework is impossible. This limits the size of the guest molecules.

Calculation of the lattice energy of hydroquinone complexes for completely or partially filled cages would be of great interest. However, such calculations have not yet been performed.

13.3.2 Clathrate Hydrates

We have already mentioned that the formation of any complex must be energetically favorable. Either additional bonds must form which decrease the overall energy of the system in relation to the energies of the crystalline components (as in ice and in clathrate hydrates), or else the density of the complex increases, i.e., the short-range interaction energy between particles decreases (as in packing complexes). Let us discuss the packing in *ice polymorphs*. The volume per water molecule is $18 \times 1.65 = 30$ Å^3. When water molecules are connected by hydrogen bonds, the molecular volume may be taken as that of a sphere of radius of 1.4 Å, which is the oxygen non-bonded (intermolecular) radius. But 2.8 Å is an ordinary distance between the oxygen atoms connected by a hydrogen bond, and is also the intermolecular non-bonded diameter of oxygen. In other words, geometrically one can view the hydrogen bond as a proton pressed between two oxygen spheres of radius 1.4 Å, and of volume 11.5 Å^3. Hence, the packing coefficient of ice is 11.5/30, or less than 0.4. Such a loose structure is retained only because of the strong hydrogen bonds: each oxygen atom participates in four hydrogen bonds.

A structure with such a great number of voids possesses a very high energy. A tendency to decrease the energy is responsible both for the high polymorphism of ice and also for the formation of clathrate hydrates.

Clathrate hydrates are space structures in which water molecules form a framework whose O‑O edges are hydrogen bonded. The protons lie on these edges or deviate only slightly from the straight line connecting the oxygen atoms. The hydrogen bond energy in clathrate frameworks is not inferior to that of ice. However, the cages formed in clathrate hydrates are larger (5 ‑ 7 Å in size). A structure with such voids cannot be retained even with the strongest hydrogen bonds. But, whereas in ice the voids are too small even for the smallest atoms to be located in them, fairly large molecules and ions can occupy the larger cavities of clathrate hydrates. Clathrate hydrates resemble the hydroquinone clathrates considered above.

Clathrate hydrates were first discovered at the beginning of the last century. In 1823 Faraday obtained yellow crystals of *chlorine clathrate hydrate*. NIKITIN [13.18] was the first to show that the gas hydrates are inclusion compounds. However, it was POWELL and PALIN [13.19] who elucidated the nature of clathrates, performed a structural investigation of *hydroquinone clathrate with sulphur dioxide*, and for the first time used the term "clathrate" (a Greek word translated as "cage"). The use of this term is especially justified when we deal with gaseous hydrates. The molecules are "imprisoned" in cages and must overcome high energy barriers to get out. Under standard conditions gas hydrates are metastable compounds.

However, packing complexes are also termed "clathrates" when the complex is thermodynamically stable. This term is often used for all packing compounds, i.e., for all complexes in which no additional forces arise between the molecular components. But we prefer to limit the term to those complexes whose cages are hydrogen bonded or are bonded by other interatomic forces stronger than the interaction between the walls of the cage and the "prisoner", or guest molecule.

We assume that in clathrates with cages of the same type, a stable compound can only be formed when all the cages are occupied. When the clathrate has cages of two or more types, as in hydrates, all large cages must be completely occupied. Only in this case can one presume that interstitial solid solutions, i.e., compounds of a variable composition, can exist. However, their existence has not yet been proved; furthermore, it is not correct, theoretically to consider all clathrates as interstitial solid solutions [13.20].

When voids exist whose size allows the incorporation of a sufficiently large molecule (6 Å, for instance), and when the cage is formed by hydrogen bonds, then filling and closing the void must produce a considerable distortion of the lattice. The existence of such a cage or a channel whose structure is retained undistorted by packing of rigid molecules does not contradict our opinion concerning the elasticity of the bonds. But then the host lattice must be stable with voids unfilled and the formation of terminal solid solutions must be possible. However, no reliable data are available on the existence of such stable solid solutions. The existence of clathrates with variable intermediate composition is more probable, but has not been definitely proved.

Now we shall briefly discuss the structure of clathrate hydrates presented in the review by JEFFREY [13.21,22] who performed a careful X-ray diffraction study of the most intricate of these complexes, and elucidated geometrical regularities of the frameworks of water molecules.

A cage of water molecules that has an optimum number of hydrogen bonds must satisfy the following requirements. It must be a polyhedron with edges of equal length (2.8 Å) and the angles between the edges close to $110°$. In an ideal case, the polyhedron should be regular. However, a small distortion (scattering of the angle values by $2-3°$ and lengths of the edges by several hundredths of Å) is energetically insignificant. The polyhedral vertices are occupied by oxygen atoms. It is impossible to fill space with general polyhedra of the same type. But the problem can be solved by adjoining polyhedra of different types by common vertices, edges, or faces.

It may be assumed that there is a very large number of frameworks that allow molecules of different shape and size to be incorporated into the cavities. Many new structures of water hydrates may yet be discovered. But even today there are many different types of clathrate hydrates. The simplest of them are the most frequent. In fact, about twenty years ago it was widely believed that there were only two types of clathrate hydrates with cubic lattices of a period 12 and 17 Å, respectively.

In the majority of clathrate hydrates we find almost regular *pentagonal dodecahedra*. This polyhedron has 30 edges and 20 vertices, with three edges meeting at each vertex. If a pentagonal dodecahedron is considered as a structural unit for water with formula $H_{40}O_{20}$, then ten protons will be directed outward so as to form hydrogen bonds with oxygen atoms belonging to other polyhedra. It is impossible to fill space completely with pentagonal dodecahedra. Filling can, however, be achieved with slight distortions if a pentagonal dodecahedron is combined with *14, 15, or 16-hedra*.

The simplest hydrate, whose crystals possess a cubic cell with edge 12 Å, is built of 12- and 14-hedra. A cell contains two dodecahedra and six 14-hedra. A framework element of this structure is shown in Fig.13.9.

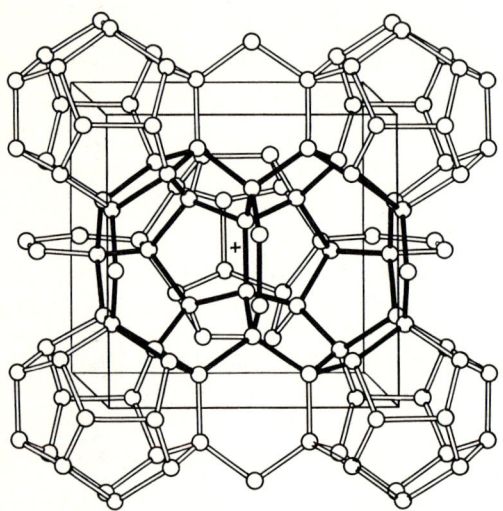

Fig.13.9. A framework element of the simplest hydrate. Two 14-hedra are shown with bold lines. The centers of 12-hedra coincide with the ends of the cubic lattice

The number of polyhedral vertices, i.e., the number of oxygen atoms (and, consequently, water molecules) per cell is 46. The cell contains 8 cages. Two of them (the dodecahedral cavities) are of approximately spherical shape 5 Å in diameter; six cages are of an ellipsoidal shape with diameters 5.3 and 6.4 Å.

Filling all the voids in this structure (8 voids per 46 water molecules) leads to the clathrate composition X · 5.75 H_2O. The molecules of inert gases, *methane, hydrogen sulphide*, and many others form *clathrate hydrates* of this type. As follows from scarce and contradictory literature data, small molecules fill all the cages of the clathrate. One should bear in mind that sometimes two molecules may fit in the same cage, which may result in a deviation from the ideal ratio 1 : 5.75.

The formation of a clathrate leads, naturally, to an increase in the structure's packing density. By way of illustration, Δ for the clathrate of methane will be calculated. The specific volume of a water molecule in a crystal of ice is 30 Å. The molar volume of liquid *methane* is 39 cm^3, and the molecular volume is 65 Å^3, and methane will fit all 8 cages of Fig.13.9. Hence, from (3.17):

$$\Delta = \frac{8 \times 65 + 46 \times 30}{12^3} - 1 = 0.1$$

is positive. Clathrates may exist in which only the large voids are occupied. As was already mentioned, there are 6 such large voids and two smaller ones per 46 water molecules.

Molecules of *ethane*, its derivatives, and halogens cannot be located in the small voids. We shall leave the small voids vacant and fill all the large voids. In this case also Δ is positive. For instance, the molar volume of liquid C_6H_5F is 60 cm^3, and its molecular volume is 100 Å^3. In the formation of clathrate Δ becomes:

$$\Delta = \frac{6 \times 100 + 46 \times 30}{12^3} - 1 = 0.14 \quad .$$

The clathrate becomes much more stable if the small voids contain small molecules, for instance, the molecules of sulphur dioxide. In this case Δ will attain large values.

It is not difficult to see that the packing coefficients acquire normal values when the cavities are filled with guest molecules. When all the clathrate voids of the x.5.75 H_2O structure are filled with methane molecules, the packing coefficient increases from 0.4 (for ice) to 0.6. If large and small voids are filled with molecules of suitable size, the packing coefficient may acquire a normal value for molecular crystals (more than 0.7).

We shall describe the other frequent structure of clathrate hydrates, which yields cubic crystals with a cell edge of 17.3 Å. It has 16 *pentagonal dodecahedra* (as in the clathrate of the first type) and 8 *16-hedra*. The 16-hedron forms spherical cages 6.6 Å in diameter and the dodecahedron forms, as above, spherical cavities of diameter 5 Å. There are 136 water molecules per cell. If all the voids are occupied, the clathrate formula is x.5.67 H_2O. If the guest molecules enter only the large cavities, the formula becomes x.17 H_2O.

Thus, one may assume that in the clathrates of the first type small voids are vacant or partially occupied (the number of small voids is one third of the number of large voids). This means that the existence of solid solutions in such clathrates is possible due to a continuous filling of the small cages.

But in the clathrates of the second type a solid solution is unlikely because now the number of small voids is twice that of the large ones. The stability of such a structure is doubtful. The following calculation strongly supports this argument. Let all the large 16-hedral cages be occupied by molecules 6 Å in diameter and 113 Å^3 in volume, and let all dodecahedral voids be vacant. Then Δ is

$$\Delta = \frac{136 \times 30 + 8 \times 113}{17.3^3} - 1 = -0.04 \quad .$$

Thus, no increase in structure's packing density takes place as compared to a mechanical mixture of the components. The formation of a clathrate without filling the small voids gives no gain in the packing density and consequently is not energetically advantageous.

At the same time the existence of double hydrates which are formed with the aid of a so-called auxiliary gas, for instance, hydrogen sulphide, which fills all small voids, seems quite natural. *Double hydrates* are known for benzene and for *halogen derivatives* of *methane, mono and disubstituted ethane*. *Benzene* is known to be disc-shaped, 7.2 Å in diameter, and 3.6 Å in cross section. Its location in an approximately spherical cavity 6.6 Å in diameter probably requires an adjustment of the molecule to the shape of the cage and a slight lattice distortion. Clathrate compounds formed with the aid of the auxiliary gas have high packing coefficients and Δ values.

The formation of clathrate hydrates is strongly affected by the conditions of crystallization. Some facts are available which unambiguously indicate that a part of small voids can remain vacant. In the case of *ethylene oxide*, which forms the hydrate of the first type (with a small number of cages of small size), an X-ray diffraction study shows that the ethylene oxide molecule occupies all large cavities and 20% of the small cavities (which are occupied only with difficulty). Tetrahydrofuran in the presence of hydrogen sulphide forms a clathrate hydrate of the second type. The large molecules occupy all the 16-hedral voids, and 47% of the small dodecahedral voids are occupied by hydrogen sulphide molecules. It is premature to draw a conclusion about the thermodynamic stability of these structures. DYADIN and ALADKO [13.23] studied in detail the *bromine hydrates*. They showed that there are four bromine hydrates of different composition. With increasing concentration of bromine in an equilibrium solution, the bromine content in a solid phase is discontinuous due to the formation of new frameworks. This refutes the approach to clathrate hydrates as solid solutions.

Evidently, the large clathrate cavities are completely occupied by guest molecules in all cases. For the smaller dodecahedral voids, the situation is not quite clear yet. Most likely, partial filling of small cavities may take place but it corresponds to the formation of metastable structures rather than of solid solutions stable in some concentration range.

Many compounds which form clathrate hydrates are hydrophobic. They are formed from a two-phase system. The solubility of these compounds in the solid state water clathrate is an order of magnitude higher than their solubility in liquid water. As a result of the lack of a specific interaction between the guest molecules and water molecules, only the molecule's shape and size play a part in the formation of a clathrate hydrate.

It should be noted that the formation of a cage from water molecules leads to a saturation of hydrogen bonds. Therefore, such molecules as *tetrahydrofuran*, *ethylene oxide*, *acetone*, and others behave just like hydrocarbons, i.e., do not form any particular bonds with the framework. In such cases we deal with a clathrate obtained from a monophase system, i.e., from an aqueous solution. Despite this apparent fundamental difference, the final result, namely, the formation of the clathrate, is the same.

The geometrical factor remains dominating even when clathrates are formed from a solution of *peralkylammonium salts*. Their structure differs considerably from that of the clathrate hydrates described above. Cations are now located in the cages whereas anions together with water molecules form the cage walls. In this case electrostatic interaction takes place between the cation in the cage and the cage walls. However, the regularities of the clathrate hydrates' structure are affected only slightly since long-range electrostatic forces are always of minor importance in the process of crystal formation. Of prime importance is the packing coefficient, i.e., the short-range interaction, which is the same for cations and for neutral molecules.

The structure of these cages is much more diversified. There is a class of compounds in which guest molecules do form hydrogen bonds with water atoms. Among them are *alkylamine hydrates*. Chemists believe that this is an essential factor and suggest another term for such cases. Clathrate hydrates in which a guest molecule forms bonds with the atoms of the cage are termed *semiclathrates*. However, the packing consideration is a deciding factor as before. The only result of a compromise between the tendency towards maximum saturation of all hydrogen bonds and a convenient arrangement of molecules in cages formed by water molecules is that almost all semiclathrates have different structures. Nevertheless, in the majority of cases clathrate semihydrates can also be considered as combinations of polyhedra. There are, however, some exceptions, for example, the hexahydrate of hexamethylenetetrammine. This interesting molecule has an approximately spherical surface. Anhydrous crystals form a body-centered cubic lattice so rare in organic crystallochemistry. The sphere has "hollows" corresponding to those sites of the molecule where four nitrogen atoms reach the surface. In a clathrate three of the four nitrogen atoms succeed in forming hydrogen bonds with hydrogen atoms of the cage, the result of which is a structure which cannot be described as a combination of polyhedra, but resembles hydroquinone clathrates. A schematic illustration of this unique structure is given in Fig.13.10.

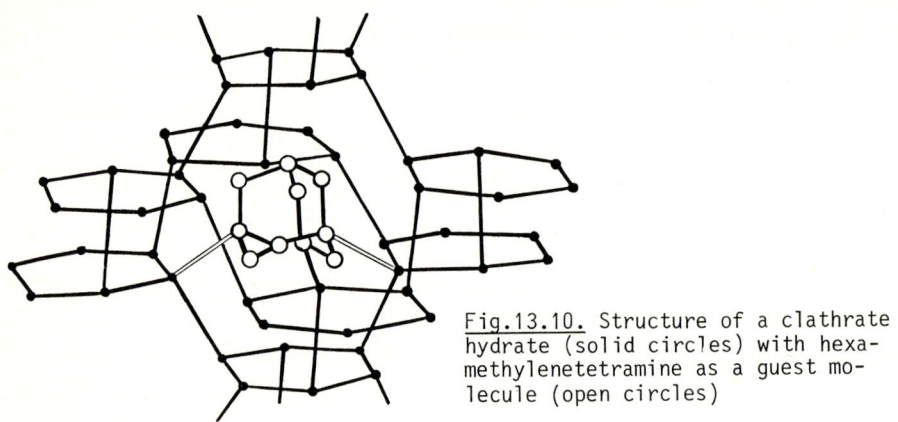

Fig.13.10. Structure of a clathrate hydrate (solid circles) with hexamethylenetetramine as a guest molecule (open circles)

It is quite clear which route we should follow in order to predict or at least to explain why in some cases clathrate hydrates are formed and in others water molecules enter the crystal only to help the hydrogen-bonded molecules form denser structures. The calculation of lattice energy by the atom-atom potential method may be quite suitable for this purpose. However, estimations of this type have not been done yet. Therefore, we shall end this section with a brief listing of the classes of molecules which form and do not form hydrates.

Carbohydrates are readily soluble in water but do not form hydrates. In those rare cases when water molecules enter the crystal, as is the case with raffinase pentahydrates, water molecules fill voids in the channels formed by the spirals of trisaccharide molecules. Very seldom do water molecules enter into the crystals of aminoacids, peptides, pyrimides, and organometallic compounds. If hydrates are still formed, they have nothing in common with clathrates. Water molecules are arranged one by one, or in chains or ribbons. Clathrates of molecules whose linear dimensions exceed 7-8 Å have not yet been observed. This, however, does not mean that such compounds will not be encountered in the future.

It should be mentioned that only *acetone*, out of all ketones, forms a clathrate hydrate. *Ethanol* does form clathrate hydrates. *Other alcohols* and *glycols* usually form hydrates with small amounts of water. But there are some exceptions. Data are available on several hexahydrates of alcohols. One may expect that we deal here with clathrates since the amount of water is large, and the size of the molecules does not exceed the permissible limits.

There is an enormous literature on clathrate hydrates and hydrates. A reader who is interested not only in the principles of structure but also in experimental material will find many references in the above-cited reviews [13.14,21,22]. One more reference can be added [13.24]. This paper is devoted to clathrate hydrates of *tetrabutylammonium, fluoride,* and *oxalate* and shows convincingly that one should think twice before treating clathrates as solid solutions. It has been shown for quite different compounds (clathrates in which neutral molecules and cations are guests) that we always deal with complexes and that hydrate numbers change discontinuously. This demonstrates that a water cage can easily be adjusted to the molecules incorporated.

General considerations related to the reasons for the formation of clathrate hydrates can also be applied to other hydrates. The looseness of ice and its low thermodynamic stability often promote the formation of different crystallohydrates. It can always be shown that inclusion of water molecules into the structure in the form of single molecules, chains, or nets increases the number of hydrogen bonds per unit volume with a simultaneous enhancement in the packing density.

A comparison of the structure of anhydrous crystal with that of its hydrate is very interesting. Such examples are scarce and we shall consider only one of the most recent examples. In 1980 the structures of *pyridine* and *pyridine trihydrate* were determined. Water molecules form not a cage, but an almost planar army of bonds. The pyridine molecules are between these water arrays. Comparing the specific volumes of the molecules of pyridine, water, and the complex, we find that Δ is 0.06. We believe the same is true for all hydrates.

Now we turn our attention to the study of *hexa-* and *heptahydrates* of α-*cyclodextrin*. In Sect.13.3.4 we shall consider in more detail these ringlike molecules. The compound crystallizes only when the voids inside the ring molecule are filled with guest molecules. In the hexa- and heptahydrate, the guests are water molecules arranged in five-membered "cycles" which give the complex a high packing coefficient [13.25].

13.3.3 Urea and Thiourea Complexes

Urea molecules form crystalline complexes with many compounds: normal paraffins, esters, acids, (halogenated substituted) aliphatic compounds. Just as in the case of hydrates and compounds of hydroquinone, this ability can be attributed to hindrances which arise in the process of urea crystallization. The structure of these molecules is such that it is almost impossible to attain dense-packed structure and at the same time enable the hydrogen atoms of amino groups to form an optimum number of hydrogen bonds. Although the packing coefficient of urea molecules in the crystal is rather high, non-bonded interatomic distances are much larger than equilibrium van der Waals distances. Besides that, the nitrogen, oxygen, and hydrogen atoms of urea which are involved in hydrogen bonds are not colinear. The energy of this crystal is undoubtedly very high.

A hexagonal crystal structure ($a = 8.23$, $c = 11.0$ Å; 6 molecules in a cell) is optimal with respect to saturation of the urea hydrogen bonds. Each oxygen atom is bonded via a hydrogen bond to four nitrogen atoms and each nitrogen atom to two oxygen atoms. The hydrogen-bond framework is optimum from the standpoint of energy. However, a hexagonal structure is impossible. The molecules are so arranged as to form channels about 5 Å in diameter along the hexagonal axis. The packing coefficient of such a structure with no guest molecules would be 0.5. If these channels fill with guest molecules, low energy crystals can form.

The complexes of *urea with normal hydrocarbons* have been studied in more detail. In crystals of paraffins the molecular cross section is 18.6 Å^2; for a cylinder its diameter becomes 4.9 Å. Molecules of such size can fill tightly the hexagonal channels of urea. Hydrocarbon molecules fill all the hexagonal urea channels. No voids remain between the ends of these long molecules.

The inner "surfaces" of the urea channels are rather smooth. Therefore, there are no preferential sites for paraffin molecules to be arranged in the channels. For this reason, stoichiometry does not play any part in filling the channels, and urea forms additional products with all even and odd paraffin molecules without exception.

As is known, in paraffin crystals the carbon atoms of the aliphatic chains form a planar zigzag array. The distance between the closest non-bonded carbon atoms in such a chain is 2×1.277 Å.

In recent papers (for instance, [13.26]) it has been shown that the carbon chain length in the complex can be expressed as

$1.26(n - 1) + 3.55$.

The fact that the formula contains 1.26 instead of 1.277 is quite natural. It means that paraffin molecules adapt themselves to the geometry of the channel: slight spiralization takes place.

As the paraffin-chain length increases the packing coefficient becomes larger since the number of contacts between the nonbonded atoms per unit length decreases. For C_{14} the packing coefficient is 0.705.

The filling of the channels with molecules of an arbitrary length results, strictly speaking, in a loss of periodicity along the hexagonal axis. If the channels are filled with molecules of the same kind, two periods will appear along the c axis: one 11 Å period formed by the urea molecules and the other equal to the length of the guest molecule in the channel. The chain of identical molecules, which strictly follow one another and fill the channel, forms a peculiar one-dimensional crystal. If bromine or iodine derivatives of normal paraffins are used as the second component, contiuous diffraction lines are observed on the X-ray rotation photographs about the hexagonal axis, just as predicted by the theory of X-ray scattering (Sect.6.2).

Urea always forms complexes with guest molecules and never forms solid solutions, the meaning of this assertion being that the channels must always be completely filled. But the channel can also be filled with different molecules. This can be elegantly proven by plotting a phase diagram after filling the channels of hexagonal urea with molecules of two different lengths, C_5 and C_{10}. The same authors obtained a diagram typical for a continuous series of solid solutions. the channel-filling, C_5 and C_{10} molecules are arranged in statistical disorder.

As was already mentioned, a wide variety of molecules can fill the channels of hexagonal urea. However, the geometry of the channel imposes rigid limitation. For instance, an additional methyl group may "distort" the nearly cylindrical shape of the molecule of normal hydrocarbon. But the channels can nevertheless be filled if the molecule is long enough (12 - 13 carbon atoms). It is quite clear that complexes are formed when we change the end groups of the chains and are not formed if the same substituents are located in the middle of the chain. Geometrical analysis and the atom-atom potential method may be used to forecast the possibility of forming complex as successfully as by the methods described for solid solutions. However, no such calculations have been performed yet.

Thiourea behaves similarly to urea. The crystals of pure thiourea also show poor agreement between the requirements for dense packing of the molecules and those for saturation of the hydrogen bonds. A suitable framework of hydrogen bonds is accomplished in a crystal of hexagonal symmetry, but cannot exist independently because of large channels whose diameter is much greater than that of the urea channels.

The inner surface of the channels in thiourea is not as "smooth" as in urea. Therefore the agreement between the shape and size of the guest molecule and the shape of the channel is much more important here than in the case of urea. The channels of thiourea can be filled with different molecules (branched paraffins, cyclic molecules). A complex is formed even with such a relatively large molecule as decahydronaphthalene.

Since the channel has varying widths, incorporated guest molecules may occupy "suitable sites" and will not pack, i.e., touch one another. Thus in the case of thiourea we may expect that complexes with a stoichiometric ratio between the components will be formed. Most probably, thiourea does not form solid solutions with the guest molecules entering into its channels, any more than does urea.

If the host molecules' framework cannot exist without molecules filling the voids, one may doubt the existence of solid solutions (i.e., the existence of such crystals with noticeable vacancies along the channels). Such vacancies would have resulted in substantial lattice distortions and consequently in great energy losses. Therefore, an "all or nothing" principle seems to hold: either a complex is not formed at all, or it is formed with all the voids occupied.

Let us repeat once more: in our opinion, solid interstitial solutions may exist only when the framework, i.e., the crystal of the host molecules without the "guests" is thermodynamically stable. In [13.26] the thiourea-chloroform complex was studied, and the data on the complexes of thiourea with a number of other molecules of small size were summarized. The authors point out that the integral thiourea : guest ratios obtained in previous papers can be accidental. In reality, for the following guests the ratios are rather close to 3: 2.96 for $C_2H_2Cl_4$, 2.94 for C_2HCl_5, and 2.95 for C_2Cl_6. However, the experimental host : guest ratio for the urea : benzene is 2.40 and for thiourea : chloroform it is 2.25. Even for these complexes, the obtained experimental host : guest ratios were substantiated theoretically when the channels are filled with molecules in mutual contact in accordance with the most dense packing principle, i.e., *hollow-to-recess packing*. Thermodynamic study has also shown that thiourea, as well as urea, forms complexes and not solid solutions with all the guest molecules studied.

Although these interesting complexes have been studied for a long time, a number of questions still remain open. In particular, it is of great interest to study the degree of azimuthal ordering of the molecules which fill the channel. A detailed study by X-ray diffraction and by nuclear magnetic resonance may elucidate this interesting question. At the same time, quantitative calculations of the structure and dynamics of the lattice of the complexes by the atom-atom potential method may be successful.

13.3.4 Inclusion of Molecules into Intramolecular Cavities: Tri-o-Thymotide and Cyclodextrin Complexes

Chemists can synthesize molecules of almost any shape: ring, toroid, cage, cup, etc. Molecules can be rigid or flexible. If parts of a molecule are connected with the aid of ordinary bonds about which free or slightly hindered rotation is possible, then upon crystallization such a molecule will almost certainly roll up into a dense ball. But among molecules with a complex shape there can exist rigid molecules. We can divide them, somewhat conditionally, into two classes: molecules with an inner cavity and molecules unsuitable for creating a dense-packed structure. Molecules of both types have a strong tendency to form complexes with other kinds of molecules and, by themselves, either will not crystallize (an amorphous compound is formed) or will yield a crystal with low stability.

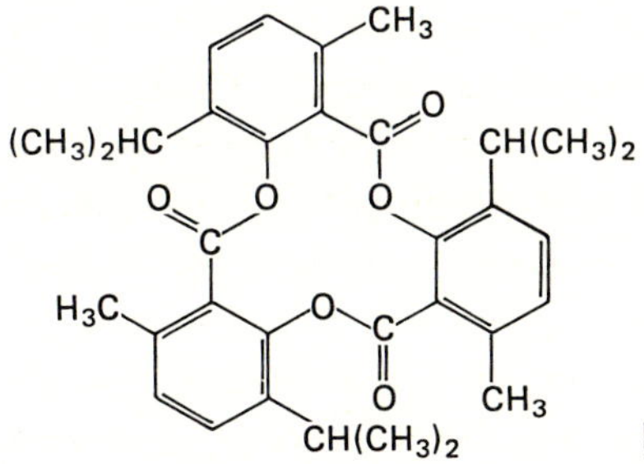

Fig.13.11. Structure of tri-o-thymotide

In this section we shall consider rigid molecules shaped as a ring with an inner cavity which, for the crystal to be formed, must be filled with guest molecules of a suitable size.

Of the known examples, we shall consider only two, namely, the binary systems based on tri-o-thymotide and cyclodextrin.

The molecule of *tri-o-thymotide* is shown in Fig.13.11. It possesses a three-fold symmetry axis. A planar conformation of this molecule is unfavorable since the oxygen atom of the carbonyl group would be sterically hindered by the adjacent nonbonded atoms. Since rotation about the single C-O bonds is possible, a conformation forms in which the three central oxygen atoms go "downward" and the phenyl rings rotate. The shape of the molecule resembles a cup with a hole in the bottom. The molecular conformation can be slightly modified to fit the molecule of the second component which uses the inner cavity in order to create a dense-packed structure. It is easy to calculate the radius of the circle tangent to the three central oxygen atoms; it equals 2.5 Å, which is quite sufficient for the incorporation of long molecules of normal hydrocarbons and their derivatives (provided the molecules of the tri-o-thymotide framework are superimposed to form a channel of voids). If we build a trihedral prism on two sets of three central oxygen atoms from two adjacent tri-o-thymotide molecules, it will become possible to inscribe a spherelike molecule about 7 Å in diameter into this prism.

It is very difficult to use molecules with large cavities and form a crystal in which molecules are packed with a reasonable density; experience in organic crystal chemistry requires that the packing coefficient must always be more than 0.6 [13.27].

The structure of a tri-o-thymotide crystal is unknown. But there is evidence that the compound crystallizes in the orthorhombic system. The crystals of packing complexes based on these molecules have a quite different structure. The molecules of the second (guest) component make use of the cavities of the main (host) molecules. As a result, the compounds that do form have a density typical for organic compounds and a high packing coefficient.

Two types of binary compounds based on tri-o-thymotide are known. One group of complexes gives crystals with a hexagonal cell about 4800 $Å^3$ in volume. The cell parameters vary slightly depending on the size of the guest molecule. With ethanol, the cell parameters are a = 13.45, c = 13.15. Normal pentanol is the largest guest molecule in complexes of this type: the cell parameters increase up to a = 13.7 and c = 30.7. More than a dozen such complexes have been studied [13.28]. Guest molecules are of almost an ellipsoidal shape and the cell parameters grow with the increasing size of the guest molecule. Tri-o-thymotide is a very large molecule. Therefore, the cavities which exist in this molecule have no strong effect on the density of the compound. Without the second component the density would be 1.09 g cm^{-3}; when all the cavities are filled with the incorporated molecules the density increases by 0.05 - 0.06 g cm^{-3}, provided these molecules do not contain heavy atoms.

A unit cell comprises 6 host and 3 guest molecules. We assume that in this case as well we are dealing with a complex and not with a solid solution.

We discuss the complexes based on tri-o-thymotide mainly because of the interesting fact that they can also form another type of complex. Frameworks are possible with channels which possess the same peculiarities as urea. Whereas in the complexes of the first type the number of incorporated molecules is fixed and the composition of the complex is 2T·M, in the second case the cavities which form the channel can be filled with molecules of any length, as a result of which compounds of a nonstoichiometric composition can be formed.

Undoubtedly, it is energetically advantageous when all the cells contain the same number of molecules. In this case, all the molecules will choose the same optimum way to pack the cavities. However, in [13.16] it was reported that the authors discovered a great number of compounds in which the cells contain different fractional numbers of incorporated molecules from 0.95 ($C_{16}H_{33}M$) to 2.67 ($C_7H_{15}OH$). Knowing the molecular length and composition, the authors have shown that the number of molecules per cell is precisely such as to enable the molecule to fill up the channel and be in close contact with one another.

Channel compounds of tri-o-thymotide form crystals with a somewhat different hexagonal cell than the complexes of the first type. Here a = 14.3 and c = 29.0. Evidently, this cell is due to the need for all the cavities of the main molecules to be strictly arranged one above another, which is not necessary in the complexes of the first type. A recent investigation is reported in [13.29].

Consider another interesting example of molecular complexes formed by cyclic polymers of glucose known as *cyclodextrins*. They comprise cycles containing 6, 7, and 8 glucose structural units. Figure 13.12 illustrates a ring-shaped six-membered molecule of dextrin. Some parts of the molecules are connected by ordinary bonds. However, rotation about these bonds is sterically hindered. Therefore, a dextrin molecule possesses limited flexibility. In the most suitable conformation the inner diameter of the cavities in these molecules is about 8 Å. Pure cyclodextrin is an amorphous powder with a density of 1.04 g cm^{-3}

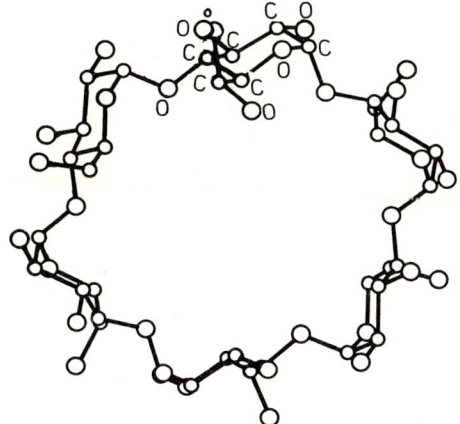

Fig.13.12. The structure of a six-membered cyclodextrin

at room temperature. The situation is quite natural, since no regular or dense-packed structure can be created from molecules of such a shape.

Cyclodextrin molecules form quite different complexes with inorganic and organic molecules. The complexes exist in solution. But in many cases it is possible to obtain perfect crystals of cyclodextrin molecules whose cavities are occupied by guest molecules.

In Sect.13.3.2 we mentioned the possibility of obtaining complexes of cyclodextrin with water. Crystalline compounds with iodine and potassium iodide have been known for a long time. The structure of orange crystals has been determined in which one formula unit consists of one cyclodextrin molecule, one iodine molecule, and fourteen water molecules. Complexes are well known in which a cyclodextrin molecule involves one or two iodine molecules and one molecule of potassium iodide.

When the guest molecules are arranged in channels, the same phenomena can be observed as in urea complexes. Guest molecules form one-dimensional periodic sequences which are seen as solid diffraction lines on X-ray rotation photographs. The decisive role of the geometric factor in the formation of complexes is evident even in complexes with halides. Thus, for instance, six-membered dextrin readily forms complexes with iodine, bromine, and chlorine; seven-membered dextrin with bromine and iodine; and eight-membered dextrin only with iodine. Even this example indicates that no complexes are formed if the cavity of the molecule is too large to keep small molecules.

Compounds with which cyclodextrins can form complexes are rather numerous. As guest molecules one can use unbranched alcohols, fatty acids, benzaldehyde, nitrobenzene, aromatic compounds, anthracene, naphthalene, and many others. There is some indirect evidence that aromatic compounds prefer to occupy the cavities in cyclodextrin molecules in such a way that the axes of host and guest molecules coincide.

Here is another interesting example which illustrates the role of the geometrical factor in the formation of a complex. Seven-membered cycles of dextrin readily form crystalline complexes with benzoic acid and all its iodine derivatives, whereas six-membered cyclodextrin forms complexes only with benzoic and paraiodobenzoic acids. There are many such examples, and the reader can find them in the above-cited reviews [13.14,16].

The structure of cyclodextrin complexes with dye molecules is a fine illustration of the decisive role played by the packing coefficient in the formation of complexes. In these cases, only one of the three phenyl rings of the dye molecule can reach the interior of the molecular cavity of cyclodextrin. A compact structure is formed which can be thought of as a toroid into whose hollow a mushroom with a cap is inserted.

In recent years the complexes of cyclodextrin with a large number of benzene derivatives have been studied in detail. Some crystals are so highly ordered that 5000 independent X-ray intensities can be measured, so that the structure can be determined with high accuracy. The isomorphism of cyclodextrin crystals in which different benzene derivatives serve as guest molecules has been studied by HARATA [13.30]. The 1:1 complexes of six-membered cyclodextrin with para-nitrophenol and para-hydroxybenzoic acid form trihydrates, i.e., one molecule of cyclodextrin and one molecule of a benzene derivative crystallize with three water molecules in space group $P2_12_12_1$, with four formula units per cell. The cell parameters differ only in their third significant digit. The volumes of the cells are practically the same (5100 $Å^3$) as are their densities (1.53 g cm^{-3}). In both cases the guest molecules are arranged in the center of the ring cavity. A dense-packed structure is produced in which all the distances without exception between the atoms of the guest molecules and the inner atoms of the ring host molecule differ only slightly from equilibrium intermolecular distances. The three water molecules fill the vacant sites of the lattice and participate in the creation of hydrogen bonds which additionally link the guest and host molecules. Such compounds are very important for chemical technology since they make it possible to obtain new catalysts and compounds capable of separating molecules close in their chemical properties but differing in size and shape.

13.3.5 Inclusion of Molecules into Intermolecular Voids

More than thirty years ago the author of the book reported that crystallization of molecules whose shape is unsuitable for creating a dense-packed structure is hindered. I formulated a principle according to which the minimum packing coefficient of molecules in a crystal must be about 0.6. If it is impossible to obtain a denser packing, no crystallization will take place and the compound can be prepared only in the form of an amorphous powder. Sometimes crystallization does take place but with great difficulty as, for instance, in the case of diocrylnaphthalene.

A great number of molecules in a cell usually indicates that the molecules pack with difficulty. Nevertheless, many molecules which have unsuitable shapes such as a cup or clover, a combination of a spherical region with long chain substituents, or some ring-shaped molecules, do succeed in forming crystals. A structure appears with a low packing coefficient and a great number of small voids. The atomic pairs which border these voids are at distances only slightly exceeding the equilibrium ones. Since the atom-atom potential curve has a small slope towards large distances, the losses in lattice energy are not very large. However, such crystals possess high energy and precisely for this reason form complexes with quite different molecules.

Even the smallest guest molecules cannot be arranged in small voids formed during the crystallization of molecules with unsuitable shapes. The molecules whose shape is unsuitable for packing can form another type of crystal containing a small number of large voids. In this case a dense-packed structure, usual for molecular crystals, is produced, in which atoms of the adjacent molecules can form a framework according to the "key-to-the-look" principle; however, large voids are formed. The pairs of atoms bordering these large voids are at such great distances from one another that the lattice energy becomes impermissibly high. The cavities will be spoiled and the framework destroyed. The crystal is unstable and can exist only if the voids are filled with guest molecules of suitable shape and size. When a compound is crystallized from a solvent whose molecules, because of their size, do not fit the cavities of the framework, either no crystal is obtained at all, or a stable crystal with a low packing coefficient and a great number of small voids is obtained. If the molecules of the solvent suit the shape of the cages of an unstable framework, a complex is formed.

Large cavities are inherently unstable. Therefore, they are all filled with guest molecules. It is most likely that solid interstitial solutions never form. Only complexes are formed whose composition is determined by the ratio of the number of the host molecules forming the framework to that of the cavities.

Many complexes of this type are known. They are thoroughly studied by chemists since they can be used to separate substances which can enter the framework cavities from those which are unsuitable for the formation of a complex. Bearing in mind these practical purposes, the chemists pay little attention to crystallographic problems. Therefore we can seldom find literature data which would make it possible to compare the crystalline structure of the complex with that of the host crystal.

Many examples of complexes, which are formed by large molecules with a shape unsuitable for packing and different guests molecules stabilizing the framework, are given in the review by MAC NICOL [13.14] who intensively studies this class of compounds. We shall consider these complexes, beginning with the compounds formed by *triphenylmethane*. In this molecule the bonds diverge from the central carbon atom at tetrahedral angles and there are "hollows" in the molecule between three phenyl rings and near the hydrogen atom. It is rather difficult to pack the molecules of such a shape with a sufficiently high density. But a crystal is, nevertheless, formed. Its structure is unknown, but the cell parameters and the number of molecules in the cell have been determined. These are the essential data which we need to explain why the complexes based on this compound can be formed.

Triphenylmethane forms orthorhombic crystals with cell parameters a = 14.71, b = 25.72, and c = 7.55, space group Pbn2, 8 molecules per cell and a density of 1.37 g cm^{-3}. There are two molecules in independent positions, so that the number of degrees of freedom doubles and the molecules have good opportunities to form dense-packed structures. The volume per molecule is 357 Å^3. The packing coefficient is almost the minimum allowable: it is just less than 0.6. Thus even 8 molecules in a cell cannot form a dense-packed structure.

Recently, ALLEMAND and GERDIL [13.31] have determined the structure of the complexes formed by *triphenylmethane* with *benzene, thiophene, pyrrole,* and *aniline*. The structure of the complexes is quite different. They are all isomorphous, space group R$\bar{3}$ with six molecules per cell. The cell parameters of the 1:1 complex with benzene are a = 10.982, c = 27.475 Å, whence the volume of a formula unit is 479 Å^3; for the 1:1 complex of triphenylmethane with thiophene a = 10.880, c = 27.475 Å, the volume of a formula unit being 471 Å^3. The cell parameters of the complexes with pyrrole and aniline differ only in their third significant figure.

The packing coefficient for the complexes can be easily calculated. It increases sharply and becomes slightly less than 0.7. Does compaction take place during the formation of the complex? The cell parameters of benzene and thiophene are known. The volume of a benzene molecule is 122 Å^3. Hence,

$$\Delta = \frac{357 + 122}{479} - 1 \approx 0 \quad .$$

Thus the formation of the complex increases sharply the packing coefficient but does not result in an increase in packing density. This is due to the fact that benzene molecules are packed very densely in the pure benzene crystal.

Turning our attention to *thiophene*, three crystal modifications of thiophene are known. In two of them the molecular volume is 116 Å^3. The cell parameters of the third modification are too large. Using the figure given above, we obtain approximately the same result:

$$\Delta = \frac{357 + 116}{471} - 1 = 0.02 \quad .$$

Thus, the packing coefficient increases sharply and the packing becomes slightly denser despite a high packing density of the small molecules in the complex and high symmetry of the complex.

The structure of the complex is very interesting. The molecules of triphenylmethane are centered along the three-fold axis and face one another with their hydrogen atoms. The hydrogen atoms probably lie on the symmetry axis. If this assumption is valid, the distance between them must be 1.92 Å. Maybe the C-H bonds may bend slightly and move away from the symmetry axis. The guest molecules are arranged along the three-fold axis so that their plane is normal to the axis. The small molecules very closely occupy a hollow formed by three phenyl rings.

Figure 13.13 illustrates structural formulas of several *chroman* derivatives. As far back as 1914, DIANIN [13.32] discovered that the compound designated as I in Fig.13.13 can create a complex. A systematic investigation of packing complexes produced by molecules whose form is inconvenient for dense packing and whose small voids are filled with guest molecules is being carried out by MAC NICOL et al. [13.14]. The ability to form inclusion compounds by different derivatives of Dianin compounds has been studied. The substances were crystallized from different solvents. In some cases complexes were formed; in others

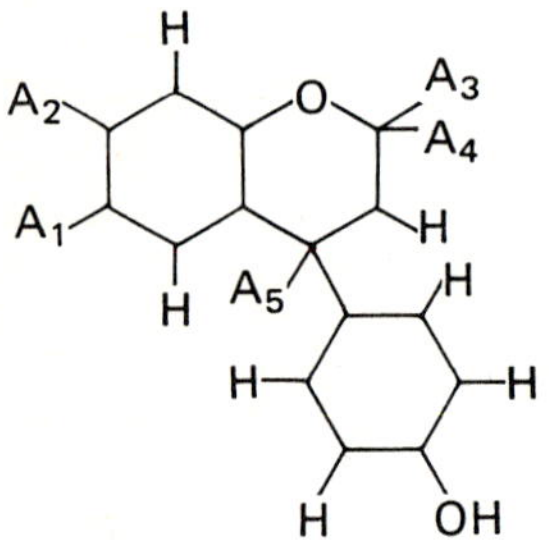

Fig.13.13. Structural formulas and packing coefficients for several chroman derivatives

	I	II	III	IV	V
A_1	H	H	H	H	CH_3
A_2	H	H	H	CH_3	H
A_3	CH_3	CH_3	CH_3	CH_3	CH_3
A_4	CH_3	H	CH_3	CH_3	CH_3
A_5	CH_3	CH_3	H	CH_3	CH_3
V_o	224	203	203	246	246
V_c	389	375	344	398	383
k	0.64	0.60	0.59	0.62	0.64

I   H + 1/6 $CHCl_3$
II H + 1/6 CCl_4
III, IV, V   none

no solvates were detected. By thorough X-ray diffraction analysis of some of the resulting packing complexes it was ascertained that complexes are formed if the host molecules create a framework with large voids and the guest molecules, by their form and size, fit the cavities. The cavities have diverse sizes and forms, usually rotation bodies of 3 - 4 Å by 7 - 10 Å. The geometric reasons for the selectivity of complex formation with any one guest molecule are quite obvious from the investigations carried out.

Unfortunately, it is not always clear from the cited works whether the crystallization of a number of the Dianin compound derivatives is possible in the absence of a solvent. Probably, frameworks with large voids cannot exist without guest molecules. And in those cases where the host molecules crystallize, the structure of the complex is distinct from that of the pure host for the reasons discussed above.

Now let us consider the latest studies of this cyclic system. The compounds designated II and III in Fig.13.13 have been studied [13.33,34]. They differ from the initial Dianin compound I by the substitution of a hydrogen atom for one of the methyl groups. Consequently, the approximate intrinsic volume V_0 of molecules II and III is less than that of I by 21.5 Å^3. The compounds IV and V in Fig.13.13, on the contrary, differ from I by the substitution of the methyl group for the hydrogen atom, i.e., they have V_0 greater than I by 21.5 Å^3.

Only derivatives I and II (Fig.13.13) yield inclusion compounds. The structure of the complex formed by the host molecules of II with carbon tetrachloride (host:guest ratio 6:1) has been determined. The structure is very close to that of the complex which forms from compound I and chloroform molecules.

The cell volume per formula unit V_c is also given precisely in Fig.13.13. Although the intrinsic molecular volumes V_0 are approximate, the packing coefficients k of Fig.13.13 are valid ±0.01. Figure 13.13 demonstrates that the packing coefficients k are small and differ little from one another. The inclusion of the guest molecules yields an insignificant gain in density. How can one account for this result? Unfortunately, the structures of the three compounds which do not form complexes (III, IV, V) have not yet been determined. However, the cell sizes and the space groups have been found. It turned out that crystals representing packing complexes (I and II) are isomorphic (space group R$\bar{3}$, similar cell sizes). As to the crystals of the substances which do not accommodate guest molecules, all the three substances possess completely different crystalline structures: type III yields an orthorhombic cell, IV and V, monoclinic cells of diverse forms. We can suppose that all these structures are loose and contain numerous voids, but all the voids are too small in size to form inclusion compounds. It would be interesting to check the validity of this prediction.

Let us dwell upon another interesting class of packing complexes whose host molecules are the *hexa substituted derivatives of benzene* with *six identical bulky substituents*. The neighboring substituents look up and down, in turn, with respect to the benzene nucleus. Of course, molecules of such a form are extremely inconvenient for creating a dense-packed structure.

The structural host molecule formulas of the investigated inclusion compounds are represented in Fig.13.14. MAC NICOL and coworkers, who studied these complexes extensively, suggest calling them "*hexa-host*" inclusion compounds for the following reason:"The term hexa-host is especially suitable for this type of host molecules since it gives not only the symmetrical picture of the central ring, but also reflects the infinity for hexamers connected by hydrogen bonds" [13.35].

We believe that the principal significance of Mac Nicol's investigations is the demonstration that the presence of hydrogen bonds is not a necessary condition for the formation of packing complexes. The selection of molecules

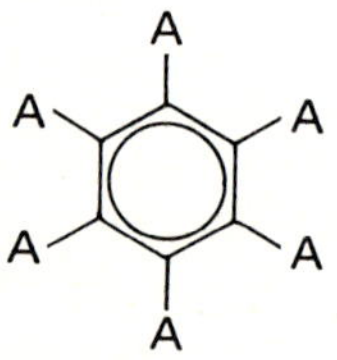

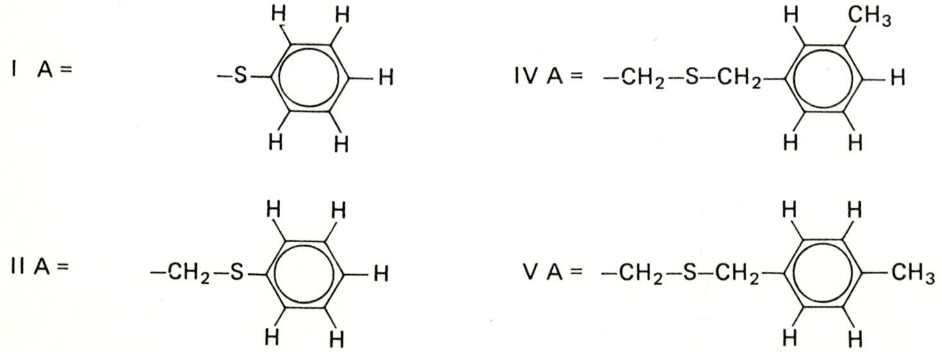

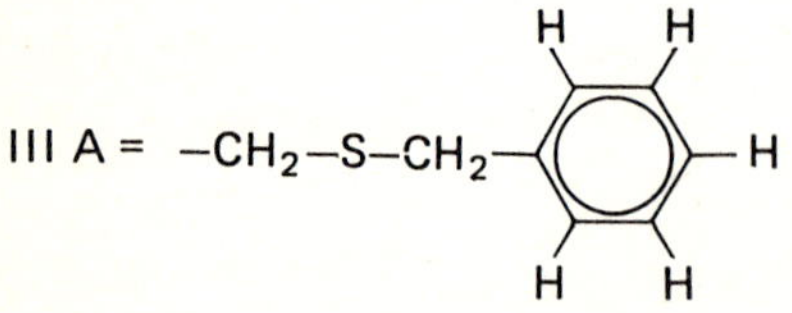

Fig.13.14. Structural formulas of five hexa-derivatives of benzene, behaving as host molecules in packing complexes

possessing a specific geometry is the main finding of these investigations. The host molecules form a rigid framework bound by the usual van der Waals forces. But this framework has large voids and hence a structure with empty cavities is not realized. When the voids are filled by small molecules suitable in shape and size, excellent highly ordered crystals are formed. The cavities possess a definite geometry and therefore the host molecules are capable of forming crystals with those molecules which fit the cavities. As a result, the intermolecular distances are brought to their equilibrium values.

Especially elegant are the experiments on solvent mixture crystallization. For instance, when crystallizing from an equimolecular ortho- and para-xylol solution, compound II absorbs 90% of o-xylol and 10% of p-xylol from the solution. In certain cases, it turned out that these insignificant differences between the forms of the guest molecules are sufficient to make the formation of the packing complex impossible.

We are interested in the crystallographic approach. Certainly, the decisive role of the packing factor in creating these complexes is obvious from comparing the form of the cavity and the character of arrangement of the molecules in them. In all cases, without exception, the guest and host molecules are located at the usual equilibrium distance from each other. Nevertheless, it would be desirable to demonstrate the idea which runs through our book, viz., that packing complexes are formed, provided the increase in density is positive. We have little information at our disposal to achieve this aim.

In [13.35] we find evidence of a crystal structure whose host framework is constructed from molecule I of Fig.13.14. One unit cell accommodates one host molecule and two guest molecules of carbon tetrachloride. The crystals have high symmetry: the space group is R 3, the cell volume 1217 $Å^3$. In the absence of solvent the substance yields a triclinic crystal. Unfortunately, the cell sizes are not given. The volume per single CCl_4 molecule equals 160 $Å^3$. Two molecules fill in 26% of the cell volume. It is clear that the framework constructed from molecule I cannot exist if the voids are not occupied.

In [13.36] we find crystallographic data on compounds III, IV, and V. These data are summarized in Table 13.1. We are able to calculate the factor Δ for compounds V. In the absence of a crystallization solvent the volume of one molecule is 1383 $Å^3$. This substance, when crystallizing with dioxane, produces an orthorhombic cell. One host and one guest molecule have a volume of 1535 $Å^3$. From data on dioxane density we find that the volume of one solvent molecule is 135 $Å^3$. The factor Δ is 0.02. The increase in packing density is not large. This can explain why perfect crystals which do not contain a solvent can form in this case. However, it may be that the crystal structures are absolutely different for filled-in or empty cavities.

Table 13.1. Some selected crystal data for three solvated hexaderivatives of benzene (III, IV, V) and for one unsolvated derivative (V) (for notation see Fig.13.14)

Host	Host:guest ratio	Space group	Lattice parameters
III	1.4 - dioxane (1:1)	$P2_1/c$ monoclinic	$a = 10.542$, $b = 20.863$, $c = 12.496$ Å, $\beta = 95.48°$, $Z = 2$ (host), $V = 2735.7$ Å^3
IV	p-o xylene (1:1)	$P2_1/c$ monoclinic	$a = 9.62$, $b = 15.45$, $c = 22.72$ Å, $\beta = 111.0°$, $Z = 2$ (host), $V = 3152.6$ Å^3
V	none	$P2_1/c$ monoclinic	$a = 8.66$, $b = 23.69$, $c = 13.50$ Å, $\beta = 92.9°$, $Z = 2$ (host), $V = 2766.1$ Å^3
V	1.4 - dioxane (1:1)	Pcab orthorhombic	$a = 18.67$, $b = 14.18$, $c = 23.22$ Å, $Z = 4$ (host), $V = 6147.3$ Å^3

It is interesting to compare the volumes of the complexes made from compounds III or V and dioxane. The corresponding figures are 1368 Å^3 and 1535 Å^3, respectively. The substitution of one methyl group for one hydrogen atom brings about an increase in the intrinsic host molecule volume by 21.5 Å^3. There are six such groups. Hence the intrinsic molecular volume of compound V is greater by 129 Å^3, and therefore the volume in the crystal should increase by about 185 Å^3. The 1535-1368 Å^3 difference is slightly less than this figure. The molecular density packing in both cases is practically the same.

Benzene hexaderivatives of the type discussed may form various packings adapting to the form of the solvent molecules. In turn, the solvent molecules "fit" their conformation of the sizes and forms of the cavities. An elegant example of this kind is given in [13.37]. A packing complex with a linear guest molecule has been found.

Consider now the packing complexes investigated by ALLCOCK et al. [13.38] formed by rigid molecules whose central six-membered ring consists of phosphorus and nitrogen atoms. Chemical formulas are shown in Fig.13.15.

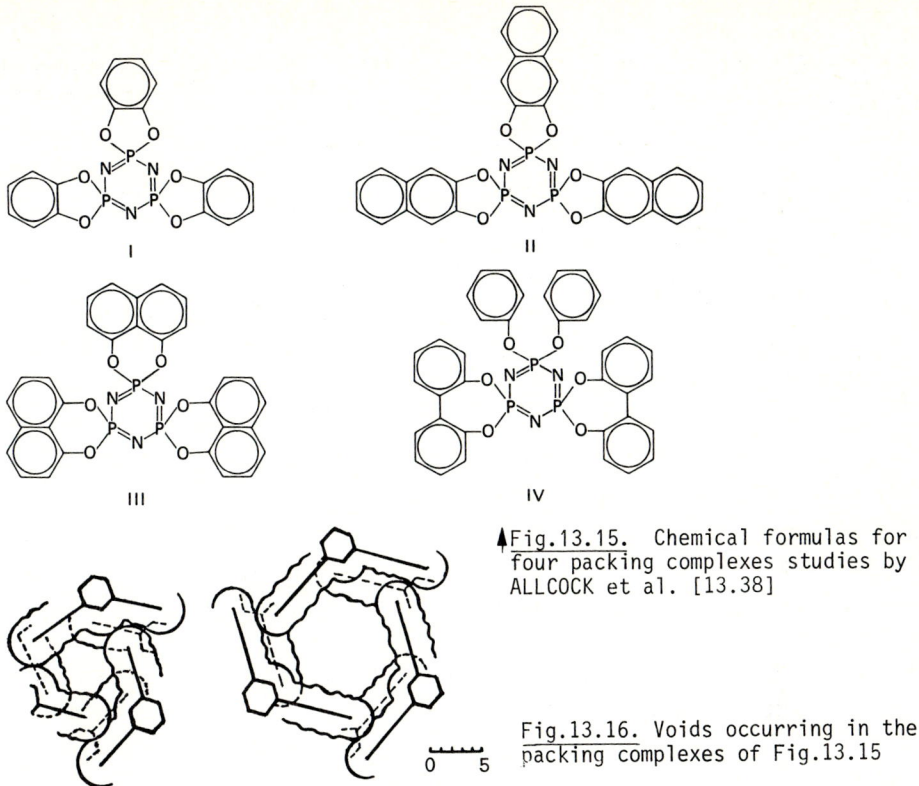

Fig.13.15. Chemical formulas for four packing complexes studies by ALLCOCK et al. [13.38]

Fig.13.16. Voids occurring in the packing complexes of Fig.13.15

No data on the structure of the crystal built from only the host molecules has been found. It cannot be ruled out that in this interesting case such crystals can be formed, since the host framework is very rigid. The framework and the resultant hexagonal cavities are shown in Fig.13.16. Different molecules from those presented in Fig.13.15 form cavities whose diameters are from 5 to 10 Å. The authors indicate that crystals with the same framework can yield complexes with such diverse compounds as tetralin and ethanol, molecules that differ considerably in size. Small molecules fill an insignificant part of the voids and large ones, up to 60% of the voids. If this is the case, the framework is very rigid and the possibility that solid solutions exist in this class of compounds is not ruled out.

Another interesting specific feature of these complexes is the manner of their formation. They can be prepared in the usual fashion, namely, by crystallization from the solution, but guest molecules can also be captured by the pure host crystals. It is not quite clear whether recrystallization, i.e., rearrangement of the host molecules, takes place in this case, or the crystal with cavities permits the matrix to be filled with guest molecules via diffusion.

In the above-cited paper an exhaustive X-ray structural analysis was carried out for the complexes of the molecules denoted as I in Fig.13.15 with benzene, mesitylene, and xylene derivatives. All these complexes are isomorphous and differ only slightly in cell parameters. The space group is $P6_3/m$. The cell parameters of the complex with benzene are $a = 11.915$ and $c = 10.054$.

For the most part the guest molecules can undergo a relatively free rotation inside the cavity. The removal of a guest molecule (to a certain extent) without destruction of the framework is also observed. The guest molecules can replace one another; small molecules can yield crystals in which one cavity can be occupied by two molecules. The authors point out that if the cavities of a hexagonal framework cannot be filled with guests, the framework is destroyed and a triclinic crystal is formed. Further study of these systems will be undoubtedly of great interest.

Quite recently, a symposium was held on clathrates and inclusion compounds, the main communications being reported in [13.39].

13.3.6 Non-Inclusion Complexes

There are a great variety of packing complexes, i.e., complexes where the intermolecular interactions are of the same nature as in the pure components, which cannot be classified as inclusions. This is because we cannot say with confidence which molecules form the matrix lattice and which must be treated as guests. Both molecules involved can be equal in size. The structures of the pure components substantially differ from the complex.

A great variety of complexes of this kind can be found among the molecules forming hydrogen bonds. The reason is oborous. The formation of the crystal structure in such cases is governed by two factors; namely the more hydrogen bonds, the better, but, at the same time, the molecules must pack as closely as possible. These two principles can interfere with one another. To satisfy them both, the complexes must have a simple stoichiometric composition, namely 1:1 or 1:2 [13.40].

13.4 Layered Complexes

There are numerous crystals packed in layers. The layers can represent planar nets of atoms, as in graphite. Bulky layers of a considerable width are encountered in silicates, the largest group of such compounds. Pyrophillite, a typical representative of this group, has a layer about 10 Å wide. Among layered crystals are mica, talc, and muscovite. The structures of layered silicates can swell when the compound is immersed into a liquid. The molecules of the liquid can be incorporated in different amounts between the layers of

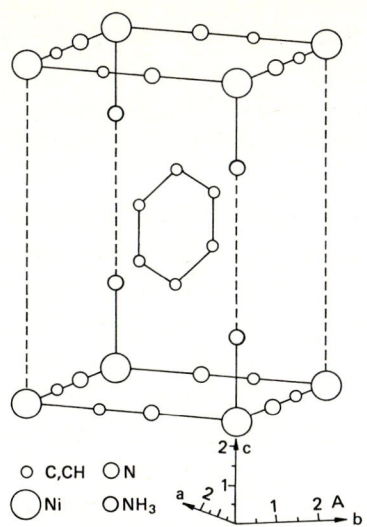

Fig.13.17. Structure of Ni(CN)$_2$NH$_3$ · C$_6$H$_6$ nickel cyanide complex layered with benzene

the minerals. Not only inorganic compounds but also organic clays are members of this family (for instance, a well-known class of alkylammonium clays).

The swelling of a compound due to penetration of a liquid between the layers leads to the formation of systems which, strictly speaking, cannot be considered as homogeneous complexes. Therefore we shall not dwell on specific problems of the structure of layered minerals capable of swelling but shall restrict ourselves to considering several examples where the molecules or atoms incorporated between the layers are ordered, and the complex has definite crystalline unit cell. We shall consider the complexes which seem to be most typical: complexes of cyanides and complexes formed by graphite with various metals and salts.

Figure 13.17 illustrates the structure of the 1:1 complex of diamminonickel (o), tetracyanonickelate (IV) with benzene Ni(NH$_3$)$_2$ Ni(CN)$_4$ C$_6$H$_6$. The Ni^{+4} cations, the cyanide anions, and the zerovalent nickel atoms form a square planar lattice; the bonds between the zerovalent Ni atoms and the ammonia molecules are normal to that plane. All these atoms form the "inorganic net". To our knowledge, nickel cyanides having such a structure give crystals only when a guest molecule (such as *benzene*, Fig.13.17) can be arranged between the nets. Without molecules incorporated between the nets, it is impossible to create densely packed layers since the NH$_3$ molecules would have to face one another. The structure of the complex of this layered cyanide with benzene molecules (Fig.13.17) was determined 30 years ago and is described in detail in [13.40]. We shall give here only the data necessary for calculating the packing density. The crystals of the complex are tetragonal space group P4/m with the cell parameters a = 7.24 and c = 8.28 Å. The benzene molecule is packed between four NH$_3$ molecules. Compounds with the formula Ni(CN)$_2$NH$_3$ · M are known, where M is a thiophene, pyrrole, furane, pyridine, or aniline molecule. All these complexes are isomorphous.

The volume of the cavity formed between two nets of this complex is such that only aromatic molecules with a molar volume less than 90 cm^3/mole may be incorporated into the cavity. And in fact *toluene* and xylene molecules with a volume of 106 and 123 cm^3/mole, respectively, do not form inclusion compounds. However, LEICHESTER and BRADLEY [13.41] found that an inclusion compound does form with biphenyl molecules whose volume is 130 cm^3/mole. It seemed that the structure of this complex differed from that of the complexes already known. IWAMOTO et al. [13.42] determined the structure of the 2:1 *biphenyl* inclusion compound of diamminonickel (o) tetracyanonickelate (IV). They grew single crystals of composition $Ni(NH_3)_2Ni(CN)_4 \cdot 2C_{12}H_{10}$ by mixing equimolar solutions. The crystals were grown for half a year. The crystal has a tetragonal cell with space group 1422 and two formula units in a cell (m.w. 564). The cell parameters are a = 7.42 and c = 25.30; the density is 1.40 g/cm^3.

Despite expectations, for the biphenyl complex the inorganic nets are very similar to those in complexes with benzene. The biphenyl molecule is arranged between the nets in such a way that its long axis is normal to the planar nets. The interatomic distances between the atoms of biphenyl and the net as well as between the atoms of the adjacent biphenyl molecules exceed slightly 3.5 Å. The environment of each of the phenyl rings in the biphenyl molecule is about the same as that of benzene in the corresponding inclusion compound. The size of the crystal axis in the complex with biphenyl is much greater than in the complex with benzene molecules. Not only are the inorganic nets moved apart, but they are arranged in a way suitable for forming a compound of sufficient density.

The interlayer distances in the inorganic nets differ to a great extent. The shortest distance (7.98 Å) was observed in a compound with pyrrole. In a compound with biphenyl this distance increases up to 12.65 Å.

As is known, in the crystal pure biphenyl is planar and in a gaseous phase the rings are twisted by an angle of 41.6°. In the present complex the phenyl rings form an angle of 33.2°. The reason for this rotation is similar to that responsible for the rotation of the rings in a gaseous phase. As in the case of a gas, the rotation of the phenyl rings does not alter the crystal symmetry and at the same time it decreases the steric hindrances between the ortho hydrogen atoms.

The *inclusion compounds of graphite* have been studied for many years, and reviews devoted to the structure of these interesting compounds are numerous [13.43,44]. Graphite reacts vigorously with gaseous or liquid alkaline metals. The metal atoms move the graphite net apart and always lie above the centers of the hexagons formed by the two-dimensional sheets of graphite C atoms. Compounds of many carbon-to-metal stoichiometric ratios have been prepared. The different compositions result from differing numbers of interlayer voids being filled with metal atoms. Thus, for instance, in the case of *lithium*, compounds with 6, 12, 18, 36, and 72 carbon atoms have been obtained. In LiC_6 the Li atoms intercalate between every two adjacent layers ("stage 1 compound"); in LiC_{12}, between every other two layers ("stage 2 compounds"); and so on. In the simplest compound (LiC_6) the layers are moved apart from 3.35 Å, which is an ordinary distance for pure graphite, to 3.73 Å. In LiC_{12} the lattice period is 7.03 Å, which equals the sum of the distance between two superimposed graphite layers and the distance between the graphite layers including a metal layer. The compounds of graphite with other alkaline metals behave in a similar way. The compound with *sodium* is an exception. The only complex obtained in this case is that with 64 carbon atoms per sodium atom, i.e., each eighth layer of carbon atoms shows intercalation.

The structure of graphite compounds with *alkali metals* is reviewed by NOVIKOV and VOLPIN [13.45].

METZ and HOHLWEIN [13.46] studied in detail the compounds of graphite with *ferric chloride*. The authors have shown that the lattice periods, normal to the layers, equal 9.38, 12.73, 16.08, and 19.43 Å when one, two, three, or four graphite layers, respectively, are arranged between two layers of ferric chloride. The structure of layered complexes is only partially ordered although there is a tendency towards the formation of strictly ordered crystals. The authors proposed that the sequence of the graphite and ferric chloride layers be characterized by the ratio s, the number of graphite layers per ferric chloride molecule. It is possible to obtain a compound with any s. If the structure is completely ordered, s is an integer. The authors have shown that the structure where two layers of ferric chloride are located next to each other is impossible.

CHUNG [13.47] has compared the results of studying the inclusion compounds of graphite with metals and molecular compounds. Even if the molecules have simple structure as, for example, *halogens*, a complex structure arises between the graphite layers. A simple arrangement of atoms in the centers of hexagons of carbon atoms becomes impossible. Defects of all kinds in the main structure play a great part in the formation of graphite inclusion compounds. Some authors observed the existence of *domains* with different degrees of ordering for ferric chloride.

BUSCARLET et al. [13.48] conducted an interesting study of the inclusion compounds of *titanium halide* derivatives. The X-ray structural data indicate that an ordered crystal may form. The distance between two graphite layers upon incorporation of a *titanium tetrafluoride* molecule must be 4.77 Å. The authors failed to prepare the inclusion compounds with titanium tetrachloride.

At the same time it is known that *zirconium* and *hafnium tetrachlorides* form layered complexes.

It should be noted that the results obtained for layered graphite complexes can be interpreted in a variety of ways. It is not quite clear yet which cases are thermodynamically stable compounds and in which cases the included molecules will be gradually removed from the complex. It is sure that the complexes cannot be classified as packing complexes although the role of van der Waals interactions in the process of their formation is far from insignificant. The fact that graphite layers can act either as *donors* or as *acceptors of electrons* complicates the interpretation of the existence of the complexes. All possible additional bonds may arise and cannot be ruled out a priori. In the case of complexes with alkaline metals we may speak of both ionic and metal bonds. Chemists often consider layered complexes with graphite as complexes with *charge transfer*. In some studies their formation is explained by analogy with molecules of the ferrocene type.

14. Polymers

Molecules which contain at least several hundred atoms are called *macromolecules*. We encounter them in nature — cellulose, proteins — and synthesize them in laboratory — polyethylene, nylon. A molecule is called *polymeric* if it is constructed of a small number of similar links (*monomers*), which are repeated in an orderly random fashion. The monomeric links can be small, e.g., the CH_2 group is the elementary link of a polyethylene molecule, but they can also be rather large, containing scores of atoms.

The links of a polymer are connected by covalent bonds. In the simplest case the links are connected linearly. But arbitrarily branched polymeric molecules and three-dimensional frameworks can also be synthesized. Due to the great variety of schemes, according to which the links can be connected, polymers can be found with quite different properties.

Solid polymers have extremely wide application in industry, and so they should not be ignored when problems in solid-state physics are studied. It often happens, however, that they are neglected because a solid polymer fits poorly within the conventional theory of phase equilibria and the statistical physics of solids. A solid polymer can be crystalline to a certain degree. Sometimes its structure approaches a perfect single crystal, but in other cases it should unconditionally be considered as an amorphous body. The individual polymer molecules within a given polymeric substance usually do not have an identical number of links, i.e., do not have a definite molecular weight. This leads to the fact that transformation points are blurred and turn into transformation regions. The mechanical and several other properties of such polymers as rubber have no analogs in the world of low-molecular-weight crystals. Nevertheless, the structural features of polymer molecules enable one to extend to solid polymers the main laws which govern the behavior of other solids, particularly organic solids. The analogies between the behavior of polymers and low-molecular-weight organic substances, and polymers and metallic alloys, inspire new experiments and the extension of theoretical concepts developed in one field of science to the other.

The physics of polymers and principles of their structure have been considered in hundreds of books. Referring the reader for details to certain fundamental monographs [14.1-6], we shall say only a few words about the structure of polymers and then choose several problems that are directly related to the subject of our book, the structure of mixed systems.

14.1 The Structure of Polymeric Materials

14.1.1 A Polymer Molecule

The properties of an individual polymer molecule

-A-A-A-A-A-A-

are determined by the structure of the link A and the way the links are connected to one another. If the bar between the links means an ordinary single (sigma) bond between two atoms (and only in this case), the molecule possesses *flexibility*. One part of the molecule can *freely rotate*, relative to the other part, about the ordinary bond.

The possible rotations of a molecule about thousands of bonds permit the molecule to take on very fanciful shapes. It can form planar arrangements, spirals, coils .... However, only in rare cases is the rotation about an ordinary bond completely free. The configuration of an elementary link may hinder this rotation, because atoms of adjacent molecules may come up against one another while rotating *(steric hindrance)*.

The repulsive forces between nonbonded atoms increase rapidly as the atoms approach one another. The rotation of one part of a molecule relative to another part becomes impossible if this requires energies considerably in excess of 10 kcal/mole. The repulsion energy attains such a value when the distance between the atoms is about 0.6 of the equilibrium distance.

The flexibility of a molecule and, correspondingly, its ability to take various shapes *(conformations)* will increase as the allowed range in angles of rotation about ordinary bonds increases. The molecules whose parts can rotate by 360° possess quite specific properties.

In the case where a group of atoms -A-B is the polymeric link, one may write the formula of a polymer as

-A-B-A-B-A-B-A-B- .

If the groups A and B (or a larger number of links) follow one another in disorder,

-A-B-B-A-B-A-A-A-B-A-B-

this affects the flexibility of the molecule as well as the properties of a solid formed from such molecules.

14.1.2 Packing of Polymer Molecules

A condensed system is formed according to the close-packing rule of particles. The most favorable is the formation of molecular packets. If the molecules are identical in their structure, i.e., the groups A and B alternate regularly, *highly ordered segments* will appear which can tentatively be called *crystalline*. The formation of a perfect crystal is hindered by the difference in the molecules' lengths and, of course, by the kinetic factor. Actually, when parts of two molecules fit each other according to the projection-to-hollow principle (key to the lock), they "do not know" the distance between their ends. This is probably the reason for rather frequent occurrence of crystals formed by folded molecules.

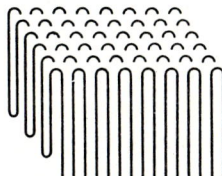

Fig.14.1. Packing of linear polymer chains

Figure 14.1 demonstrates the packing of linear polymer chains, clearly seen under a microscope, that form single crystals. Several links of the chain are involved in the folding of molecules that takes place at the crystal boundary. Naturally, such folds cannot be formed if the chains are bulky. The simplest polymers: *polyethylene, polypropylene,* and various *polyamides*, form faceted crystals. The molecules are folded every 100 - 150 Å. An elementary layer of the crystal is formed by a large number of molecules. In the course of such an arrangement a molecule fits itself to the preceding one.

It does not follow from what was said above that crystals must be planar in shape. The mechanism of *spiral growth*, well known for low-molecular-weight substances [14.1], permits the formation of bulk crystals faceted with *steps*, whose height corresponds to the length of a rectilinear segment of the fold.

If a polymeric crystal is heated, the length of the *rectilinear segment* increases. This shows the possibility of longitudinal *sliding* of chains relative to one another, an argument in favor of a ribbon structure of a crystalline lamella. The crystal grows by the wrapping of the ribbons around its side faces.

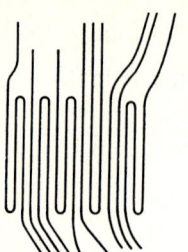
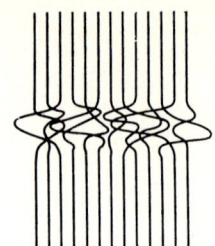

Fig.14.3

Fig.14.2

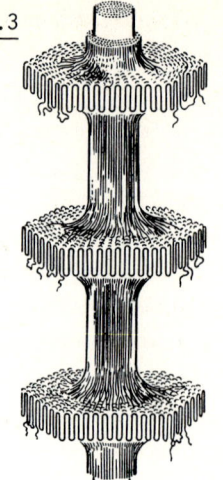

Fig.14.2. Possible structure of linear polymers crystallized from solutions

Fig.14.3. Packet of macromolecules with alternating crystalline and amorphous regions

Fig.14.4. Schematic representation of the "shish-kebab" structure

Fig.14.4

Some polymer crystals are obtained by crystallization from solutions. These can be represented by the model of Fig.14.2. A polymer consists of more or less extended linear segments passing into folded formations. The ordered segments inevitably alternate with low-ordered domains (*amorphous regions*). On pulling, folded chains are likely to unfold to give molecular packets with alternating crystalline and amorphous domains (Fig.14.3). The tendency to order is manifested in that the alternation of crystalline and amorphous domains has a fairly definite *period*. Such structures also arise upon crystallization from a flowing melt.

The structure's periodicity is determined by small-angle X-ray scattering and electron microscopy. The electron microscope made it possible to observe peculiar systems in which the same long-chain molecules participate in the formation of alternating linear and folded segments. The diagram depicting such a packing is shown in Fig.14.4. These formations are known in the literature as the "*shish-kebab*" structures.

The characterization of a polymer can be greatly simplified by using only one numerical parameter, the *degree of crystallinity*, which indicates the volume fraction occupied by ordered segments. To describe in detail the structure of a partially ordered polymer, one has to indicate the mutual arrangement of crystallites, the way they are fitted to one another, and the shapes of the crystalline and amorphous regions. It is also essential to know the fraction of molecules that pass from a crystalline region to an amorphous region and then to the next crystalline region.

Crystallites can be organized in very different ways. They can be arranged in the form of a *step-ladder*, but most frequently they form *spherulites*. Spherulites are observed under a polarizing microscope in shapes similar to a *Maltese cross*. It is difficult to suggest an exact model for a spherulite structure. One may say, at any rate, that they consist of lamellas interconnected by common chains. The space between the lamellas is filled with an amorphous material.

A crystallite can also be internally ordered in different ways. We have already said that crystalline segments form polymer chains constructed of regularly repeated similar links. But a certain order can also be obtained for chains, in which links follow one another randomly. This is *azimuthal disorder*. In some cases any azimuthal position is equally probable. This leads to the structure which can be described as *hexagonal packing of cylinders*, which represents an averaged figure of rotation of a chain about its axis. Such hexagonal packings are widespread and can be encountered in the case of very complicated polymers constructed of bulky links, containing dozens of atoms (rotation-crystalline state; see [14.7]). Besides the azimuthal disorder, other types of disorder are also possible, namely, large displacements of chain axes from positions corresponding to a perfect lattice, inclinations of chains, etc.

The low order of polymer crystals is clearly seen in an X-ray pattern of a *stretched polymer*. Although such an X-ray pattern consists of more or less distinct reflections, their number is very small. An X-ray pattern of 10 - 20 unique reflections is considered excellent. From the geometry of the diffraction spots one can (using packing arguments or, better still, calculations by the atom-atom potential method) not only determine the size of a unit cell, but also roughly define the polymer structure.

The most important parameter determined from an X-ray pattern of a stretched polymer is the *repeat period* along the molecular chain. If the chemical structure of a polymer is known, we can determine the *degree of spirality* of a polymer chain, using interatomic distance data.

What is an *amorphous polymer?* Like any system of particles devoid of order it can only be characterized by parameters, governing the material homogeneity. An amorphous polymer may be assumed to contain regions where chains are coiled, form spirals, distorted lamellas, ribbons, curved packets of molecules, etc. The disordered microstructure of an amorphous polymer can be observed under an electron microscope. Yet, one is usually restricted to qualitative structural characterization.

At high temperatures an amorphous polymer possesses a high elasticity, which is close to that of rubberlike materials. As the temperature lowers, the polymer undergoes a *glass transition*, i.e., becomes a solid. The transition occurs in a narrow temperature interval, so one can speak, conditionally, about the temperature of the polymer glass transition. The homogeneity of an amorphous polymer can be judged by a gradual change in its properites with temperature and sometimes by its transparency.

14.2 The Structure of Polymer Blends

A homopolymer is also, in a certain sense, a two-phase system because it consists only partially of crystallites. However, it does not possess most of the properties typical of a low-molecular-weight two-phase system.

This will be demonstrated quite clearly in Sect.14.3, which deals with block copolymers. There the reader will see that the terminology of the physics of low-molecular-weight systems cannot be applied unambiguously to polymers. It will be quite evident that a polymer, consisting of various links, can be considered as two-component system both with respect to the molecules and with respect to the links.

Now, primary attention will be given to two-component systems constructed of two kinds of molecules: -A-A-A-A- and -B-B-B-B-. The possible types of polymer blending can be divided into the following categories:

    1) both polymers are amorphous,
    2) one of the components can crystallize,
    3) both components can form crystals.

14.2.1 Both Polymers are Amorphous

Three cases are possible. In the first case, molecules of the components are *mixed randomly*. If we consider an amorphous structure, the similarity of molecular configurations may not in this case play any important role. It may seem at first sight that almost all amorphous polymers should form solid solutions, i.e., exhibit mixing of molecules; this, however, seldom occurs [14.8-10]. This may be explained as follows. For amorphous polymers, as well as for crystalline ones, the packet model is valid. If it is so, only in rare cases will very small gain in entropy lead to lowering of the free energy upon mixing of even slightly different molecules. In the rare cases when it does occur, how can one prove that an amorphous two-component polymer is a solid solution? Several ways can be suggested. First of all, the glass transition temperature is a characteristic function of the polymer composition. In the case of a solid solution we should obtain a smooth curve, the glass transition temperature satisfying the FOX equation [14.11]

$$T_g^{-1} = w_1 T_{g_1}^{-1} + w_2 T_{g_2}^{-1} \quad . \tag{14.1}$$

Here T_g, T_{g_1}, and T_{g_2} are the glass transition temperatures for the mixed crystal and for the components.

The second way lies in measuring small-angle X-ray diffusion scattering. For a uniformly arranged solid, DEBYE et al. [14.12] have shown that the reciprocal of the square root of the scattered X-ray intensity should be proportional to the squared sine of the scattering angle. Finally, light scattering is also sensitive to the system homogeneity.

STEIN et al. [14.13] present convincing evidence that in a particular concentration range a homogeneous solution occurs in an amorphously mixed polyvinyl chloride-polycaprolactam polymer.

The second type of structure that occurs on mixing two amorphous polymers is a *two-phase mixture of amorphous domains*, each containing molecules of one kind. This type of structure also includes polymers formed according to the matrix-with-inclusions principle. Such a morphology can be revealed by electron microscopy. Interconnection of the domains or interconnection of the domains with the matrix is most likely to occur via partial penetration of the chains of one component into a domain constructed of molecules of the second kind.

Finally, the third case is a *complete incompatibility* of polymers, i.e., the impossibility of obtaining a domain structure in a natural way (say, cooling the melt of a mixture). This is the case where polymers do not wet one another. Evaporation of a solution of two polymers will result in separate segregation of the two polymers.

14.2.2 One Component Crystallizes

Here we can also list a priori three possible variants. The first variant corresponds to *crystalline formations*, lamellas, large crystallites, and spherulites, included into an amorphous matrix. The second variant consists of *spherulites*, the amorphous component *filling the space* between crystalline lamellas of the spherulite. Finally, in the third case the sample, as in the preceding variant, consists of *spherulites*, but the amorphous component forms *elongated domains* inside the spherulite.

Electron microscopy proves to be very useful in deciding to which variant a given mixture belongs. But only by measuring large periods on X-ray patterns and using TSVANKIN's theory [14.14,15] can one determine whether amorphous material is included between crystalline lamellas. The theory permits measurement of the linear crystallinity of a spherulite, i.e., the ratio of the crystal thickness to the repetition period (to a large period). A linear variation of a large period with the thickness of the crystalline lamella remaining unchanged was shown for one of the systems studied [14.13]. This proves unreservedly that amorphous material is included between the lamellas. WARNER et al. [14.16] showed for another system that the addition of an amorphous component did not change the X-ray pattern at small angles; both the repetition period and the size of the crystalline region remained unchanged. Thus the second variant of the structure should be rejected. The last question is whether segments of the amorphous polymer are located inside or outside the regions occupied by the crystalline component. The answer to this question was obtained by determining the percentage of the crystallinity of the mixed system and was confirmed independently by microscopic observations.

KEITH and PADDEN [14.17] showed that the structural variant of a given mixed system could be predicted if the diffusion coefficient of the amorphous component into the crystalline region of a polymer was known. The ratio of the diffusion coefficient to the growth rate of a crystal governs the fate of the amorphous material: at large diffusion coefficients and low growth rates the amorphous material remains inside the spherulite.

14.2.3 Both Components Crystallize

In this case the number of possibilities naturally increases because the system contains an amorphous phase which may have two different structures (Sect.14.2.1), and crystals of two components.

The formation of true solid solutions is possible, in principle, i.e., the formation of two-component crystals. Such situations are rare, however, and can be expected only when polymer molecules differ insignificantly; for instance, the fluorine atom in one molecule is at the same site as the chlorine atom in a molecule of the second component. The possibility of forming true solid solutions in the crystalline region of a polymer should be analyzed by using packing arguments or by the atom-atom potential method, as was done in the analysis of solid solutions of low-molecular-weight organic substances.

Even after excluding the formation of true solid solutions in crystalline lamellas, we must choose one structure out of a large number of possible structures. Crystals of both components can be dispersed in either a homogeneous or an inhomogeneous amorphous matrix. One or both components can form superstructures of the spherulite type. A spherulite of each of the components can represent an individual formation. Yet, there can be spherulites containing lamellas of both components.

The problem is a complicated one, but it can be solved again by using X-ray scattering, optical scattering, and electron microscopy. References to particular studies can be found in [14.13].

14.2.4 Some Remarks on the Technology of Polymer Blends

In recent years the focus in the search for new polymeric materials has shifted. Formerly, technologists expected from chemists new molecules, homopolymers and copolymers, but at present physicists are asked to produce new materials by means of *mechanical mixing* of components.

Such new materials may be *new composites of polymers*, which do not form true blends. A certain amount of progress has been achieved in combining various elastomers; for example, fibers woven of cotton and polyesters are utilized in the textile industry. Mechanical mixtures of various elastomers and plastics can be obtained. Such systems are, naturally, thermodynamically unstable, so a researcher should help a technologist in developing a manufacturing process that will lead to a final product which has no tendency toward delamination. Interweaving of fibers, dispersion of particles of one component in the continuous medium of the second component, the use of adhesion properties of materials, ... , the list of possible techniques could be continued.

But better new mixed materials should be the *thermodynamically stable mixtures*. This explains the extensive search for conditions under which a polymer blend possesses a *negative free energy of mixing*. It has been admitted rather unanimously that the main role in the formation of stable mixtures is played by energy, rather than by entropy. But in what case should one expect a decrease in energy when two polymeric substances are mixed? Almost in all studies that seek to predict miscibility, the interaction of polar groups is taken to be the main factor. Few researchers have paid attention to the fact that in some cases miscibility is attained at the expense of a change in stereoregularity of a polymer chain. For instance, isotactic polymethyl methacrylate does not mix with polyvinyl chloride, while syndiotactic polymethyl methacrylate does.

It is quite probable that the idea we consistently propound throughout the book also works in the case of polymer blends. If one could find two polymers with the geometries satisfying the key-to-the-lock principle, one would perhaps expect the formation of a system with negative energy of mixing. As usual the magnitude of the short-range part of molecular interaction energy is decisive.

14.3 Block Copolymers

14.3.1 Packing of Molecules

The name *"block copolymers"* is assigned to macromolecules consisting of two links connected end to end as follows:

$$-A_n-B_m-A_n-B_m- \quad .$$

The numbers n and m of links in A and B blocks can be rather large. Chain segments constructed of blocks, containing links of the same sort, have a length of the order of hundreds of angströms.

The configurations of the A and B links may differ considerably. One wonders how close-packed structures of such molecules can be formed. It is obvious a priori that the molecules should have a tendency to pack in such a way that the segments -A-A-A- and -B-B-B- be packed separately. This has proven to be the case. Though the links A and B belong to the same molecule, they occupy different regions in the formation of a solid material. As a result, a very peculiar two-phase system is formed. Each phase is created by the same segments of molecules rather than by individual molecules.

Such a two-phase system can be studied by electron microscopy and X-ray diffraction techniques. The structure of a copolymer whose blocks are polysterene and polybutadiene has been studied in great detail. The results obtained when studying such a solid do not, in principle, differ from those obtained for other systems. Therefore the packing schemes of all block copolymers can be described rather reliably. Our discussion is based on the summary paper by GALLOT [14.18].

There are three types of block copolymers, which form *layered*, *cylindrical*, and *cubic* structures.

a) Layered Structures

The possibility of dyeing polybutadiene with osmium tetroxide permits easy identification of regions occupied by different blocks by electron photomicrographs. The electron micrograph reveals a strictly periodic alternation of dark and bright bands, whose thickness can readily be measured.

The same distances are found from X-ray patterns, which consist of sharply outlined bands with uniform spacings related to integer multiples of the basic interplanar distances. These bands are the reflections of different orders from the same system of planes.

The band widths are on the order of two hundred angströms. They can be measured rather accurately or calculated from the simple formula

$$d_A = d \left\{ 1 + \frac{c(1 - x_A)v_B + (1 - c)\varphi_B v_s}{cx_A v_A + (1 - c)\varphi_A v_s} \right\}^{-1} , \qquad (14.2)$$

where c is the polymer concentration in the solution, x_A is the concentration of the A block in the copolymer, v_A and v_B are specific volumes of the blocks, v_s is the specific volume of the solvent, and φ_A and φ_B are the partition coefficients of the solvent. The formula is simplified if the solvent acts only on the A block (then $\varphi_B = 0$).

The applicability of these formulas and the data of electron microscopy and X-ray diffraction analysis permit the reliable representation of layered polymers by the diagram shown in Fig.14.5. Open circles are the solvent; full circles mark the transition from one block to another. Since the blocks A and B have, in general, different length (d_A, d_B), a strictly periodic structure can be formed only when the molecules are arranged randomly inside the layer. The amorphous layers are alternated with the crystal periodicity.

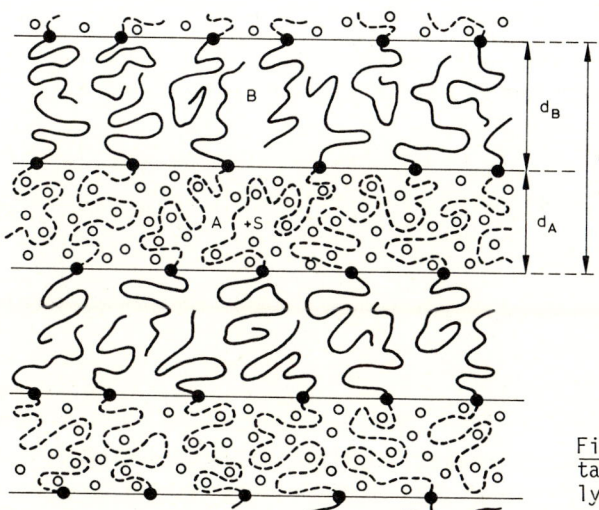

Fig.14.5. Schematic representation of layered block-copolymer structure

b) Hexagonal Structures

X-ray patterns of these block copolymers reveal five distinct lines. The ratio of the squares of reciprocal interplanar distances is 1, 3, 4, 7, and 9. This is typical of hexagonal packing. When dyeing a polymer with osmium tetraoxide, one can obtain two types of electron micrographs: dark circles against a bright background and light circles against a dark background. The distance between the circle centers, which form a perfect hexagonal network, is several hundred Å. The sizes of the circles constitute various fractions of the network period, depending on some block's content in the copolymer. Longitudinal micrographs of these materials exhibit systems of bands. Thus there is no doubt that the structure represents a system of parallel cylindrical formations.

The section perpendicular to the "infinitely" long axes of the cylinders is shown in Fig.14.6. The notations are the same as in the previous diagram. By varying the conditions for forming a block copolymer, one can obtain a structure in which cylinders are filled with B parts of a molecule, but the reverse case is also possible. The solvent (s) may be inside a cylinder but may also occupy the space between the cylinders.

If D is the distance between the cylinder axes and R is the cylinder radius, the following relation holds:

$$R^2 = \frac{D^2 \sqrt{2}}{2\pi} \left[1 + \left(\frac{1 - cx_B}{cx_B} \cdot \frac{v_B}{v_{As}} \right)^{\pm 1} \right]^{-1} . \tag{14.3}$$

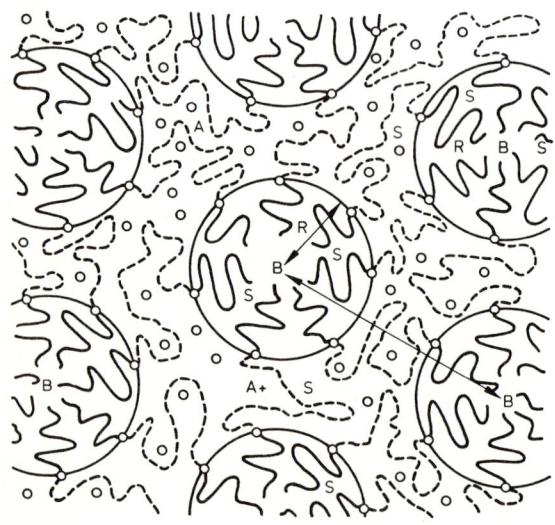

Fig.14.6. Schematic representation of hexagonal block-copolymer structure

The plus exponent after the parentheses is for the case where the cylinders are formed by insoluble B links, and the minus exponent is for the case where A links and the solvent are inside the cylinders.

c) Cubic Structure

The presence of a *cubic structure* — formed according to the same principles as the structures discussed before, constructed of spheres, filled with links of one kind, and implemented into a matrix of links of the other kind — is revealed by electron microscopy and by the ratio of interplanar distances measured on an X-ray pattern. Here we can also encounter two types of cubic structures that differ in the kind of links filling the spheres and in the location of the solvent. Relations for sphere radii, similar to (14.2,3), can easily be derived.

The reference [14.18] reports the values of the structure parameters for various polymers. The same polymer may form one of three structural types, depending on the block-copolymer composition, chemical formula and concentration of the solvent, and temperature.

The regions of structure stability and variation of the parameters as a function of the above three factors have been extensively studied. Structures of the same type can also be obtained for polymers which do not contain a solvent.

Attempts have been made to develop a theory for the types of block-copolymer structures. However, the calculation of free energy in an amorphous region is hardly possible at present. Attention has not been given to the development of rough packing models, which could help in understanding the relative advantage in forming the above-described very interesting structures.

d) Segment Copolymers

There is of course a continuous transition between block copolymers and copolymers consisting of alternating monomeric links. Technologists consider it to be reasonable to consider polymers consisting of two short blocks, one rigid and one soft, as a special class. The degree of polymerization of rigid blocks in this class of materials does not exceed 10. The soft blocks (flexible ones, to be exact) have the degree of polymerization from 15 to 30.

A remarkable fact is that a heterophase system appears in this case too. Short rigid blocks conglomerate to form microcrystals (which cannot always be easily detected because of their small size). The soft blocks manage to become packed rather closely. Among these systems are thermoplastic and polyurethane elastomers, remarkable materials that combine strength and plasticity and find wide application in industry.

We believe that the formation of these very interesting structures, which differ from the above-mentioned block copolymers in random orientation of rigid microcrystallites, may be given the same explanation: a tendency of the polymer chains to pack most closely. Block copolymers are discussed extensively in [14.19].

14.3.2 Deformation of Block Copolymers

To demonstrate the power of X-ray diffraction method in the structural studies of systems with a perfect lattice constructed of large amorphous particles (regularly placed sacks with potatoes), we shall consider in detail a part of the comprehensive work by TARASOV et al. [14.20] concerning the deformation mechanism of single crystals which are hexagonal systems of butadiene-styrene block copolymers.

Investigations were performed with a linear block copolymer (molecular mass $M \approx 8 \cdot 10^4$, weight content of bound styrene 28.3%). The samples were prepared by pressing at 140°C under low pressure. A heated polymer placed at the center of the mold spreads in radial directions, forming a *circular plate* 70 mm in diameter and 1 - 1.5 mm thick. The samples for investigation in the form of rectangular plates (5 - 10 mm long and 1 - 2 mm wide) were cut exactly parallel or perpendicular to the cylinder radius in the region far apart from the center.

A system of cylinders forms a supermolecular crystal, whose dimensions are rather large, and exceed the cross section of the incident beam. The geometry of the X-ray experiment and of the system of cylinders in the supermolecular crystal are explained in Fig.14.7. On the X-ray pattern for the unstretched

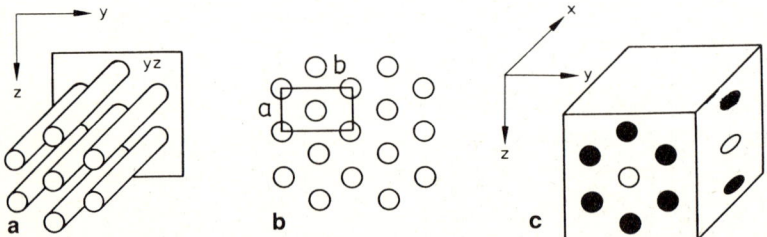

Fig.14.7a-c. Geometry of the small-angle X-ray diffraction study of butadiene-styrene block copolymers, from [14.19]. (a) Hexagonal pattern of circular cylinders of block copolymers, and orientation of sample coordinate system (xyz). The cylinder axes are along x, and the polymer stretching direction (see text) is along z. (b) The unit cell sides a and b (along z and y respectiveley) for a centered orthorhombic cell equivalent to the hexagonal cell. (c) Scheme of the X-ray pattern in the yz plane (hexagonal symmetry, primary X-ray beam along x) and in the xz plane (linear symmetry, primary beam along y)

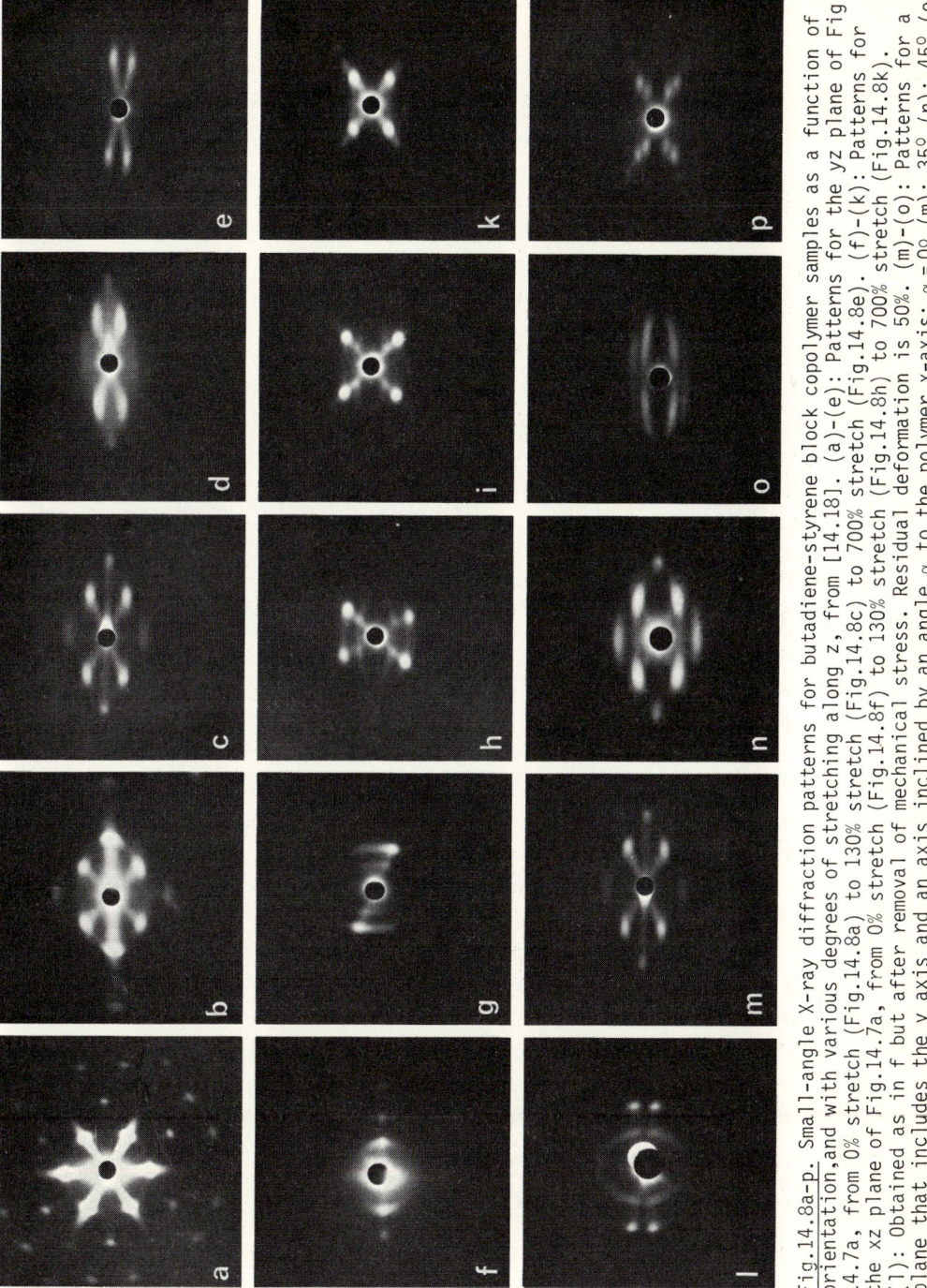

Fig.14.8a-p. Small-angle X-ray diffraction patterns for butadiene-styrene block copolymer samples as a function of orientation, and with various degrees of stretching along z, from [14.18]. (a)-(e): Patterns for the yz plane of Fig. 14.7a, from 0% stretch (Fig.14.8a) to 130% stretch (Fig.14.8c) to 700% stretch (Fig.14.8e). (f)-(k): Patterns for the xz plane of Fig.14.7a, from 0% stretch (Fig.14.8f) to 130% stretch (Fig.14.8h) to 700% stretch (Fig.14.8k). (l): Obtained as in f but after removal of mechanical stress. Residual deformation is 50%. (m)-(o): Patterns for a plane that includes the y axis and an axis inclined by an angle α to the polymer x-axis; $\alpha = 0°$ (m); 35° (n); 45° (o). (p): Patterns for the yz plane inclined by 45° angle to the y axis, $\varepsilon = 700\%$.

333

sample ($\varepsilon = 0\,\%$) of Fig.14.8a the primary X-ray beam is parallel to the base radius of the cylindrical pressed plate. The most intensive reflection corresponds to $d = 220$ Å, and the other reflections correspond to $d = 127$, 110, and 83 Å. The X-ray pattern of the unstretched sample taken along the y axis in the direction normal to the plate (Fig.14.8f) reveals reflections with the same values of d as in the X-ray pattern of Fig.14.8a, but located along one straight line.

The system of cylinders can be seen on electron micrographs of ultrathin slices of the pressed samples (Fig.14.9). The slices are made in the planes xz (Fig.14.9a) and yz (Fig.14.9b). White bands and white circles correspond to polystyrene.

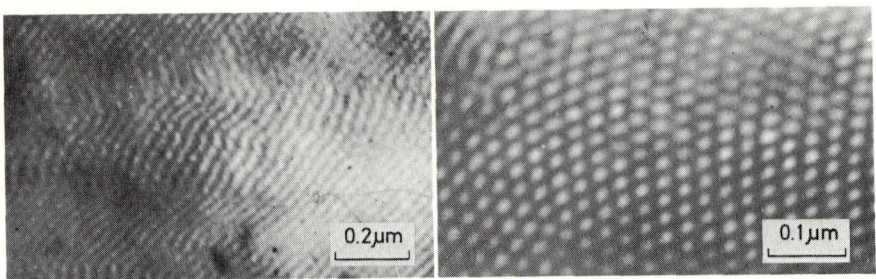

Fig.14.9a,b. Electron micrographs of ultrathin slices of pressed butadienestyrene block copolymer samples in the xz plane (a) and the yz plane (b)

Calculations show that for the unstretched sample the location of the sections of cylinders in the yz equatorial plane of Fig.14.7 corresponds to a plane hexagonal cell with the lattice parameter $a = 255$ Å. Since the hexagonal cell becomes distorted due to deformation, it is more convenient to describe such a structure by a plane centered orthorhombic cell with the parameters $a = 255$ and $b = 440$ Å (Fig.14.7b). It should be noted that X-ray patterns of a polymer obtained at large angles contain only amorphous reflections.

Deformation of samples is accompanied by considerable structural changes. Figure 14.8 also shows a series of small-angle X-ray patterns obtained for different degrees of stretching of the sample in the direction z of Fig.14.7a, i.e., perpendicular to the cylinder axes, and for various orientations of the sample relative to the incident beam. When X-ray patterns are taken along the cylinder axes (primary beam along the x-axis, yz diffraction patterns), the hexagonal network of reflexes becomes more and more distorted with the increasing deformation, and at the same time the intensity of all reflections

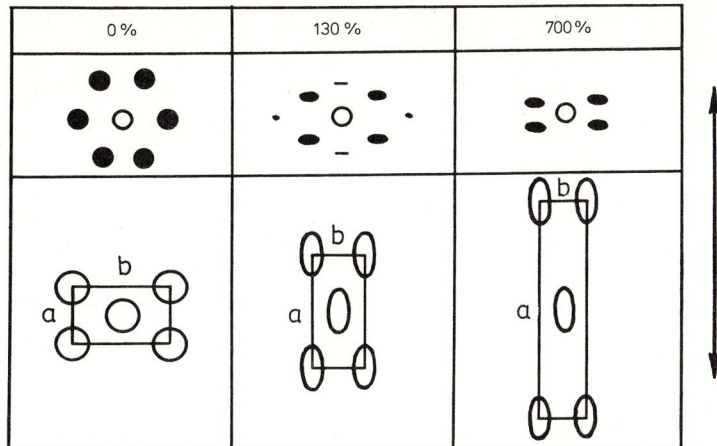

Fig.14.10. The diffraction spots and the corresponding unit cells for butadiene-styrene copolamer samples with different degrees of stretching

decreases significantly ($\varepsilon = 0 \rightarrow 700\%$, Fig.14.8a-e). The periods d along the deformation axis z gradually increase and the reflections corresponding to this direction approach the center of the X-ray pattern. In the other direction (y) the values of d decrease and the spots recede from the center. From the position of reflections in these X-ray patterns one may calculate the variation of the unit cell with increasing deformation. Figure 14.10 shows the diagrams of the X-ray patterns and the corresponding unit cells. The b axis of Fig.14.7b is slightly compressed while the a axis, parallel to the stretching direction z, increases dramatically as the elongation goes from $\varepsilon = 0\%$ to $\varepsilon = 700\%$.

X-ray patterns of a new type which have probably not been observed so far in block copolymers were obtained in the direction perpendicular to the cylinder axes (in the sample xz plane, Fig.14.8f-k). As the stretching is increased from 0% (Fig.14.8f) to 700% (Fig.14.8k), peculiar four-point X-ray patterns arise, the angle α between the cylinder direction x and the line connecting the reflections gradually increasing (Fig.14.11). These patterns show that the general character of the structure changes significantly under stretching, because the diffraction pattern is distributed not in the equatorial (yz) plane alone, as was the case with the initial sample containing a system of parallel cylinders, but in the entire reciprocal space. To calculate the structural changes due to stretching, X-ray patterns in directions intermediate between the xz and xy planes were also obtained (Fig.14.8l-p).

As an example, let us consider in more detail the X-ray patterns corresponding to deformation $\varepsilon = 130$ and 700%. For $\varepsilon = 130\%$ the most intense X-ray pattern of the distorted hexagonal type in the plane xz arises when the pattern is recorded at an angle α = 35° to the initial direction (x) of the cylinder axes (Fig.14.8n). X-ray patterns for other directions in the xz plane (Fig.14.8m,o) are much less intense and distinct, though they do contain distorted hexagonal reflections. This direction (35°) coincides well with angle between the x-axis and the direction towards an intensive reflection on the X-ray pattern at ε = 130% of Fig.14.8h as is shown in Fig.14.11a. A comparison of the orientations shows that the main reflection on the X-ray pattern of Fig.14.8h is the same as in Fig.14.11. X-ray patterns recorded in the yz and xy planes for ε = 700% (Fig.14.8e,k) reveal a loss of the hexagonal reflection network. This loss can be duplicated at ε = 700% by finding a maximum intensity X-ray pattern at an angle α = 55° from the xy plane (Fig.14.8p).

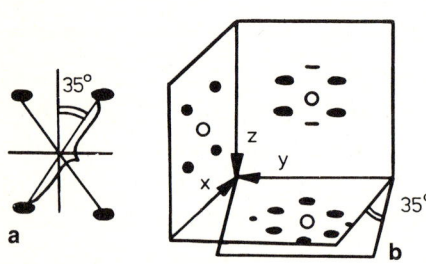

Fig.14.11. Analysis of small-angle x-ray scattering pattern for ε = 130% along z axis. (a) The xz pattern (Fig.14.8h) has orthorhombic symmetry. The two diffraction sports of Fig.14.8f (xz pattern at ε = 0% stretch) have become four spots. The angle 35° corresponds to a tilt of the cylinders of Fig.14.7 from the x direction. (b) Summary of patterns for 130% stretch. Fig.14.8c (yz pattern), Fig.14.8h (xz pattern), Fig.14.8n (patterns declined by α = 35° from xy pattern)

When comparing all the X-ray patterns for ε = 130% (Fig.14.8c,h,n and Fig.14.11), one may conclude that as a result of stretching, the main part of the cylinders has rotated through an angle of 35° from their initial position towards the deformation direction z, the mutual packing in the form of a distorted hexagonal structure remaining the same. The rotation has occurred in the xz plane. As distinct from this main portion of the cylinders, some of them retain the initial orientation with the rotation angles distributed in the range of 0 - 35°. This conclusion is confirmed by the fact that the distorted hexagonal X-ray pattern in the equatorial plane is retained (Fig.14.8c) by similar X-ray patterns obtained between the angles of 0 - 35°, and by continuous distribution of the intensities along a horizontal line passing between the main reflections on the X-ray pattern (Fig.14.8h).

An exactly similar series of X-ray patterns are obtained for other degrees of stretching. In all cases the analysis of X-ray patterns obtained for various directions shows that the main portion of the cylinders is rotated by a certain angle α. The remaining small portion has orientations distributed between this direction and the initial one, perpendicular to the deformation direction.

This distribution takes place only in one plane (xz in Fig.14.7), which passes through the deformation direction (z) and the direction of the initial orientation of the cylinder axes (x). The orientation angle α of the main portion of the cylinders increases with deformation from $\alpha = 0°$ at $\varepsilon = 0\%$ through $\alpha = 35°$ for $\varepsilon = 130\%$ to $\alpha = 55°$ for $\varepsilon = 700\%$.

The most obvious interpretation of such a distribution of cylinder group orientations is that as a result of deformation, the crystal, constructed of closely packed cylinders, is gradually bent in the deformation plane (the xz plane of Fig.14.7), taking the shape of a planar zig-zag.

It is interesting that some irreversibility is characteristic of such a transition of the domains into a zig-zag form. The X-ray pattern recorded in the Y direction after the removal of mechanical stress (Fig.14.8l) reveals some residual fracture of polystyrene cylinders.

The transformation of a linear crystal into a planar zig-zag formation is the first result of deforming a specimen that possesses the structure of a supermolecular single crystal. Another result of the deformation is the distortion of a regular hexagonal packing of cylinders inside this zig-zag type single crystal. In the preceding sections we have determined the changes in the size of a plane orthorhombic cell from X-ray patterns recorded in the x-direction (Fig.14.10). These X-ray patterns and the corresponding unit cells refer to a section in the bent part of the zig-zag formations. The cell deformation in this section reaches considerable values. Denote a' and b' the parameters of the deformed cell. The parameter a' can be found from the X-ray patterns of Fig.14.8f-k. Unfortunately, the other parameter b' is difficult to determine experimentally. Thus structural changes under deformation can be characterized by changes in the parameters a and b of a cell in the section of the bent part (Fig.14.8a-e), by the parameter a' in the section intersecting the straight part (Fig.14.8f-k), and by a change in the angle α of the inclination of the cylinder axis to the initial direction. All these data are listed in Table 14.1.

The percent microdeformation λ_a, λ_b, $\lambda_{a'}$ of the unit cell have been calculated from the parameters a, b, and a', respectively. The results presented in Table 14.1 show that the deformation of the initial hexagonal packing in the straight parts is considerably less than in the bent parts, the difference increasing for $\varepsilon > 130\%$. The maximum deformation along a reaches 330%, while it is only 160% along a'. This result confirms the deformation scheme suggested above, because it is obvious that in bent regions the deformation of the lattice should be much larger than in the weakly deformed straight regions. For

Table 14.1. Cell deformation on stretching the sample perpendicular to the axis of cylindrical domains. $\lambda = [(1-l_0)/l_0] \cdot 100\%$ ($l = a, b, a'$; $l_0 = 1$ at $\varepsilon = 0$)

ε[%]	a[Å]	λ_a[%]	b[Å]	λ_b[%]	a'[Å]	$\lambda_{a'}$[%]	α[deg]
0	255	0	440	0	255	0	0
33	330	30	330	25	-	-	10
130	500	100	250	40	440	75	35
200	720	185	220	50	500	100	46
250	850	235	220	50	-	-	-
500	950	275	220	50	540	110	48
700	1100	330	220	50	660	160	55

deformations not exceeding 100% the cylindrical groups are not bent, and in both sections the deformation is almost the same, amounting to approximately 100% along the axis of stretching (the values of a and a'). When the stretching is large, the cylinders are bent and the lattice deformation in straight regions for the stretching values up to 700% increases only by a factor of 2 ($\lambda_{a'}$), while in bent regions the lattice deformation λ_a increases more than threefold. This, naturally, is related to an increase in the angle α and to a larger deformation in the regions where orientation changes.

The b axis of the cell (Fig.14.7b) is perpendicular to the deformation direction z (Fig.14.7). The compression in this direction is relatively small and reaches a maximum of 1/2 (Table 14.1). Probably, even at such a deformation the compression of the cylinders is very high, and its further increase requires much higher stresses. The data of Table 14.1 permit the determination of the lattice deformation coefficient $k = \lambda_b/\lambda_a$. The parameter k gradually decreases from 0.83 for $\varepsilon = 33\%$ to 0.15 for $\varepsilon = 700\%$.

Thus the stretching of a single-crystal specimen in the direction normal to the cylinder axes proceeds in several stages. At the first stage, under low deformations (< 100%), the orientation of all the cylinders remains unchanged, and only the cell undergoes deformation, which can probably be followed by a distortion of the cylinder cross section. A circular cross section may turn into an ellipse with the major axis located along the deformation direction. It should be noted that at this initial stage of deformation, the recording in the Y direction often yields asymmetrical small-angle X-ray patterns of the two-point type (Fig.14.8g). They show that prior to the second stage of deformation, the orientation may be disturbed and certain parts of the crystal may be deflected to one side or another.

At the second stage of deformation (exceeding 100%) cracks appear in the weakest places of the crystal along its full length, and the crystal acquires the shape of a planar zig-zag. The crystal is bent in the xz plane passing through the cylinder axes (x) and in the stretching direction (z). The straight sections of the zig-zag are directed uniformly to both sides, this being confirmed by a symmetrical shape of X-ray patterns in Fig.14.8h-k. The inclination angle of the zig-zag increases with stretching. The cell distortions increase considerably at the points of bending, but in the straight parts the cell deformation grows very slowly. It is significant that even at very large values of stretching the bent crystal remains a single whole and does not break into several crystallites of smaller size. This is confirmed by the fact that all the X-ray patterns reveal the same set of reflections corresponding to an orthorhombic cell. If the specimen ceased to be single crystalline, the X-ray pattern would exhibit reflections corresponding to several unit cells or to a texture (rings in the case of isotropic structure). Unfortunately, data on the length of the straight parts, and hence on the zig-zag period, are not available so far. The period can be assumed to be rather large, of the order of several thousand angströms.

14.4 Solid Solutions

As has been mentioned above, solid solutions of polymer are seldom encountered and are of no special interest, because a large similarity in the polymer chain configuration is required for the formation of such solutions. Crystalline regions constructed of macromolecules that differ only in that some atom in the molecule is replaced by another atom (close in size) have not been considered by researchers. We can hardly expect any considerable changes in polymer properties in this case. Crystalline regions of polymers usually contain defects and therefore a slight distortion of the lattice which will occur if a bromine atom or a methyl group replaces a chlorine atom is of little interest. If the distinctions in the structure of polymer molecules are more significant, then different polymer molecules will crystallize separately and not form solid solutions.

The formation of solid solutions of polymer molecules in a polymer matrix can be expected, in our opinion, for those macromolecules which form a hexagonal lattice. Rotational-crystalline structures, mentioned in Sect.14.1.2, are formed if molecular axes are packed parallel to one another, but an *azimuthal disorder* exists. An "average" molecule is a cylinder. It seems quite possible that different polymer molecules, which form cylinders of the same radius when they rotate about their axes, can form solid solutions. Molecules of biological origin are likely to form such systems. It would be interesting from this standpoint to investigate all the systems (though their number is fairly large) in which macromolecules form cylindrical packing.

We do not know of such investigations. That is why we shall consider two cases of solid solutions: solutions of polymer molecules in monomers and solutions of low-molecular-weight substances in polymers.

14.4.1 Solid Solutions of a Polymer in a Monomer

Since in a solid the definite positions occupied by molecules severely constrains any conceivable chemical reaction due to heat, light, or radiactivity, it is of great interest to carry out a solid state reaction. Whereas the result of a reaction in solution or gas is ambiguous, in a crystal one of the possible ways is usually presupposed. This is due to a simple reason, namely there is only one possibility for the formation of a new compound.

THOMAS [14.21], one of the leading researchers in the field of chemical reactions in organic crystals, justly points out that the vast variety of close packings of organic molecules enables one to introduce the concept of *molecular engineering*.

A reaction in solution proceeds in a.way which minimizes the reaction barrier. In the cases where there are several reaction routes the reaction will result in several, rather than one, reaction products. This is not the case with a crystal. The reaction will proceed only if the reaction centers are close to one another, and if this condition is satisfied the chemical process will develop in the direction dictated by the molecular packing. THOMAS calls this type of reactions *topochemical* [14.22].

Reactions proceeding in a solid may lead to its destruction, i.e., to the formation of a polycrystalline product. There are, however, interesting examples where as a result of a reaction a single crystal of a compound is transformed into a single crystal of another compound, the shape of the object being retained. Finally, there are cases where a reaction does not go to completion. If the reaction centers are distributed over the entire crystal, a two-phase system can be obtained, but the transformation of a single-component system into a solid solution of two **components** is also possible.

The work done by WEGNER [14.23] has attracted great attention among researchers. He showed that for an interesting class of organic crystals, it was possible to obtain a solid solution of a polymer in a monomer by means of UV irradiation. Two classes of such substances are known at the present time. There is no doubt that if the character of the molecular packing capable of being polymerized in a solid state is thoroughly investigated, other classes of such substances will be found. Moreover, employing the laws of packing organic molecules in a crystal, one may synthesize substances with the hope of obtaining a molecular packing appropriate for a topochemical polymerization reaction (this hope, if not always realized, will be realized rather frequently).

We shall consider this phenomenon using the example of *diacetylene* polymerization. The reactions under discusssion were first discovered in these substances. About a hundred diacetylenes with various radicals are known at present. These diacetylenes behave as follows.

In a crystal of monomer diacetylenes, the molecules in some cases are packed as is shown in Fig.14.12a. The radicals can rotate freely about single covalent bonds. Because of this, the central carbon atoms can be packed closely, i.e., approach one another to within typical non-bonded distances of 3.4 - 3.6 Å. Figure 14.12b shows the original crystal and the polymer formed. The reaction will proceed topochemically if the carbon atoms of methylene groups remain in their places under such a transformation. If this is possible, a polymer molecule will be quite freely packed between monomers. In this case new impermissibly close contacts between the molecules will not arise, and undesirable voids will not be formed. The polymer chain will most probably break at the site of some defect in the crystal. That will result in the ap-

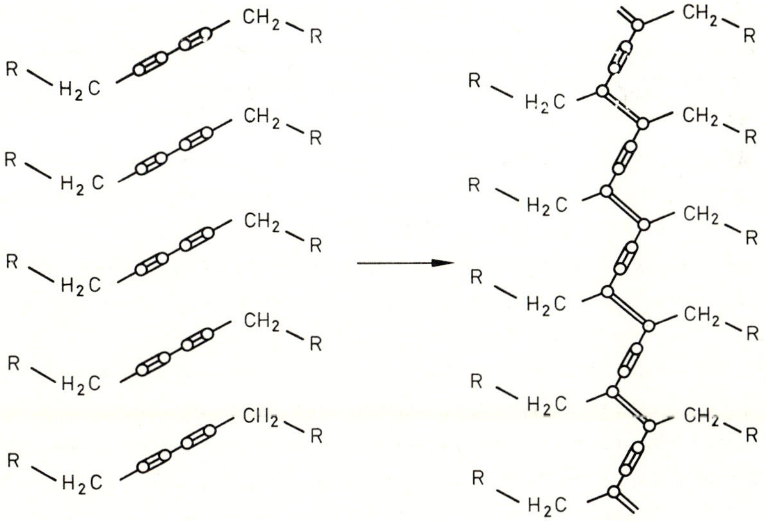

Fig.14.12. Example of the topochemical reaction for a derivative of diacetylene

pearance of polymer chains of various lengths. Polymerization may also cease due to the accumulation of slight differences in the packing geometry of a polymer in a monomer. In the latter case one would expect only a slight deviation of the polymer molecule length from some average value.

The density of such a solid solution should not differ from that of the original monomer, since in any volume the number of the atoms does not change, only the arrangement of the central part of the molecule is varried.

In contrast to the idea suggested by Wegner, we think it is possible to predict changes in the geometry of the atomic arrangement that will take place upon polymerization. Accordingly, we shall be able to decide whether polymerization will or will not proceed, and if it proceeds, what will be the molecular lengths formed.

It is obvious that a column of monomers cannot be transformed into a polymer molecule by the displacement of atoms in one plane. It is easy to understand that a planar polymer molecule, shown in Fig.14.12b, brings the carbon atoms of the methylene groups to a distance of about 5 Å from each other. This will not hinder the polymerization of a monomer even when for the monomer crystal the lattice period in the direction of the monomer column is 4 Å. In fact, the carbon atoms of the methylene groups can be left at a distance of 4 Å in the polymer crystal provided that the central part of a polymer molecule is made spiral. Such a geometrical transformation encounters little steric hindrance.

However, if the lattice period in a monomer crystal is too large, the formation of a polymer chain becomes impossible. A very high energy is required to stretch a planar polymer molecule into a chain with the identity period exceeding 5 Å even by only a few tenths of an angström.

A decisive role of molecular packing in a crystal during the process of polymerization was shown by GERASIMOV et al. [14.24]. These researchers have shown that the chemical process does not require thermal activation. They also estimated the changes in the molecular interaction energy which accompany the transition of a monomer crystal into a solid solution of polymer and monomer molecules.

It is well established that a solid solution forms upon polymerization of diacetylene [14.25]. Cases are possible, however, where polymerization will lead to a heterophase system. In certain cases the new phase is embedded into the monomer matrix in the form of elongated regions strictly oriented relative to the host crystal. Undoubtedly, the formation of a heterophase system without destruction of the object undergoing polymerization (or any other chemical reaction) requires the exact geometrical correspondence between the old phase and the new phase. Yet, the formation of a solid solution is possible with a slight deviation of the geometry of the matrix molecules from the geometry of the newly formed molecules. From this standpoint, homogeneous polymerization is more probable than a heterogeneous one.

Solid solutions of a polymer in a monomer are evidently the only case of solid solutions in a low-molecular-weight substance. It is impossible to imagine a situation where the joint crystallization from a melt and from a solution would lead to the formation of a single crystal, inside which polymer molecules would be randomly distributed.

14.4.2 Limit Solubility of Low-Molecular-Weight Substances in a Solid Polymer

Without resorting to experiment, one may state that a continuous series of solid solutions of low-molecular-weight substances in a polymer is impossible, and that substitutional solid solutions are equally impossible. It is unlikely that small molecules and crystallites of a low-molecular-weight component can be embedded into the crystalline regions of polymers.

Dissolved molecules should find some room in the amorphous regions of a polymer. In the case of a heterophase system, crystallites of a low-molecular-weight component are either in the amorphous regions or are mixed with polymer domains in a way similar to the eutectic mixture of low-molecular-weight substances.

Cases are rather frequent where polymer molecules do not interact with a low-molecular-weight substance. In this case both the solution and the eutectic system are thermodynamically unstable. The substances do not mix either in a liquid or in a solid state and crystallize separately.

Since a solid polymer is almost always imperfect, one may assume that there is always a limited solubility of small molecules in a solid state whenever these substances are mixed in a liquid state. In fact, it is very difficult to make sure that molecules of a low-molecular-weight substance do not dissolve in a polymer but only form a eutectic mixture. Neither X-ray diffraction analysis, nor electron microscopy, nor other physical methods can distinguish reliably true molecular mixing from penetration into a polymer by a low-molecular-weight substance in the form of small crystallites.

However, for a 3 - 5% low-molecular-weight impurity the crystallites, if any, can be detected reliably enough by physical methods. If a low-molecular-weight substance has been detected in a polymer, but the X-ray patterns do not reveal diffraction strips of the impurity and the electron micrographs do not exhibit crystallites, then we can assert with confidence that the system is a solid solution.

Solidification of a homogeneous mixture of a low-molecular-weight substance and a polymer most likely leads to the formation of a mixture of low-molecular-weight crystals and solid solutions of small molecules in the amorphous regions of the polymer. To prove this statement and to predict the phase diagrams of the system polymer — low-molecular-weight substance by theoretical calculations is almost impossible. The behavior of each system should be studied experimentally.

Two examples of the formation of thermodynamically stable mixtures can be investigated. First, one can study objects obtained from crystallizing a polymer from a solvent (the melting point of the low-molecular-weight component is lower than that of the polymer). Second, one may study specimens obtained by cooling a mixed melt of a polymer and a low-molecular-weight substance (the melting point of the low-molecular-weight component is higher than that of the polymer).

Let us first consider the dissolution of a polymer in a low-molecular-weight solvent. Dissolution of a polymer has its peculiar features related to the different sizes of the polymer and the solvent molecules. As the solid polymer comes in contact with the solvent, small molecules penetrate into the amorphous regions of the polymer. The polymer swells because of a peculiar, unidirectional mixing. The swollen polymer, being a solution of a low-molecular-weight liquid in a polymer, coexists for some time with a layer of pure low-molecular-weight liquid. When the polymer chains are separated far apart, the macromolecules can penetrate into the liquid and a homogeneous solution is formed. In some cases the swelling is limited, i.e., an equilibrium system represents a two-phase mixture of a swollen polymer and pure solvent. The swelling of polymers, which is of interest to experimenters who introduce a low-molecular-weight substance to plasticize the material, has been studied thoroughly.

The solidification of a swollen polymer has been studied to a much lesser extent. Probably the solvent molecules do not leave the polymer material as the temperature lowers. However, no data have so far been available to determine which system is formed after solidification, a solution or a eutectic mixture. It is most probable that the plasticizer molecules do not penetrate into the crystalline regions. What is still unknown is the maximum percentage of a low-molecular-weight impurity that can be present in a polymer in molecular-dispersion state, and at what impurity percentage the crystallites of the low-molecular-weight impurity begin to form.

As the plasticizer percentage increases, the glass transition temperature naturally decreases. Experimental correlations between the chemical structure and geometry of the plasticizer molecule and of the polymer molecule are unlikely to be generalizable. The existing theories of plasticization are not very suitable for predicting the behavior of a system polymer—low-molecular-weight substance.

The interaction between macromolecules and small molecules is simpler to investigate when the solvent melting point is higher than that of the polymer. One may assume that the laws found for such cases can be extended to the dissolution of a polymer (solid at room temperature) in a liquid solvent. There is no essential difference between these two cases, but the patterns of phase diagrams are probably different. Polymer crystallization from solution is possible for weak solutions. And this means that the eutectic point is very close to the melting point of the pure solvent. Experimental data also show that the liquidus curve runs sharply upward starting from a small percentage of polymer, rapidly reaches a temperature close to the melting point of the polymer, and then runs almost parallel to the abscissa axis to the polymer melting point.

When the polymer melting point is lower than that of a low-molecular-weight component, the phase diagrams of the polymer—low-molecular-weight-substance system resemble those for two low-molecular-weight substances.

One of the few studies of the phase diagrams of the polymer—low-molecular-weight-substance system where the polymer melting point is lower than that of a low-molecular-weight component was performed by CHERDABAEV and TSVANKIN [14.26]. Mixtures of polymers with low-molecular-weight substances were prepared from powders. The powders, taken in appropriate amounts, were thoroughly mixed in a grinder. For X-ray diffraction analysis of phase transformations, the powder mixtures were put into special capillaries 0.7 - 1.0 mm in diameter. A *ferrocene* system was studied in detail as an example of a binary system consisting of a crystalline polymer and a low-molecular-weight substance.

Investigations performed by means of X-ray diffractometry as a function of temperature have shown that melting of *polyethylene* (PE) proceeds almost in the same manner as in pure PE, whether ferrocene is present or not. In a mixture with a small content of ferrocene (no more than 16 wt.%), the intensity of the main ferrocene X-ray reflections gradually decreases with increasing temperature, starting from 120°C. This means that ferrocene gradually dissolves in PE, as the temperature increases and PE becomes amorphous. All the PE melts at 136° and the entire system becomes completely amorphous. In the course of the second heating-cooling cycle, the diffraction patterns exhibit the same variations, but after the first cycle a portion of ferrocene remains dissolved in PE. The higher the ferrocene content, the higher is the temperature at which all the ferrocene is dissolved in PE. With a low ferrocene content (no more than 7 - 8%), the ferrocene lines completely disappear from the diffraction pattern after the first heating-cooling cycle. In this case a solid solution of ferrocene in PE is formed. A solubility diagram of the PE-ferrocene system was constructed from the results obtained for a large number of samples (Fig.14.13). The points on the diagram show the amount of ferrocene that can be dissolved in PE under the given conditions.

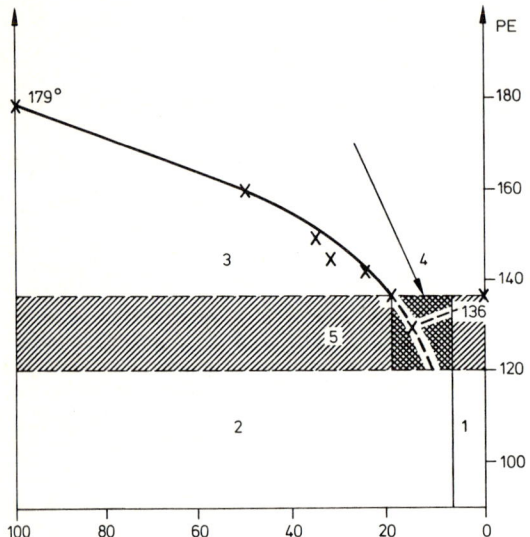

Fig.14.13. Polyethylene-ferrocene solubility diagram

In Fig.14.13 region 1 is the region of solid solutions of ferrocene in PE; in 2 there is a mixture of ferrocene and solid solutions of ferrocene in PE; 3 is a two-phase region of ferrocene and a solution of ferrocene in the PE melt; 4 is a one-phase region of this solution. In transition region 5 ferrocene melts at all concentrations. Eutectic or peritectic points may be present in this region, but they cannot be singled out because complicated processes take place in this interval due to a gradual melting and recrystallization of PE.

Small-angle X-ray scattering studies of ferrocene:PE mixture revealed a gradual decrease in the intensity, with the increasing ferrocene content corresponding to a large PE period. This means that the density of amorphous PE regions increases and approaches the density of its crystallites. This equalization of the densities is obviously caused by the penetration of heavy ferrocene molecules into amorphous PE regions. Since the degree of PE crystallinity is 60%, the actual ferrocene concentration in the amorphous regions at 20°C is 17%. In this case the density of the amorphous regions increases from 0.86 g/cm^3 to 0.96 g/cm^3 and becomes almost equal to the density of the crystallites, 1.00 g/cm^3. The maximum concentration of ferrocene in amorphous PE is 17% at room temperature, 25% at 140°C, and 50% at 160°C.

Thus upon the formation of solid solutions, ferrocene molecules are located in amorphous PE regions. An increase in the ferrocene content, at constant temperature, leads to the saturation of amorphous PE regions with ferrocene and when its content becomes even higher, ferrocene crystallizes out. At higher temperatures the number of amorphous PE regions increases, and as a consequence, the amount of ferrocene dissolved in PE grows.

Besides the ferrocene-PE system, similar investigations were performed with other binary systems consisting of crystalline polymers and low-molecular-weight substances. Investigations in the mixing of polymers with benzanilide has shown that *benzanilide* is not dissolved in *PE* but is well dissolved in POE (*polyoxyethylene*). On the other hand, *anthracene*, like ferrocene, is dissolved both in PE and in POE. The solubility diagram for the anthracene-PE and benzanilide-POE systems does not differ from the ferrocene-PE diagram. It has been established that in the pentaerythrite-PE, pentaerythrite-POE, hydroquinone-PE, and urea-PE system the components do not interact with each other and the dissolution of a low-molecular-weight substance in the polymer melt is not observed.

The solubility of low-molecular-weight substances in amorphous polymers at T > 105°C was studied in detail for the system *anthracene-polymethyl metacrylate* (PMMA). The diffraction patterns for the anthracene-PMMA mixtures, and correspondingly their structure, remain unchanged up to 140°C, the glass transition point of PMMA. At higher temperatures the intensity of anthracene lines decreases. This is an indication of the gradual dissolution of anthracene in PMMA above the glass transition point where the mobility of PMMA macromolecules increases. Figure 14.14 shows the solubility curve of anthracene in PMMA. The diagram was obtained from measurements of the integral intensity of the main anthracene X-ray reflections at different temperatures. The diffraction patterns corresponding to regions 1 do not reveal lines of crystalline anthracene, so it should be a one-phase region of anthracene solution in PMMA. In the two-phase region 2, pure anthracene is present besides a solid solution of anthracene in PMMA.

Thus, in the process of forming solid solutions, molecules of low molecular weight penetrate into the polymer's amorphous regions, thereby increasing their density; in the case of a crystalline polymer, the degree of crystallinity, the structure of polymer crystallites, and their size and shape remain unchanged.

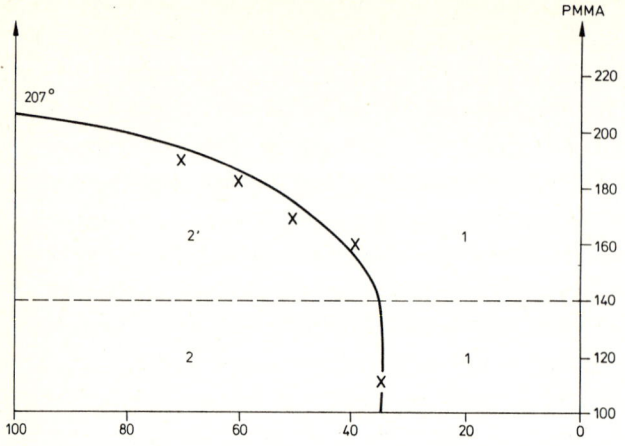

Fig.14.14. Solubility curve of anthracene in polymethyl metacrylate

The solubility of low-molecular-weight substances in polypropylene was studied by MYASNIKOVA [14.27]. The samples were obtained by pressing the mixtures at a temperature of 180°C and a pressure of 50 atm with a subsequent slow cooling under pressure. X-ray diffraction analysis has shown that there is only one phase present up to 0.12 diphenyl molecules per polypropylene polymer link. At higher diphenyl content X-ray lines due to pure diphenyl appear. The sample deformation was also measured as a function of temperature under constant loading. The results agree with the X-ray analysis. If the diphenyl content does not exceed 12 - 15%, the system is an interstitial solid solution. The eutectic melts at a temperature of 65 - 70°C, the solid solution at 130 - 140°C. Electron micrographs also confirmed the above findings. Besides diphenyl, the solubility in propylene of other low-molecular-weight admixtures (tolane and resorcinol) was also investigated.

It would be premature to judge a polymer structure and its lability, from the solubility of low-molecular-weight admixtures in it. The solubility limit depends not only on the porosity of the amorphous regions, where the dissolution or crystallization of a low-molecular-weight mixture takes place, but also on the relationship between the interaction energy between admixture molecules and the interaction energy of admixture molecules with the polymer. One may assume, in particular, that molecules bound in "their own" crystal by hydrogen bonds will not dissolve in a polymer like polyethylene.

14.5 Deformation of an Eutectic Mixture

CHERDABEYEV and TSVANKIN [14.28] investigated the mutual orientation of a polymer (high-density polyethylene) and a low-molecular-weight substance (ferrocene). X-ray patterns of the samples show that an isotropic structure consisting of crystalline ferrocene and PE appears after compressing. The new effect lies in the fact that as the mixture is stretched both PE and ferrocene are both oriented. The sample with maximum orientation exhibits a double texture: one is PE crystallites, the other is ferrocene crystallites.

Ferrocene crystals were found to form growth habits with axes [001] and [110]. It is known that pure ferrocene may crystallize to form needles [001] and prisms with long axes [110]. One may assume that the orientation of the mixtures of ferrocene with polymers leads to the formation inside the polymer matrix of ferrocene crystallites consisting of oriented needles [001] and prisms [110].

Oriented double systems were obtained not only for ferrocene, but also for some polymers with anthracene, naphthalene, caprolactam, and others. For each pair it was found that above a certain concentration of the low-molecular-weight component the mixture could not be oriented.

A new type of texture, a plane texture of low-molecular-weight crystals, has been obtained in oriented anthracene-PP and naphthalene-POE systems. Such a plane texture is especially typical of the orientation of substances with a plate-like structure, such as graphite or molybdenum disulphide. No double texture was obtained for such pairs as PE-pentaerythrite, POE-pentaerythrite, PE-hydroquinone, and PE-urea, though preparation conditions and mixture composition were widely varied. In this case, a stretched sample of the mixture contains only one texture of polymer crystallites, while the crystals of the low-molecular-weight substance remain unoriented. A direct test showed that for all the systems in which double-oriented samples were obtained, complete dissolution of the low-molecular-weight substance in the polymer melt was observed at a temperature exceeding the melting point. In contrast, in systems where no interaction between the components was observed, samples with a double texture of crystallites could not be obtained.

The first stage in the preparation of samples with a double texture is the *compressing* of the mixture at temperatures exceeding the polymer melting point. The result of the joint crystallization after *compressing* is the formation of fairly small and uniformly shaped crystals of the low-molecular-weight substance in the form of needles or plates, as is usual for eutectic crystallization.

During the second stage of sample preparation, the pressed samples are stretched. Processes taking place during this step are difficult to describe in detail at the present time. It is possible that under stretching (as during the first stage under compressing) the low-molecular-weight organic substance dissolves and recrystallizes due to a local temperature fluctuation caused by the rearrangement of an isotropic polymer structure into an oriented one.

If the low-molecular-weight substance is not dissolved in the polymer and its crystallization point is higher than that of polymer melting, then the compression forces a crystallization of the low-molecular-weight component and forms fairly large crystals of an arbitrary shape which cannot orient in the later process of stretching the polymer matrix.

The reversible orientation of an organic low-molecular-weight substance in elastomers has been monitored. A mixture of anthracene and natural rubber, NR (smoked sheet), was prepared by rolling with subsequent irradiation. It was found that under elastic stretching and crystallization of NR, anthracene crystals also become oriented and form their own independent axial texture. After contraction, the NR sample becomes amorphous, and anthracene is disoriented, i.e., the texture in the arrangement of its crystallites disappears. Thus it is possible to observe this reversible process.

In the work cited oriented mixtures of PE and POE with a molecular mass of 6000 were also studied. In the oriented PE a texture of POE is formed in which the macromolecular axes are oriented at an angle of $45°$ to the texture axis. The POE texture seems to consist of individual spherulitic POE radii directed along the axis of PE orientation. When the mixtures of PE with some thermoelastoplastics are oriented, the joint orientation of the mixture components is observed, thermoelastoplastics being oriented to much lesser degree than PE.

14.6 Crystalline Complexes of a Polymer and a Low-Molecular-Weight Compound

A small molecule can often be packed together with polymer molecules to form a regular crystalline structure. Such complexes are possible only for a definite stoichiometric relation between polymer links and the small-mass molecules. However, one may hypothesize that an intermediate phase may exist in some concentration range, while the stoichiometric ratio, the number of links per one small-mass molecule, remains unchanged. This would arise if the degree of crystallinity changes with a change in composition; however, this intermediate phase has not yet been observed.

MYASNIKOVA et al. [14.29] described the structure of a complex formed by polyethylene oxide molecules and resorcinol. A phase diagram was constructed that reveals a distinct maximum corresponding to the ratio two links of polyethylene oxide per resorcinol molecule. Polymers of different molecular weights were investigated. They all gave the same phase diagram. The crystallinity percentage is probably the same in all cases. The phase diagram was constructed by the contact method described in Sect.2.4.4. Mixed systems were studied by X-ray diffraction and electron microscopy. Thermomechanical curves were also plotted, thermogravimetric analysis was performed, and calorimetric measurements were carried out.

On the phase diagram of the POE-resorcinol system (Fig.14.15) the bell-shaped region demonstrating the existence of a molecular compound (MC) is separated from the pure components by two eutectic points, the MC melting point being about 30°C higher than that for pure POE. The center of the bell corresponds to the stoichiometric ratio 1:2, i.e., one resorcinol molecule per two monomer units of POE.

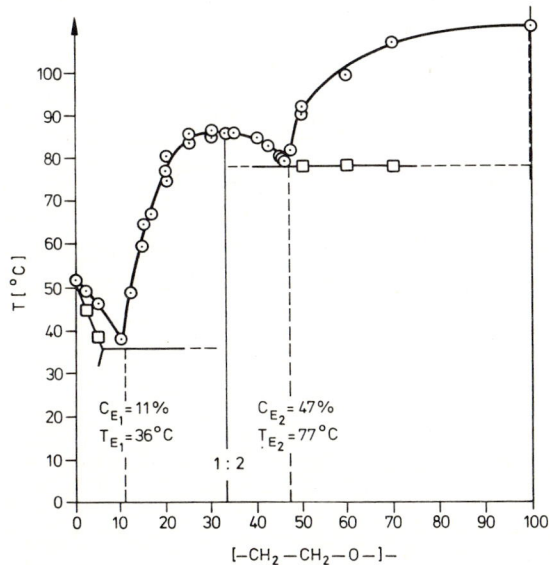

Fig.14.15. POE-resorcinol phase diagram

The formation of MC was assumed to proceed via an "embracing" of the POE chain with resorcinol molecules using hydrogen bonds as shown in Fig.14.16. This scheme clearly exhibits a periodicity along the chain; the translation contains two resorcinol molecules and 4 POE links. Two C atoms following each O are in a special position, since they are beyond the "embracing" zone. It is clear that if the "embracing" takes place, the X-ray patterns should reveal such a well-defined periodicity.

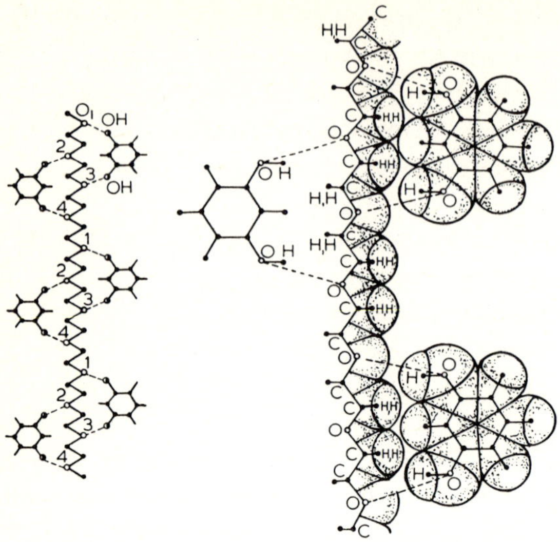

Fig.14.16. "Wrapping" of the POE chain by resorcinol molecules

This model for the formation of MC is also confirmed by the following facts.

1) POE samples of different molecular masses (2000, 6000, 20 000, 40 000, and $2 - 4 \times 10^6$) form with resorcinol the same type of phase diagram, containing the 2:1 ratio at the center of a bell-shaped region and with variation of the MC melting point in the range 86 - 99°C.

2) Pure resorcinol sublimed easily, but almost no MC sublimes from the 2:1 melt in air; an excess of resorcinol (beyond the stoichiometric ratio 2:1) sublimes easily as the molten mixture is stirred.

Geometrical analysis has shown that hydrogen bonding is accompanied by close fitting of resorcine molecules to a POE chain according to the "projection-to-the-hollow" principle. In the case of other rigid molecules (α- and β-naphthol, phenol, benzoic acid) such fitting proves to be impossible.

To investigate the system POE-resorcine, the authors used X-ray diffraction, electron-microscopic, and physio-mechanical techniques.

An oriented MC film was obtained only upon the mixing of resorcinol with high-molecular-weight POE (molecular mass $2 - 4 \times 10^6$). After thorough stirring the mixture of a definite concentration was pressed almost in vacuo at 120°C (the melting point of resorcinol is 111°C) to a pressure of 50 atm. The mixture was then slowly cooled. The MC film obtained by this techique could be stretched manually.

MC yields a much more informative diffraction pattern than POE. The $d_{nk\ell}$-values were measured for 49 reflections. The presence of axial reflections 00ℓ $\ell = 2,3,4,\ldots$ shows that the symmetry of an MC crystal cannot be higher than monoclinic. MC was found to crystallize in the monoclinic space group

$P2_1/a$ with the parameters a = 16.05, b = 14.25, c = 9.84 Å, β = 112°, and Z = 4. This means that four MC chains pass through the plane ab, each chain containing 4 POE links and 2 resorcinol molecules. The density ρ_{calc} = 1.261 g/cm³; ρ_{exp} = 1.241 g/cm³ (pycnometry). The difference in the densities determined experimentally and calculated from X-ray data indicates that intercrystalline amorphous layers exist in MC, as in any crystalline polymer.

It is known that in the structure of pure POE [a = 8.16, b = 12.99, c = 19.30 Å (fibril axis), β = 126.5°, Z = 4] a POE molecule represents a 7/2 helix, i.e., there are links making 2 revolutions in the period c = 19.30 Å. Hence, each elementary link has a height of 2.76 Å, while in a planar zig-zag the height of a link is 3.5 Å.

In the MC structure the shortened period c = 9.84 Å contains 4 POE links, that is, the height of each link is 2.46 Å. One may therefore assume that in the formation of MC the conformation of a POE chain varies little, and only a small contraction of the links takes place due to the formation of hydrogen bonds.

If an entire complex two-component chain is twisted into a spiral, resorcinol molecules will follow along the chain without intervals (Fig.14.16), i.e., the resorcinol molecules will also be closely packed with their end-faces. Calculation of the volume of such a complicated molecule from "increments" [14.30] yields a fairly high coefficient of molecular packing in a crystal, 0.73 (neglecting the approach of the groups in the formation of hydrogen bonds).

Let us calculate the increase in density. The specific volume per one link of a polymer molecule in the POE structure is 58.7 Å³, the specific volume of a resorcinol molecule is 141.8 Å³, the specific volume of the complex is 266 Å³. Hence,

$$\Delta = \frac{2 \cdot 58.7 + 141.8}{266} = -0.025 \quad .$$

Thus the formation of a complex results in a slight loosening of the structure. Other physical techniques confirmed that a molecular complex forms in the resorcinol:POE system.

15. Biopolymers

At the present time a great deal of attention is devoted to the study of biological polymer molecules and supermolecular systems formed by them. Naturally, biologists are primarily interested in the mechanism by which these molecules and systems function in living organisms. But the study of this mechanism remains incomplete until more detailed information has been compiled on the structure.

Here again, the principal methods for investigating condensed systems are electron microscopy and X-ray diffraction analysis. As a rule, biological macromolecules form multicomponent condensed systems. In the condensed state the macromolecules are usually ordered. However, this order is often imperfect, valid only in small volumes. This makes it extremely difficult to conduct structural investigations. Nevertheless, we encounter biological substances whose molecular organization, despite its complexity, is amazingly regular. They form quite perfect crystals. Numerous viruses can serve as an example.

There are many supramolecular structures with macroscopic order. The macromolecules form a regular lattice, but the internal arrangement of the atoms in these molecules cannot be described by the lattice. We have already come across such structures in the previous chapter and can liken them to accurately arranged bags of potatoes or apples. Muscle tissues are an example of regular structure of this type.

To discuss ordered systems created by the many kinds of biological macromolecules would require a separate monograph on structural molecular biology. In this book we have set ourselves to the task of illustrating by examples taken from all fields in the study of condensed-media structure, the universality of the principle of dense packing, the fertility of the idea "the key to the lock", that is, to show the possible adaption of molecules and parts of molecules to one another whereby "projections" fit into "hollows". The decreasing free energy of the system is due, first of all, to the filling of space to maximum density. This principle is particularly important in molecular biology, for it allows one to understand not only the principle of constructing systems containing hundreds of thousands or millions of atoms, but also to understand the mechanism of such important biological processes as DNA replication, the formation of proteins on matrices, transfer of biological molecules, muscle contraction, etc. This does not mean, of course, that a simple geometrical approach is sufficient for explaining all the details of biological processes. But, in any case, we cannot manage without employing the idea of close-packed structures.

We could illustrate the heuristic value of this principle by many examples, but this is not our purpose. Suffice it to remember that the DNA double helix was discovered in a search for the densest assembly of two chain molecules of nucleic acids.

In this concluding chapter we shall confine ourselves to several examples of biological systems which seem most suitable for illustrating the dense-packing principle. Dense-packed structure in biological systems (as well as in synthetic polymers) are realized at two levels. First, biological macromolecules stretch out into a chain or curl up into a coil in such a fashion that the atoms occupy a volume as small as possible, and second, the macromolecules themselves are densely packed.

15.1 Proteins

A huge number of protein crystals have been investigated and many of their chemical structures have been established. The crystallography of proteins has developed intensively over the past two decades despite many technical complications. For instance, it is very difficult to produce single crystals and perform an X-ray diffraction analysis of the crystals whose unit cell accommodates thousands of atoms. Because molecular biology is such a young branch of science, we consider it extremely important to carry out such demanding investigations in order to establish the conformation of protein molecules, whose geometry plays a crucial role in many biological processes.

Data on the structure of protein molecules can be found in [15.1]. We will not consider the structure of protein molecules in this book since we are only interested in protein crystals made up of the binary system protein-water (or some other solvent), specifically a mixed crystal.

Before we begin to describe the peculiarities of protein crystal structures, we would like to dwell briefly on *globular proteins*, especially since their structure is a remarkable illustration of our main idea — a system of atoms that fills up space as densely as possible exhibits stability.

15.1.1 Globular Proteins

Protein molecules are polymers constructed from disordered combinations of 20 kinds of amino acids. The macromolecules contain from a thousand to a hundred thousand atoms. The polymer chain is unusual in its ability to acquire an almost spherical form. The backbone of the protein molecule is a so-called polypeptide chain (Fig.15.1). The *protein globule* has a rough surface with rather deep hollows and projections.

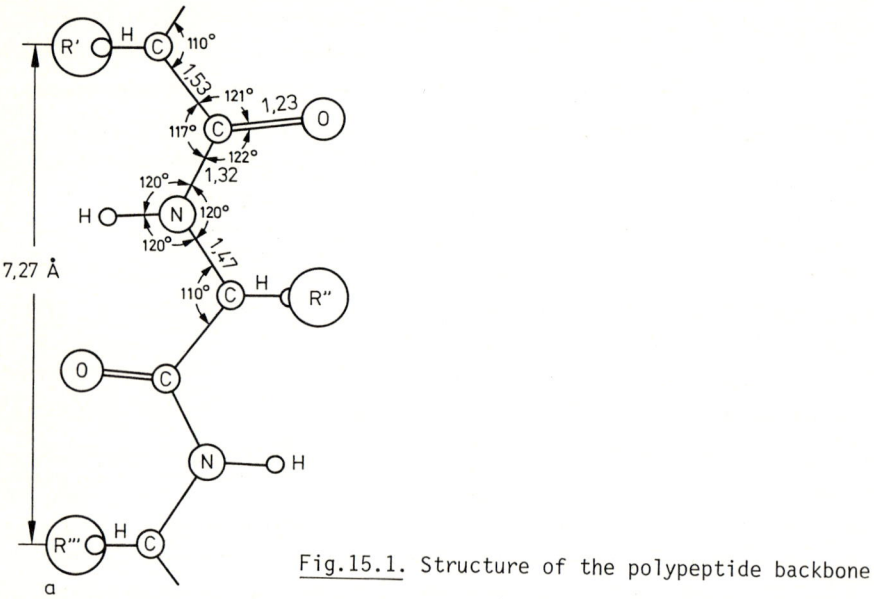

Fig.15.1. Structure of the polypeptide backbone

Unlike synthetic polymer molecules, protein molecules from animals of the same species are entirely identical in the number, kind, and sequence of amino acid residues. In addition, they possess exactly similar architecture, i.e., similar *conformations*.

Molecules of the same name (e.g., hemoglobin, ribonuclease, or cytochrome) in all biological species differ by composition and sequence of their amino acids in polymer chains. Remarkable however is the circumstance that they all have practically the same conformation.

Numerous examples of the protein globules have made it possible to demonstrate the general principle of dense packing in naturally ordered systems.

When studying proteins with familiar structures, we detect regions of close-packed lateral chains in all cases. The lateral chains curl into spheres. The close-packing principle can be applied to the arrangement of these spheres. The packing stereochemistry of atoms of amino acid residues into spheres and of those spheres into dense formations was considered in detail by EFIMOV [15.2]. The close-packing principle has allowed LIM and PTITSYN [15.3] to develop the theory of protein globule construction on the basis of their chemical structure, i.e., the sequence of amino acid residues.

The possibility of applying the close-packing principle to a free protein globule is important and very convenient for investigators, since in nature globules are found only in solutions. Thanks to the fact that it is much easier to closely pack the aliphatic or so-called hydrophobic parts of amino acids, the process of organizing a globule brings about an optimization of the free energy in the protein-water system. The gain in the free energy is due to the fact that the contacts between water and the nonpolar parts of the globule prevent the formation of a maximum number of hydrogen bonds.

An elegant calculation by SARKISOV and DASHEVSKY [15.4] showed that two methane molecules in water were attracted to each other, and their contact decreases the system's energy. The so-called hydrophobic interaction boils down precisely to that. Of course, the packing of atoms in a globule is not as elementary as the packing of close-packed spheres. Therefore, there may exist voids in the globules which do not affect the rigidity of the molecular skeleton. In certain cases, when the polymer chain is curled into a globule, the capture of several water molecules inside the globule may prove to be beneficial.

Knowing the chemical composition of a protein molecule, one can calculate, by using the increments of the volumes of atomic groups [15.5-7], the intrinsic volume of the protein molecule and compare it with the globule volume determined empirically. By dividing one volume by the other we obtain the packing coefficient of the atoms located inside the globule from which we can then calculate its density.

The intrinsic molecular volumes for the three proteins *ribonuclease, lysocine*, and *thermolysine* (12000, 12550, and 36740 $Å^3$, respectively) were calculated using the volume increments indicated in [15.7]. The molecular weight of thermolysine is 34600 daltons; its density is 1.18 g/cm^3. The volume per single molecule in a dry protein equals $34600 \times 1.65/1.18 = 48500$ $Å^3$. Hence the packing coefficient of the atoms inside the globule equals 0.76.

The volume per single molecule of ribonuclease is known to equal 15800 $Å^3$. The packing coefficient is practically the same, 0.77. We shall dare to generalize this result by assuming that the atoms in the protein globule are packed with the same density as in low-molecular-weight organic crystals. Probably, the number of voids and water molecules located inside the globule is not very large.

15.1.2 Protein Crystals

Practically no attention has been paid to the study of the protein crystal, an exclusively original structure. Protein crystals contain a very large percent of solvents which promote the growth of crystals. The solvent can be water, salt solutions, or aqueous alcohol solutions. A liquid protein solution is a thermodynamically unstable system. Crystals start to grow in them at any protein concentration and any temperature. Naturally, from purely practical considerations, the protein concentration amounts to several per cent. Nevertheless, crystals of the size suitable for investigation will slowly grow for a very long time, as long as several months.

The resulting crystal is a rigid skeleton of protein globules. The globules contact one another at a small number of sites. Large voids between the protein molecules are filled with molecules of water or some other solvent. The weight fraction of the solvent in the crystal is definite. Thus a protein crystal should be considered to be a molecular compound of protein and water. In other words, a protein crystal is a *pseudocrystallohydrate*. We are obliged to add the prefix "pseudo" because the water molecules in the voids are not strictly ordered. It is true that a significant part of the water molecules bonded directly (by hydrogen bonds) with the globule surface are fixed at well-defined positions; however, the degree of ordering of the solvent molecules located far from the surface of protein molecules has not been sufficiently investigated.

The advantage of forming a protein-water system is obvious enough. Only a relatively insignificant part of the atomic groups capable of forming a hydrogen bond are concealed inside the protein globule. A greater part of these groups are located on the outer surface of the globule and are capable of forming hydrogen bonds with the water molecules. These groups, protruding from the outside, allow the protein molecule to acquire an envelope of water molecules.

As has been mentioned earlier, the liquid solution containing the protein globules is a *thermodynamically unstable system*. At a certain protein concentration and under certain external conditions the pseudocrystallohydrate containing a definite amount of solvent starts to grow.

The amount of water (or some other solvent) is really large. In most cases, the cell volume is found to contain 30 - 60% water. In some cases this percentage is as low as 20%; the tropomyosine crystals are known to hold a record content of 95% water.

The pseudocrystallohydrate lattice skeleton is formed with a small number of macromolecular contacts. For instance, there are four such contacts in cytochrome. A small number of amino acid residues is involved in the protein-protein contacts.

The protein molecules form an absolutely regular lattice. One can judge the degree of regularity by their X-ray data sets which sometimes consist of hundreds of thousands of reflections. Thanks to the strict ordering of the protein skeleton, the pseudocrystallohydrate takes the form of a faceted crystal. At a given concentration and temperature the growing crystals are always of the same structure. However, varying the solvent and other crystallization conditions may give rise to quite a different lattice. The pseudocrystallohydrates of one and the same protein can be obtained in a great number of modifications. It is better to omit the adjective "polymorphic" since different structures arise from different solvents, and these solvents enter into the crystalline lattice. Therefore, strictly speaking, different crystallohydrates of one and the same protein represent different crystals.

There does not exist a thermodynamically stable protein liquid solution. Neither is there a stable protein crystal. A crystal removed from the mother liquor will degrade. With a decrease of water the crystal changes its form. The molecules change orientation. However, the most important circumstance is the constant loss of ordering with a decrease in the water content. Evidence of this fact is the drastic decrease in the number of X-ray reflections. If the most distant reflections in the ordered crystal correspond to an interplanar distance of up to 1.5 Å, the most distant reflexes in a "dry" crystal close to destruction correspond to 10 - 15 Å. In the absence of water, the protein molecules cannot be packed into a regular lattice because of their complicated form. The water serves as a kind of cement which allows the voids to be filled up and the ordering to take place.

Upon further drying, an *amorphous powder* is obtained which, in all probability, consists of the same globules, but in random orientations. Such a powder still contains a substantial amount of water (sometimes as much as 30%) which most likely is closely bound to the protein molecules by hydrogen bonds. *Fully dried proteins* can apparently be obtained when the globules are uncoiled, i.e., in the process which chemists call *denaturation*.

Of interest for the theory of mixed crystals is the system composed of protein pseudocrystallohydrate and a very weak protein liquid solution (evidently, an equilibrium system).

If we could write the expression for the free energy of these two systems, it would be possible to predict the protein crystal lattice, provided the protein globule form is given. How do the two systems differ from each other? Let us first of all consider the configuration difference. In a disordered solution the globules possess an arbitrary orientation at each moment of time and are located at different distances from one another; in the crystallohydrate the orientation and the distances are the same. However, these differences do not lead to any appreciable differences in free energy. As far as the voids filled up with water molecules are concerned, it is quite another matter. In a disordered solution they are disordered both in form and size, while in crystals they are identical. A usual procedure of probability theory may, by simplifying the model, give an estimate of the increment of the thermodynamic probability arising upon the transition from order to disorder.

Let us now consider the differences in the molecular interaction energy. The ordering leads to a gain in the interaction. We may assume that the small number of contacts which create the protein crystal skeleton are the contacts between the nonpolar parts of the molecule which happen to be on its outer surface. In other words, the protein crystal skeleton is created by hydrophobic interaction. The gain in ordering consists in the possibility of forming additional hydrogen bonds between the water molecules. The validity of the hypothesis expounded can be verified by considering the contacts between the globules and calculating the energy gain that occurs at the expense of the hydrophobic interactions, using the same scheme used in the above-cited model calculation [15.4]. By calculating the entropy increment that occurs when water molecules are distributed within "boxes" of different sizes and equating it to the hydrophobic interaction energy of the protein molecules in the crystal, we can find the solvent concentration in the crystal and mother liquor.

ESIPOVA et al. [15.8,9] suggest another treatment of the protein crystallization problem. Thermodynamic measurements have led the authors to the conclusion that no changes take place in the intermolecular interaction energies, but that all the variations in the free energy come from the absolute growth of the entropy during crystallization. Here follows a citation from [15.9].

The difference in the value of ΔS may be related only to the change in the hydrate protein molecule envelope which becomes partially disordered upon

the crystallization of the system, which ensures a decrease in the entropy of
the transition in the crystal. When being organized into crystals, the protein
molecules turn out to be surrounded by the protein and water molecules. The
number of intermolecular protein-protein contacts is not large enough to retard the rotational diffusion or, at least, to retard the fluctuations around
the contact areas. This fact and the small amount of protein molecules as compared to the amount of hydrate water molecules in the protein-hydrate water
system imply that the decrease in the entropy of the crystalline part of the
system at the expense of the protein molecule orientation in the crystal should
be less than the increase in entropy at the expense of the disordering of the
hydrate water. Thus the contribution of the protein part to the change in the
entropy of the protein-water system during crystallization must be small. Everything that has been said means that there is an absolute rise in the entropy of the crystal-water system during crystal growth at the expense of disordering of part of the hydrate water during crystallization.

The construction of the phase equilibrium theory for protein crystallohydrate-water is a complicated task, for the gain in the free energy occurring
upon spontaneous crystallization of proteins from a hermetically sealed vessel
at a constant temperature is, apparently, very insignificant. Indeed, the mutual interaction of protein molecules bound by a small number of bridges is
weak, as is the mutual interaction of water molecules in the mother liquor
and inside the protein crystal. There is no particular difference in the interactions between the water and protein molecules either. We can speak of very
small changes in the free energy during crystallization because slight variations in the crystallization conditions lead to quite different crystallohydrate structures.

Let us now consider a classical study carried out under the direction of
HARKER [15.10]. The researchers set themselves to the task of obtaining as
many ribonuclease crystal modifications as possible. By varying the solvents,
the authors obtained 14 forms of ribonuclease. Some of them are obviously unstable. But it is quite impossible to decide which modifications possess the
lowest free energy, because the differences in the energy of some forms are
undoubtedly very negligible.

Eleven modifications crystallize in space groups $P2_1$ and $P2_12_12_1$, most convenient for crystallization of asymmetrical molecules. In one case a hexagonal
cell was detected, and in two cases the cells turned out to be centered.

The intrinsic volume of the ribonuclease molecules was estimated by the
authors from the data on partial specific volume and molecular weight. The
value 15800 $Å^3$ was accepted. From the information on the cell dimensions it

was possible to calculate the volume per molecule. Then it was easy to calculate the percentage of the solvent content in the crystal.

Thirteen modifications out of 14 yield very close values: the solvent content in the crystal varies from 46 to 55%. One modification obtained from the solvent stands out; it contains 55 - 60% methyl alcohol. In this case the water content equals 37%. When this crystal is dried, the diffraction pattern does not change and the moist and "dry" crystals give reflections to 2.8 Å resolution.

At the end of Sect.15.1.1 we determined the packing coefficients of a protein globule. We compared the molecular volumes calculated with the aid of increments to the specific volumes determined from the experimental dry protein density.

As was emphasized earlier in this chapter, researchers engaged in structural investigation of proteins are interested in the structure of molecules rather than that of crystals. Therefore, only in separate cases was attention paid to the independent determination of the volume of water filling the space between the globules in a protein crystal.

The accuracy of these measurements does not exceed 10% in most cases. It is therefore impossible to draw a comparison between the protein globule volume calculated earlier and the volume determined from the data on the unit-cell dimensions and the share of the solvent.

We can only make a supposition that a packing coefficient of the atoms inside a globule of the order of 0.75 is valid for a real molecule surrounded by water molecules.

Details of the structure of protein crystals can be found in [15.11].

15.2 Systems Based on DNA Molecules

15.2.1 DNA Crystals

Deoxyribonucleic acid (DNA) molecules are located in the cell nucleic of animal and plant organisms. They are responsible for the growth of organisms since they produce proteins. They carry genetic information that unambigously chacaterizes the organism; they are responsible for the transmission of heredity.

A DNA molecule is a chain whose sugar-phosphate backbone is the same in all DNA molecules of various organisms. Four bases — adenine, thymine, guanine, and cytosine — are connected to the backbone.

The formation of a double helix consisting of two such strands is energetically profitable. The two strands are bound by hydrogen bonds between the

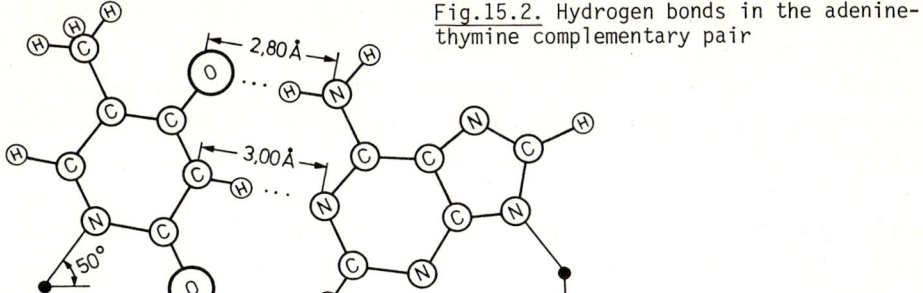

Fig.15.2. Hydrogen bonds in the adenine-thymine complementary pair

so-called complementary pairs (or base pairs) (Fig.15.2) adenine-thymine (the two-ring compound in Fig.15.2 is adenine, the other is thymine) and guanine-cytosine. The double helix promotes the formation of hydrogen bonds and the close-packing of atoms. Approximate calculations have shown that the packing coefficient inside the double helix, as in any organic substance, is close to 0.7.

The sequence order of the bases up and down the backbone is unique for each organism. Thus, strictly speaking, the DNA molecule representing both single and double helices is not a periodic sequence. However, the close resemblance of the base pairs connected by hydrogen bonds can create a quasi-periodic structure, if not along the whole length of the molecule, then at least after large segments of it. Here identical molecular groups are arranged in such a manner that the structure of these segments corresponds to one and the same periodicity and symmetry. Thus in a rough approximation the double helix of DNA consists of one-dimensional "crystallites" of comparatively small dimensions. Parallel molecules of DNA can be packed relative to one another in a regular fashion, so that the entire array of these molecules forms a three-dimensional periodical lattice.

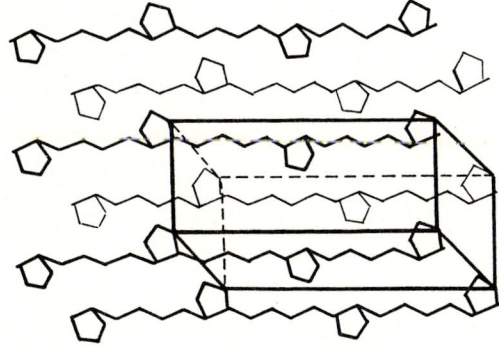

Fig.15.3. Lattice of the sugar-phosphate backbones occurring in DNA crystals

In crystalline preparations of DNA, as is seen in Fig.15.3, the sugar-phosphate backbones of the chains make up a regular space lattice, while the "rough" lattice is constituted by the base pairs having the same dimensions. There exist several different structural modifications of DNA crystals which result from different salt solutions, absorbing in this process various portions of water.

When the humidity is above 92%, irrespective of the nature of the cation, DNA crystals possess the so-called B form. The diameter of the double helix in the B form is about 20 Å. The height of the turn is 33.7 Å each turn of the helix has 10 base pairs. Molecules of DNA in the B form can form a hexagonal or a rhombic lattice, depending on the content of salt in the solution. According to experimental data, there are 20 water molecules per monomeric unit (nucleotide) in a sodium salt of DNA in the B form.

With a 75% relative humidity, the sodium of DNA forms A form crystals and the content of water decreases twofold; each nucleotide accommodates 10 water molecules. The molecular packing is much more ordered. The diffraction pattern reveals distinctly clear and well-resolved reflections. In the A form the DNA molecules are packed in a monoclinic lattice with two molecules per unit cell. In the case of the A form one complete helical turn of DNA accommodates 11 pairs of bases.

With a 44% humidity, the lithium salt of DNA forms C form crystals. The C form molecules are similar to those of the B form in many of their structural parameters. One complete helical turn accommodates 9.3 pairs of bases, the distance between the neighboring pairs being 3.32 Å. Depending on the cation content in the solution, C form DNA can produce an orthorhombic or a hexagonal lattice.

Natural molecules of DNA residing in a cell at a high relative humidity exist in the B form, according to much experimental data.

Hence it follows that the situation observed during the formation of mixed DNA-water systems is very similar to that of protein crystallization. No DNA crystals can form in the absence of water, because this would lead to systems having very low packing coefficients, i.e., to systems having large intermolecular voids. Water fills the voids between the molecules. Again a pseudocrystallohydrate is formed.

A large number of degrees of freedom allows the double helix of DNA to change its form. When the water content is high, different forms of the double helix prove to be more favored. The interaction between the water molecules and the nucleic acid plays an essential role in this system.

Unfortunately, numerous DNA conformation calculations have been made without taking into account the interaction of the water molecules. However, the ability of the atom-atom potentials method to predict that the forms of the DNA molecules detected experimentally in the DNA-water binary systems are associated with the minima on the multidimensional surfaces of the potential energy [15.11] is already a great success for these calculations.

15.2.2 Chromosomes

Complexes of nucleic acids and proteins are often observed in the most important biological aggregates; it is sufficient to mention chromosomes and viruses as examples. In chromosomes and virus heads, the DNA molecules are packed particularly densely.

In the bacterial cell of Escherichia coli, the DNA molecule which consists of several million pairs of bases is packed in a small space of $1 \times 0.5 \; \mu m^3$. Such a compact ordered packing is due to proteins. According to current conceptions, the basic structural unit of a chromosome is a nucleosome. Nucleosomes are made up of nuclei containing an octamer of proteins with a DNA length consisting of 140 base pairs and with a part of DNA of varied length external in relation to the nucleus. Investigators have managed to crystallize the nuclei of nucleosomes and study the main features of their structure.

It turned out [15.12] that the double helix of DNA twists itself into a smooth superhelix having a period of 1 3/4 of a turn around a protein rod of 100 Å in diameter (Fig.15.4). As the chromosome structure is being formed, the protein-DNA (nucleosomes) complexes go through two more organization levels. The nucleosomes acquire thick chromatin fibrils which constitute a substructural element of the chromosome.

One presumes that the packing of the complexes has to satisfy two requirements: the creation of a tightly close-packed structure and the prevention of DNA molecules from tangling. The packing calculations seem not to be very precise, but approximate estimates can be made. It is known, for instance, that a molecule of DNA 700 Å long resides in a nucleosome whose diameter is close

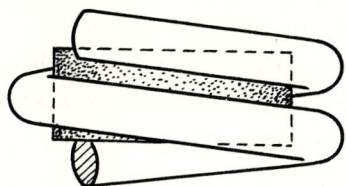

Fig.15.4. Superhelix formed by the double helix of DNA

to 100 Å. If we regard DNA as a cylinder, with a radius of 10 Å, the intrinsic molecular volume will equal 20×10^4 Å^3. Assuming that the packing coefficient of the DNA molecule entering a chromosome is 0.7, we may expect a DNA molecule to have a volume of 30×10^4 Å^3.

The volume of a sphere 100 Å in diameter equals 60×10^4 Å^3. Thus a DNA molecule occupies half the volume of a nucleosome. If there are 8 protein molecules in the nucleosome core, then each molecule will have a volume of 4×10^4 Å^3. Assuming that the protein molecules also have the same packing coefficient of 0.7, we obtain a volume of about 25000 Å^3 per protein molecule which is quite reasonable. Thus biological polymer atoms are packed in a nucleosome as close as possible.

If we know the chromosome size and the number of nucleosomes contained in it, we can establish the packing coefficient of the spherical nucleosomes contained in a chromosome. Apparently, the calculation should lead to the "usual" value, close to 0.7.

15.2.3 Viruses

Like chromosomes, viruses are complexes formed by globular proteins and nucleic acids. We do not intend to dwell on the biological functions of viruses here. We only wish to point out that a virus penetrates a cell and makes the cell apparatus synthesize protein in accordance with the assignment coded by the nucleic acid of the virus, but not by the nucleic acid of the cell.

Protein globules combine with the nucleic acid in the virus so that the globules could defend the nucleic acid, either by forming a hollow cylinder in which the ribonucleic acids hide, or by forming a hollow sphere layer within which the DNA molecules are arranged. Figures 15.5 shows a structural scheme of a virus, cylindrical in form, Fig.15.6 that of a virus, spherical in form.

It is amazing that such complicated multiatomic molecules can form strictly ordered crystals. The packing coefficient dictates the formation of these structures. Since the particles are large, the entropy term of the free energy fades into the background.

The globules in a cylindrical virus are packed very closely. The virus can be destroyed and then the protein globules can be compelled to gather into the same kind of formation, but without the DNA molecule inside the cavity.

From among the representatives of this type of viruses, most thoroughly studied was the tobacco mosaic virus. Its length is 3000 Å, the outer diameter is 170 Å, the channel diameter is 80 Å. The molecular weight of each globule is 17420. In all, 2140 units are packed in this virus. The subunit volume

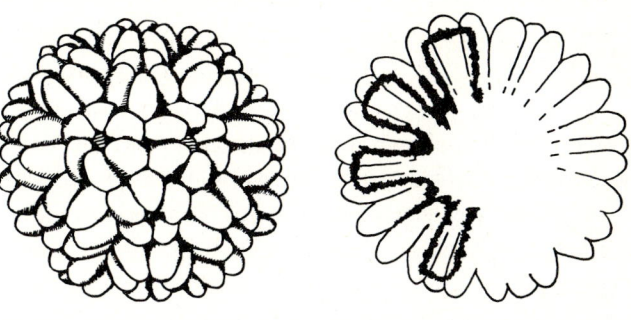

Fig.15.5. Schematic representation of the structure of a cylindrical virus

Fig.15.6. Schematic representation of the structure of a spherical virus

equals 25000 $Å^3$; the globule density is 1.15. The data are quite reasonable, showing that the atoms located inside the globule are arranged with the same packing density as in amino acid and polypeptide crystals.

Spherical viruses are bodies with a high degree of symmetry, namely, they are icosahedra. Certainly, one can approximate them as spheres. Due to their high degree of symmetry and to the fact that their outer surface is almost spherical, viruses of this kind often form cubic crystals with an enormous lattice period, sometimes up to 2000 Å.

The high ordering of virus crystals is really amazing. In some cases up to 200000 X-ray reflections have been measured.

We present approximate sizes of the spherical virus, whose structural scheme was shown in Fig.15.6. This is the virus of the yellow turnip mosaic. The outer diameter equals 300 Å, but the thickness of the protein envelope is only 35 Å. About 20000 identical globules are located in this spherical layer. Each globule has a volume of about 40000 $Å^3$. In this case, too, we have quite a reasonable value for the globule density.

Exceptionally complicated X-ray diffraction analysis of virus crystals has been carried out in recent years. The sequence of the polypeptide chain forming the dense globule has been established. The complexity of the initial elements does not prevent them from forming a structure possessing a high degree of symmetry. They are all built up according to "the key-to-the-lock" principle. Only the principle of close packing of particles, that of the maximum filling of space, makes it possible to understand the possibility of creating so many regular and highly symmetrical structures out of tangled asymmetrical chains.

However, the ordering of the virus structures occurs not only for the envelope, but for the DNA molecules residing inside the protein envelope as well.

Bacteriophages and organelles are other complicated systems which consist of many kinds of different molecules (DNA, proteins, lipids, etc.), and only in their separate elements can one find strict structural regularities. For example, bacteriophages consist of heads packed with DNA and "tails" which can attach the phages to a living cell and inject the bacteriophage DNA into it. The tail appendix consists of a rod with an inner channel and a cover whose protein subunits are monomers and dimers. The covers of some of the phages are capable of contracting. In the process the reciprocal arrangement of the subunits is reconstructed; however, the helical symmetry of their packing (with other parameters) is retained. There are two or three dozens of such protein molecules. Part of them form the head envelope which is polyhedral; there are different forms of such polyhedra. The protein molecules in the tail are packed with helical symmetry, which is fairly well established by the three-dimensional reconstruction method [15.13].

Let us now examine the packing of DNA inside the viral particles, e.g., the T2 bacteriophages, a bacterial infection virus (Fig.15.7). The DNA molecule isolated from the virus occupies a volume in solution which is dozens of times greater than that of the phage head.

Inside the protein envelope of the virus the DNA molecule has an ordered compact form and in such a form (biologically inactive state) it is stored inside the phage particles. The minimum length of the T2 virus phage head is about 1/500 of the contour length of DNA. Let us compare the volume of the

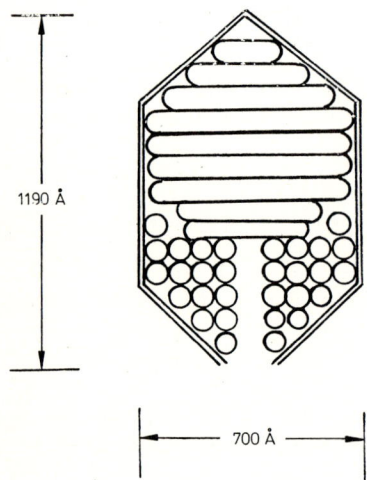

Fig.15.7. Packing of DNA inside a virus particle

virus head to that of a free DNA strand. Assuming the molecular weight of the free DNA to be to 130×10^6, which corresponds to a molecular length of 60×10^4 Å, we can determine the volume of the DNA strand by regarding it as a cylinder 20 Å in diameter. The volume of the DNA strand is then 2×10^8 Å^3. The geometrical form of the head of a bacteriophage is approximated by an elongated icosahedron. The external volume of the head was measured to be 4×10^8 Å^3. Taking account of the protein envelope thickness of about 35 Å, we find the internal volume of the head comes to 3×10^8 Å^3. When comparing the values obtained we see that the arrangement of the free DNA molecule in the phage head has a normal packing coefficient, close to 0.7.

A remarkable circumstance is the fact that the intracellular medium promotes the peculiar "crystallization" of the DNA molecule. The DNA molecule curls into a compact particle and is then covered by protein. One of the models of packing a DNA strand in the bacteriophage heads [15.14] is illustrated in Fig.15.7. The double helix of the DNA molecule in the B form curls into a secondary helix which is located in the phage head like a spooled thread. Thus the virus particles reveal a regular architecture both in the external form (envelope) and in the internal nucleus (DNA).

References

Chapter 1

1.1 J.L. Proust: Ann. Chim. (Paris) *32*, 45 (1800)
1.2 C.L. Berthollet: *Essai de statique chimique* (Paris 1803)
1.3 J.W. Gibbs: Trans. Connecticut Academy (1873-1878)
1.4 J.D. Van der Waals: *Die Kontinuität des gasförmigen und flüssigen Zustandes* (J.A. Barth, Leipzig 1881)
1.5 H.W.B. Roozeboom: *Die heterogenen Gleichgewichte vom Standpunkt der Phasenlehre* (Vieweg, Braunschweig 1918)
1.6 M.W. Lomonosov: Novi Commentarii Acad. Petropolitanae *1*, 245 (1748)
1.7 C. Blagden: Trans. R. Soc. London *78*, 277 (1788)
1.8 B. Hadfield: *Faraday and His Metallurgical Researches* (London 1931)
1.9 M. Faraday, R. Stodart: Ann. Chim. Phys. *21*, 62 (1822)
1.10 N.P. Belaev: Gorn. Zh. *3*, 313 (1914)
1.11 H.C. Sorby: Rep. British Ass. *2*, 189 (1864)
1.12 F. Rudberg: American College of Physicians, Annals (2) *48*, 353 (1831)
1.13 E. Wiedemann: Wied. Ann. NF *3*, 237 (1878)
1.14 W. Heintz: Pogg. Ann. *92 (168)*, 588 (1854)
1.15 W.C. Roberts-Austen: American College of Physicians, Annals (5) *13*, 111 (1878)
1.16 Fr. Guthrie: Philos. Mag. (5) *18*, 105 (1884)
1.17 Fr. Guthrie: Philos. Mag. (5) *17*, 462 (1884)
1.18 D.K. Chernov: Zh. Russ. Metall. Ova. *1*, 1 (1915)
1.19 F. Osmond: Ann. Mines *8*, 5 (1885)
1.20 J.W. Gibbs: *Equilibre des systemes chimiques* translated by H. Le Chatelier (Paris 1899)
1.21 G. Tammann: Z. Anorg. Chem. *37*, 303 (1903)
1.22 G. Tammann: *Lehrbuch des heterogenen Gleichgewichtes* (Vieweg, Braunschweig 1924)
1.23 G. Tammann: Z. Anorg. Chem. *47*, 289 (1905)
1.24 G. Tammann: Z. Anorg. Chem. *45*, 24 (1905)
1.25 N. Kurnakov, S. Zhemchuzhny, M. Zasedatelev: Zh. Russ. Fiz. Khim. Ova. *47*, 871 (1916)
1.26 E.S. Shepherd, G.A. Rankin, F.E. Wright: Z. Anorg. Chem. *93*, 327 (1915)
1.27 J.H. Van't Hoff: *Die Bildung ozeanischer Salzablagezungen* (Braunschweig 1905-1911)
1.28 M.E. Chevreul: *Recherches chimiques sur les corps gras* (Paris 1889)
1.29 I. Timmermans: *Les solutions concentrées* (Masson, Paris 1936)
1.30 N.N. Efremov: Zh. Russ. Fiz. Khim. Ova. *45*, 348 (1913)

Chapter 2

2.1 J.V. Gibbs: *Scientific Papers*, Vol. 1, Thermodynamics (Dover, New York 1961)
2.2 A. Findley: *The Phase Rule and Its Applications* (Dover, New York 1951)
2.3 H.W.B. Roozeboom: Z. Phys. Chem. *28*, 494 (1899)
2.4 N.S. Kurnakov: *Vvedenie v fisikokhimicheskii analiz* [Introduction to physicochemical analysis] (Izdanija Akademii Nauk SSR, Moscow 1940)
2.5 W. Wendlandt: *Thermal Methods of Analysis* (Wiley and Sons, New York 1974)
2.6 W.T. Sproull: *X-Ray in Practice* (McGraw-Hill, New York 1946)
2.7 C.N. Fenner: Am. J. Sci. *36*, 331 (1913)
2.8 A.I. Kitaigorodsky, R.M. Myasnikova: Dokl. Acad. Nauk. USSR *129*, 380-383 (1959)
2.9 N.Ya. Kolosov: Sov. Phys. Crystallography *3*, 707 (1958)
2.10 R.M. Myasnikova, I.E. Kozlova, A.I. Pertsin, L.G. Radchenko: Kristallografia *21*, 964-970 (1976)
2.11 S.N. Kipot, R.M. Myasnikova: Journ. Appl. Chem. USSR *49*, 815-820 (1976)
2.12 L. Kofler, A. Kofler: *Thermo-Mikro-Methoden* (Universitätsverlag Wagner, Innsbruck 1954)

Chapter 3

3.1 W.B. Pearson: *The Crystal Chemistry and Physics of Metals and Alloys* (Wiley-Interscience, New York 1972)
3.2 G. Schultze: *Metallphysik* (Akademie, Berlin 1967)
3.3 W.B. Pearson: *Handbook of Lattice Spacings and Structures of Metals and Alloys*, Vols. 1 and 2 (Pergamon, Oxford 1958 and 1967)
3.4 A. Taylor: *X-Ray Metallography* (Wiley, New York 1961)
3.5 A.P. Wells: *Structural Inorganic Chemistry* (Clarendon, Oxford 1950)
3.6 R.C. Evans: *Einführung in die Kristallchemie* (Barth, Leipzig 1954)
3.7 A.I. Kitaigorodsky: *Organic Chemical Crystallography* (Consultantas Bureau, New York 1961)
3.8 A.I. Kitaigorodsky: *Molecular Crystals and Molecules* (Academic, New York 1973)
3.9 J. Dunitz: *X-Ray Analysis and the Structure of Organic Molecules* (Cornell University Press, Ithaca, NY 1978)
3.10 E.S. Fedorov: *Osnounye raboty* [Fundamental Works] (Nauka, Moscow 1949)
3.11 A. Schoenflies: *Kristallsysteme und Kristallstruktur* (Leipzig 1891)
3.12 M.J. Buerger: *Elementary Crystallography* (New York 1956)
3.13 W.H. Zachariasen: *Theory of X-Ray Diffraction in Crystals* (Wiley and Sons, New York 1945)
3.14 B.K. Vainstein: *Modern Crystallography I*, Springer Ser. Solid-State Sci., Vol. 15 (Springer, Berlin, Heidelberg, New York 1981)
3.15 E.M.H. Henry, K. Lonsdale (eds.): *International Tables for X-Ray Crystallography* (Kynoch, Birmingham 1952)
3.16 N.V. Belov: *Struktura ionykh kristallov* [Structure of ionic crystals] (Nauka, Moscow 1947)
3.17 L. Pauling, M. Huggins: Z. Kristallogr. *87*, 205 (1934)
3.18 S. Geller: Acta Crystallogr. *9*, 885 (1956); *10*, 687 (1957)
3.19 I.A. Wasastjerna: Soc. Sci. Fenn., Comm. Phys. Math. *38*, 22 (1923)
3.20 V.M. Goldschmidt: Skr. Nor. Vidensk. Akad. Oslo 1 N 2 (1926)
3.21 L. Pauling: *The Nature of the Chemical Bond*. 3rd ed. (Cornell University, Ithaca 1960)

3.22   W.H. Zachariasen: Z. Kristallogr. *80*, 137 (1931)
3.23   E. Teatum, K. Geschneider, A.C. Larson, M. Prince: Trans. Metall. Soc. AIME *227*, 717 (1963)
3.24   R.D. Shannon: Acta Crystallogr. *A32*, 751 (1976)
3.25   A.I. Kitaigorodsky: Acta Physicochim. USSR *21*, 575 (1946)
3.26   B.K. Belsky, P.M. Zorky: Sov. Phys. Crystallography *15*, 607 (1970)
3.27   A.G. Khachaturyan: Prog. Mater. Sci. *22*, 1 (1978)
3.28   J.W. Christian: *The Theory of Transformations in Metals and Alloys* (Pergamon, London 1975)
3.29   H. Warlimont: *Order-Disorder Transformations in Alloys* (Springer, Berlin, Heidelberg, New York 1974)
3.30   A.G. Khachaturyan: *Structure Changes Caused by Phase Transformations in Alloys* (Wiley and Sons, New York in press)
3.31   F. Laves: In *Theory of Alloy Phases* (Am. Soc. Metals, Cleveland 1956) pp. 124-204
3.32   W. Hume-Rothery: *The Structure of Metals and Alloys* (Institute of Metals, London 1936)
3.33   L. Vegard: Z. Phys. *5*, 17 (1921)

Chapter 4

4.1   A.I. Kitaigorodsky: *Molecular Crystals and Molecules* (Academic, New York 1973)
4.2   A.G. Khachaturyan: Prog. Mater. Sci. *22*, 1 (1978)
4.3   R.H. Fouler, E.A. Guggenheim: *Statistical Thermodynamics* (Cambridge University, Cambridge 1939)
4.4   M. Born, K. Huang: *Dynamical Theory of Crystal Lattices* (Clarendon, Oxford 1954)
4.5   J.M. Ziman: *Principles of the Theory of Solids* (Cambridge University, Cambridge 1964)
4.6   C. Kittel: *Introduction to Solid State Physics* (Wiley and Sons, New York 1956)
4.7   M.A. Krivoglaz: *The Theory of X-Ray and Thermal Neutron Diffraction from Real Crystals* (Plenum Press, New York 1969)
4.8   A.E. Guggenheim: *Mixtures* (Clarendon, Oxford 1952)
4.9   D. Prigogine, A. Bellemans, V. Mathot: *The Molecular Theory of Solutions* (North-Holland, Amsterdam 1957)
4.10   A.I. Kitaigorodsky: Dokl. Akad. Nauk USSR *137*, 116 (1961)
4.11   A.A. Maradudin, E.W. Montroll, G. Weiss: *Theory of Lattice Dynamics in the Harmonic Approximation* (Academic, New York 1963)
4.12   M. Born, Th. von Karman: Phys. Z. *13*, 297 (1912)
4.13   J.M. Ziman: *Electrons and Phonons* (Oxford University, Oxford 1955)
4.14   G.S. Pawley: Phys. Status Solidi *20*, 347 (1967)
4.15   R.P. Rinaldi, G.S. Pawley: J. Phys. C *8*, 599 (1975)
4.16   A.I. Kitaigorodsky, E.I. Mukhtarov: Opt. Spektrosk. *31*, 706 (1971)
4.17   A.I. Kitaigorodsky, E.I. Mukhtarov: Kristallografia *14*, 784 (1969)
4.18   W.L. Bragg, E.J. Williams: Proc. Roy. Soc. London *A152*, 231 (1935)
4.19   V. Gorsky: Z. Phys. *50*, 64 (1928)
4.20   W.A. Harrison: Solid State Phys. *24*, N 4 (1970)
4.21   H.A. Bethe: Proc. Roy. Soc. London *A150*, 552 (1935)
4.22   R. Peierls: Proc. Roy. Soc. London *A154*, 207 (1936)
4.23   J.G. Kirkwood: J. Chem. Phys. *6*, 70 (1938)
4.24   J.W. Christian: *The Theory of Transformations in Metals and Alloys* (Pergamon, London 1975)
4.25   D.A. Badalyan, A.G. Khachaturyan: Fiz. Tverd. Tela *14*, 2270 (1973)
4.26   J.C. Fischer: Acta Metall. *6*, 13 (1958)

4.27 G.L. Hall: J. Phys. Chem. Solids *3*, 210 (1957)
4.28 E.I. Fedin, A.I. Kitaigorodsky: Sov. Phys. Crystallography *15*, 607 (1970)
4.29 R.A. Johnson: Phys. Rev. *A134*, 1329 (1964)
4.30 R.A. Johnson: J. Phys. F *3*, 295 (1973)
4.31 L.D. Landau, E.M. Lifshits: *The Theory of Elasticity* (Pergamon, London 1963)
4.32 E.A. Guggenheim: Trans. Faraday Soc. *44*, 1007 (1948)
4.33 J.F. Freedman, A.S. Nowick: Acta Metall. *6*, 176 (1958)
4.34 I. Waller: Z. Phys. *17*, 398 (1923)
4.35 G.S. Pawley: Phys. Status Solidi *20*, 347 (1967)

Chapter 5

5.1 G. Tamann: *Textbook of Metallography* (Chemical Catalog Co., New York 1925)
5.2 A.A. Bochvar: *Mechanism of Alloy Crystallization*, ONTI (Moscow 1935), in Russian
5.3 Ju.N. Taran: Dokl. Akad. Nauk. USSR *193*, 143 (1970)
5.4 I.V. Salli: *Alloy Crystallization* (Naukova Dumka, Kiev 1974), in Russian
5.5 V.S. Stubican: Croat. Chem. Acta *48*, 423 (1976)
5.6 H.E. Cline, J.L. Walter: Metall. Trans. *1*, 2907 (1970)
5.7 W.J. Minford, R.C. Bradt, V.S. Stubican: J. Am. Ceram. Soc. *62*, 154 (1979)
5.8 A.G. Khachaturyan: Prog. Mater. Sci. *22*, 1 (1978)

Chapter 6

6.1 A.I. Kitaigorodsky: *Rentgenostrukturnii analis* [X-ray structure analysis] (Bostechizdat, Moscow 1950), in Russian
6.2 A.I. Kitaigorodsky: *Rentgenostrukturnii analis melkokristallicheskilch veschestv* [X-ray structure of polycrystalline substances] (Gostechisdat, Moscow 1952)
6.3 A.I. Kitaigorodsky: *The Theory of Crystal Structure Analysis* (Consultantas Bureau, New York 1961)
6.4 R.W. James: *Optical Principles of X-Ray Diffraction* (Bell, London 1958)
6.5 C.A. Taylor, H. Lipson: *Optical Transforms* (Bell, London 1964)
6.6 G.N. Ramachandran, R. Srinwasan: *Fourier Methods in Crystallography* (Wiley-Interscience, New York 1970)
6.7 G.B. Bokij, M.A. Poraj-Koschitz: *Rentgenostrukturnij analis* [X-ray structure analysis](Iydatel'stvo Moskovskogo Universiteta, Moscow 1964)
6.8 G. Schimmel: *Elektronenmikroskopische Methodik* (Springer, Berlin, Heidelberg, New York 1969)
6.9 G. Thomas: *Transmission Electron Microscopy of Metals* (John Wiley and Sons, New York 1962)
6.10 E.M.H. Henry, K. Lonsdale (eds.): *International Tables for X-Ray Crystallography* (Kynoch, Birmingham 1952)
6.11 A.V. Beznosikova, V.I. Iveronova: Zh. Eksp. Teor. Fiz. *8*, 81 (1938)
6.12 M.A. Krivoglaz: *The Theory of X-Ray and Thermal Neutron Diffraction from Real Crystals* (Plenum Press, New York 1969)
6.13 A.G. Khachaturyan: Prog. Mater. Sci. *22*, 1 (1978)
6.14 B.K. Vainstein: *Diffraktsia x-luchey na cepnuykh molekulah* (Nauka, Moscow 1963) [English transl.: *Diffraction of X-Rays by Chain Molecules* (Elsevier, Amsterdam 1966)]
6.15 M. Kakudo, N. Kassi: *X-Ray Diffraction by Polymers* (American Elsevier, Tokyo 1972)

Chapter 7

7.1 W.A. Harrison: *Pseudopotentials in the Theory of Metals* (Benjamin Company Inc., New York 1966)
7.2 N.S. Kurnakov: *Vvedenie v fisikokhimicheskii analiz* [Introduction to physicochemical analysis] (Izdanija Akademii Nauk SSR, Moscow 1940)
7.3 W. Hume-Rothery: Acta Metall. *14*, 21 (1966)
7.4 H. Jones: Proc. Roy. Soc. London *A144*, 225 (1934)
7.5 W.B. Pearson: *The Crystal Chemistry and Physics of Metals and Alloys* (Wiley-Interscience, New York 1972)
7.6 E. Parthé: Z. Kristallogr. *115*, 52 (1961)
7.7 W.B. Pearson: *Handbook of Lattice Springs and Structures of Metals and Alloys*, Vols. 1 and 2 (Pergamon, Oxford 1958 and 1967)
7.8 G. Schultze: *Metallphysik* (Akademie, Berlin 1967)
7.9 F. Laves, H. Witte: Metallwirtschaft *14*, 645 (1935)
7.10 H. Jones: *The Theory of Brillouin Zones and Electronic States in Crystals* (North Holland, Amsterdam 1962)

Chapter 8

8.1 W. Hume-Rothery: *The Structure of Metals and Alloys* (Institute of Metals, London 1936)
8.2 G.V. Raynor: Trans. Faraday Soc. *45*, 698 (1945)
8.3 H.I. Axon, W. Hume-Rothery: Proc. Roy. Soc. *A193*
8.4 W. Hume-Rothery, G.V. Raynor: *The Structure of Metals and Alloys* (Institute of Metals, London 1954)
8.5 B. Chaumers (ed.): *Phase Stability in Metals and Alloys* (McGraw-Hill, New York 1967)
8.6 Ya.S. Umansky, B.N. Finkelstein: *Fizicheskie osnovi metallovedenia* [Physical foundation of metals] (Metallurqizdat, Moscow 1955)
8.7 (Fizmatgiz, Moscow 1958) [English transl.: *The Theory of Order-Disorder in Alloys* (Macdonald, London 1965)]
8.8 M. Hansen, R. Anderko: *Constitution of Binary Alloys* (McGraw-Hill, New York 1958)
8.9 R. Elliott: *Constitution of Binary Alloys* (McGraw-Hill, New York 1962)
8.10 L. Vegard: Z. Phys. *5*, 17 (1921)
8.11 Zen E-an: Am. Mineral. *41*, 523 (1956)
8.12 J. Friedel: Philos. Mag. *3*, 446 (1954)
8.13 P.S. Rudman, B.L. Averbach: Acta Metall. *2*, 576 (1954)
8.14 L. Darken, R.W. Gurry: *Physical Chemistry of Metals* (McGraw-Hill, New York 1953)
8.15 I.T. Waber, K. Geschneider, A. Larson: Trans. Metall. Soc. AIME *227*, 717 (1963)
8.16 N. Kurnakov, S. Zhemchuzhny, M. Zasedatelev: Zh. Russ. Fiz. Khim. Ova. *47*, 871 (1916)
8.17 L.D. Landau: Zh. Eksp. Teor. Fiz. *11*, 545 (1937)
8.18 E.M. Lifshits: Zh. Eksp. Teor. Fiz. *11*, 2 (1941)
8.19 A.G. Khachaturyan: Proq. Mater. Sci. *22*, 1 (1978)
8.20 S.C. Moss: J. Appl. Phys. *35*, 3547 (1964)
8.21 S.V. Semenovskaya: Phys. Status Solid *64(b)*, 291 (1974)
8.22 S.V. Semenovskaya, D. Umidov: Phys. Status Solidi *64(b)*, 627 (1974)
8.23 J.M. Cowley: J. Appl. Phys. *22*, 483 (1950)
8.24 [Short-range order in solid solutions] (Nauka, Moscow 1977); *Modern Problems of Short-Range Order* (Springer, Berlin, Heidelberg 1974) p.306
8.25 V.I. Iveronova, A.A. Katsnelson: Izv. Vuzov. Fiz. *8*, 43 (1976)

Chapter 9

9.1   V.S. Urusov: *Theoria isomorfnoi smesimosti* [Theory of isomorphic mixturing] (Nauka, Moscow 1977)
9.2   H. Hohn, W. Semann, H. John: Z. Anorg. Allg. Chem. *369*, 48 (1969)
9.3   V. Posypaiko (ed.): *Diagrammy plavkosti solevykj sistem* [Melting diagrams of salt systems] (Metallurgia, Moscow 1977)
9.4   N.B. Chanh: J. Chim. Phys. *61*, 1428 (1964)
9.5   N.B. Chanh, J.P. Bastide: J. Chim. Phys. *65*, 1425 (1968)
9.6   N.B. Chanh: J. Chim. Phys. *63*, 1181 (1966)
9.7   N.B. Chanh: J. Chim. Phys. *62*, 569 (1965)
9.8   N.B. Chanh, L. Pelissac: Bull. Soc. Fr. Minéral. Cristallogr. *91*, 13 (1968)
9.9   Y. Haget, N.B. Chanh, P. Garin, L. Bonput: Acta Metall. *23*, 723 (1975)
9.10 L. Bonput, N.B. Chanh, Y. Haget: J. Chim. Phys. *71*, 533 (1974)
9.11 I. A. Wasastjerna: Soc. Sci. Fenn., Comm. Phys. Math. *15*, N 3 (1949)
9.12 V. Hovi: Soc. Sci. Fenn., Comm. Phys. Math. *15*, N 12 (1950)
9.13 V. Hovi: Acta Metall. *2*, 112 (1954)
9.14 Yu. Vlasov, L. Makarov: Zh. Fiz. Khim. *41*, 320 (1967)
9.15 V. Hovi: Acta Metall. *6*, N 4 (1958)
9.16 I. Heitala: Ann. Acad. Sci. Fenn. Ser. *A6*, N 122 (1963)
9.17 V.S. Urusov: Izv. Akad. Nauk SSR, Mater. *5*, 705 (1969)
9.18 P.P.M. Groenewegen, C. Huiszoon: Acta Crystallogr. *A28*, 166 (1972)
9.19 E.W. Kellerman: Phil. Trans. R. Soc. London, Ser. *A238*, 513 (1940)
9.20 V.S. Urusov, V.B. Dudnikova, A.V. Garanın: Phys. Status Solidi *B102*, 695 (1980)
9.21 P.L. Cloke: Bull. Miner. *104*, 186 (1981)
9.22 M.J.L. Sangster, A.M. Steheman: Phil. Mag. *B43*, 597 (1981)
9.23 A.G. Khachaturyan: Prog. Mater. Sci. *22*, 1 (1978)
9.24 J.W. Christian: *The Theory of Transformations in Metals and Alloys* (Pergamon, London 1975)
9.25 R. Powell: In *Non-stoichiometric Compounds*, ed. by L. Mandelcorn (Academic, New York 1964) Chap.7, Sect.G

Chapter 10

10.1   N.B. Chanh, Y. Bouillard, P. Lencrerot: J. Chim. Phys. *67*, 1206 (1970)
10.2   R.M. Myasnikova, A.I. Kitaigorodsky: Kristallografia *3*, 160 (1958) [English transl.: Sov. Phys. Crystallogr. *3*, 157 (1958)]
10.3   N. Kolosov: Zh. Strukt. Khim. *6*, 926 (1965)
10.4   A.I. Kitaigorodsky, R.M. Myasnikova: Dokl. Akad. Nauk USSR *129*, 380 (1959)
10.5   A.I. Kitaigorodsky, Yu.V. Mnyukh: Vysokomol. Soedin. *1*, 128 (1959)
10.6   Yu.V. Mnyukh: Zh. Strukt. Khim. *1*, 370 (1960)
10.7   A.I. Kitaigorodsky, R.M. Myasnikova, S.A. Remiga: Kristallografia *12*, 900 (1967)
10.8   R.M. Myasnikova, A.I. Pertsin, I.E. Kozlova, A.I. Kitaigorodsky: Kristallografia *21*, 511 (1976)
10.9   A.I. Kitaigorodsky: *Organic Chemical Crystallography* (Consultantas Bureau, New York 1961)
10.10 N.F. Mott, F.R. Nabarro: Proc. Phys. Soc. London *52*, 86 (1940)
10.11 A.I. Kitaigorodsky: *Molecular Crystals and Molecules* (Academic, New York 1973)
10.12 N.B. Chanh, Y. Haget, J. Bedoin: Bull. Soc. Fr. Minéral. Cristallogr. *95*, 281 (1972)
10.13 Y. Bouillaud, N.B. Chanh, J. Bedoin: Bull. Soc. Fr. Minéral. Cristallogr. *93*, 300 (1970)

10.15 Y. Haget, N.B. Chanh, A. Corson, A. Meresse: Mol. Cryst. Li. Cryst. *31*, 93 (1975)
10.16 F. Baumgarth, N.B. Chanh, R. Gay, J. Lascombe: J. Chin. Phys. *66*, 862 (1969)
10.17 A.I. Kitaigorodsky, R.M. Myasnikova: Kristallografia *5*, 638 (1968) [English transl.: Sov. Phys. Crystallogr. *5*, 610 (1960)]
10.18 G. Belikova, L. Belyaev: *Sbornik Rost Kristallov* [Ctystal growth, collection], Vol. 3 (Isdatel'stvo Akademi Nauk, Moscow 1961) p.316; Isotopen Praxis *3*, 133 (1967)
10.19 E.V. Shpol'sky: Usp. Fiz. Nauk *80*, 255 (1963)
10.20 L. Pakhomov, E. Dazenko: Zh. Fiz. Khim. *54*, 1596 (1980)

Chapter 11

11.1 A.I. Kitaigorodsky: *Molecular Crystals and Molecules* (Academic, New York 1973)
11.2 J.M. Robertson, H.M.M. Shearer, G.A. Sim, D.G. Watson: Acta Crystallogr. *15*, 1 (1962)
11.3 N.B. Chanh, Y. Haget: Acta Crystallogr. *BA28*, 3400 (1972)
11.4 T.C.W. Mak, J. Trotter: Acta Crystallogr. *15*, 1078 (1962)
11.5 N.B. Chanh, Y. Bouillard, P. Lencrerot: J. Chim. Phys. *67*, 1198 (1970)
11.6 N.B. Chanh, Y. Haget, F. Hannoteaux, N. Chezeau: J. Chim. Phys. *72*, 670 (1975)
11.7 A. Fourme: C. R. Acad. Sci. *268*, 991 (1969)
11.8 A.I. Kitaigorodsky: *Chemical Organic Crystallography* (Consultantus Bureau, New York 1961) p.470

Chapter 12

12.1 I. Timmermans: *Les solutions concentrées* (Masson, Paris 1936)
12.2 A.I. Kitaigorodsky, R.M. Myasnikova, S.A. Remiga: Kristallografia *12*, 900 (1967)
12.3 A.I. Kitaigorodsky, L.Yakushevich: Fiz. Tverd. Tela *13*, 2386 (1971)
12.4 S.A. Remiga, R.M. Myasnikova, A.I. Kitaigorodsky: Kristallografia *10*, 875 (1965)
12.5 O. Osipov: *Handbook on Dipole Moments* (Vysshaya Shkola, Moscow 1971) in Russian; A.L. McClellan: *Dipole Moments* (Arnold, London 1971)
12.6 S. Remiga, R.M. Myasnikova, A.I. Kitaigorodsky: Kristallografia *10*, 1131 (1969)
12.7 S.A. Remiga, R.M. Myasnikova, A.I. Kitaigorodsky: Kristallografia *15*, 599 (1971)
12.8 R.M. Myasnikova, I.E. Kozlova, A.I. Pertsin, L.G. Radchenko: Kristallografia *21*, 964 (1976); Sov. Phys. Crystallogr. *3*, 157 (1958)
12.9 H.W.W. Ehrlich: Acta Crystallogr. *10*, 699 (1957)
12.10 V.G. Dashevsky: *Conformation of Organic Molecules* (Khimiya, Moscow 1974) in Russian
12.11 R.M. Myasnikova, A.I. Pertsin, I.E. Kozlova, A.I. Kitaigorodsky: Kristallografia *21*, 511 (1976)
12.12 N.B. Chanh, Y. Haget: Acta Crystallogr. *BA28*, 3400 (1972)
12.13 A.I. Kitaigorodsky, R.M. Myasnikova: Dokl. Akad. Nauk USSR *129*, 380 (1959)
12.14 P.M. Robinson, H.G. Scott: Mol. Cryst. Liq. Cryst. *5*, 405 (1969)
12.15 J.W. Harris: Nature (London) *206*, 1038 (1965)

12.16 R.M. Myasnikova, T.S. Davydova, V.I. Simonov: Kristallografia *18*, 720 (1973)
12.17 E. Gavuzzo, F. Mazza, E. Giglio: Acta Crystallogr. *B30*, 1351 (1974)
12.18 A.I. Kitaigorodsky: *Organic Chemical Crystallography* (Consultants Bureau, New York 1961)
12.19 A.I. Kitaigorodsky: *Molecular Crystals and Molecules* (Academic, New York 1973)
12.20 T. Penkala: Ann. Uni. Mariae Curie-Skladowska *11*, 77 (1956)
12.21 N.B. Chanh, Y. Haget, J. Bedoin: Bull. Soc. Fr. Minêral. Cristallogr. *95*, 281 (1972)
12.22 Y. Bouillaud, N.B. Chanh, J. Bedoin: Bull. Soc. Fr. Minêral. Cristtalogr. *93*, 300 (1970)
12.23 N.B. Chanh, Y. Haget, F. Hannoteaux, N. Chezeau: J. Chim. Phys. *72*, 670 (1975)
12.24 J.M. Robertson, H.M.M. Shearer, G.A. Sim, D.G. Watson: Acta Crystallogr. *15*, 1 (1962)
12.25 N.B. Chanh, Y. Bouillard, P. Lencrerot: J. Chim. Phys. *67*, 1206 (1970)
12.26 Y. Haget, N.B. Chanh, A. Corson, A. Meresse: Mol. Cryst. Liq. Cryst. *31*, 93 (1975)
12.27 N.B. Chanh, Y. Bouillard, P. Lencrerot: J. Chim. Phys. *67*, 1198 (1970)
12.28 N.B. Chanh, Y. Haget, A. Meresse, J. Housty: Mol. Cryst. Liq. Cryst. *45*, 307 (1978)
12.29 Y. Haget, N.B. Chanh, A. Meresse: Mol. Cryst. Liq. Cryst. *32*, 49 (1976)
12.30 N.B. Chanh, Y. Haget, L. Bonpunt, A. Meresse, J. Housty: Anal. Calorimetry *4*, 233 (1977)
12.31 Y. Haget, N. Chezeau, A. Meresse, J. Housty, N.B. Chanh: Mol. Cryst. Liq. Cryst. *55*, 109 (1979)
12.32 F. Baert: "Etude des structures de melange d'antipodes optiques"; Dissertation, Université de Lille, *N349* (1976)
12.33 M. Foulton, F. Baert, R. Fouret: Acta Crystallogr. *B35*, 683 (1979)
12.34 J. Kroon, P.R.E. van Gupp, N.A.J. Oonk, F. Baert, R. Fouret: Acta Crystallogr. *B32*, 2561 (1976)
12.35 F. Baert, R. Fouret, N.A.J. Oonk, J. Kroon: Acta Crystallogr. *B34*, 222 (1978)
12.36 E. Abovyan, R.M. Myasnikova, A.I. Kitaigorodsky: Kristallografia *22*, 1219 (1977); *23*, 768 (1978) [English transl.: Sov. Phys. Crystallogr. *22*, 693 (1977); *23*, 438 (1978)]
12.37 K.V. Mirskaya, I.E. Kozlava, V.F. Bereznitskaya: Phys. Status Solidi *62*, 291 (1974)
12.38 E. Burgos, N. Bonadeo: Chem. Phys. Lett. *49*, 475 (1977)
12.39 A.I. Kitaigorodsky, R. Myasnikova, V. Samarskaya: Kristallografia *8*, 393 (1963) [English transl.: Sov. Phys. Crystallogr. *8*, 308 (1963)]
12.40 V. Samarskaya, R. Myasnikova, A.I. Kitaigorodsky: Kristallografia *13*, 616 (1968) [English transl.: Sov. Phys. Crystallogr. *13*, 525 (1968)]
12.41 B.Z. Julkowska, R.M. Myasnikova, A.I. Kitaigorodsky: Zh. Strukt. Khim. *5*, 737 (1964)
12.42 S.V. Semenovskaya: Phys. Status Solid *64(b)*, 291 (1974)
12.43 S.V. Semenovskaya, D. Umidov: Phys. Status Solidi *64(b)*, 291 (1974)
12.44 S.V. Semenovskaya, A.A. Smirnova, A.I. Kitaigorodsky: Kristallografia *23*, 99 (1978) [English transl.: Sov. Phys. Crystallogr. *23*, 53 (1978)]
12.45 S.V. Semenovskaya, A.I. Kitaigorodsky, A.A. Smirnova: Kristallografia *23*, 299 (1978) [English transl.: Sov. Phys. Crystallogr. *23*, 166 (1978)]
12.46 A.B. Khachaturyan: Prog. Mater. Sci. *22*, 1 (1978)

Chapter 13

13.1 G.Y. Chao, J.D. Mc Cullough: Acta Crystallogr. *13*, 727 (1960)
13.2 G.Y. Chao, J.D. Mc Cullough: Acta Crystallogr. *14*, 940 (1961)
13.3 H. Hope, I.D. Mc Cullough: Acta Crystallogr. *17*, 712 (1964)
13.4 O.D.D. Holmesland, C.H.R. Romming: Acta Chem. *20*, 2601 (1966)
13.5 T. Bjorvatten, O. Hassel: Acta Chem. Scand. *15*, 1429 (1961)
13.6 T. Bjorvatten, O. Hassel: Acta Chem. Scand. *16*, 249 (1962)
13.7 T. Bjorvatten: Acta Chem. Scand. *20*, 1863 (1966)
13.8 P.L. Brun, C.I. Branden: Acta Crystallogr. *20*, 749 (1966)
13.9 A.I. Kitaigorodsky, A.A. Frolova: Izv. Sekt. Fiz. Khim. Anal. Inst. Obshch. Neorg. Khim. Akad. Nauk SSSR *19*, 306 (1949)
13.10 D.S. Brown, S.C. Wallwork, A. Wilson: Acta Crystallogr. *17*, 168 (1964)
13.11 A.W. Hanson: Acta Crystallogr. *17*, 559 (1964)
13.12 D.S. Brown, S.C. Wallwork: Acta Crystallogr. *19*, 149 (1965)
13.13 A. Damiani, E. Giglio, A.M. Liquori, A. Ripamonti: Acta Crystallogr. *23*, 675 (1967)
13.14 D. Mac Nicol, J. Mc Kendrick, D.R. Wilson: Chem. Soc. Rev. *7*, 65 (1978)
13.15 G.N. Chekhova, T.M. Polyanskaya, Yu.A. Dyadin, V.I. Alekseev: Zh. Strukt. Khim. *16*, 1054 (1975)
13.16 L. Mandelcorn (ed.): *Non-stoichiometric Compounds* (Academic, New York 1964)
13.17 G.N. Chekhova, Yu.A. Dyadin: Izv. Sib. Otd. Akad. Nauk SSSR *12*, 75 (1978)
13.18 B.A. Nikitin: Nature *140*, 643 (1937)
13.19 D.E. Palin, H.M. Powell: Nature *156*, 334 (1945)
13.20 J.C. Platteuw, J.H. van der Waals: Mol. Phys. *1*, 91 (1958)
13.21 G.A. Jeffrey, R.K. Mc Mullan: *Progress in Inorganic Chemistry*, Vol.8 (Interscience, New York 1967)
13.22 G.A. Jeffrey: Acc. Chem. Res. *2*, 344 (1969)
13.23 Yu.A. Dyadin, L. Aladko: Zh. Strukt. Khim. *18*, 1 (1977)
13.24 Yu.A. Dyadin, T.S. Terekhova: Zh. Strukt. Khim. *17*, 655 (1976)
13.25 W. Saenger: Nature *280*, 848 (1979)
13.26 G.N. Chekhova, Yu.A. Dyadin: Izv. Sib. Otd, Akad. Nauk SSSR *N5*, 78 (1979)
13.27 A.I. Kitaigorodsky: *Molecular Crystals and Molecules* (Academic, New York 1973)
13.28 D. Lawton, H.M. Powell: J. Chem. Soc. 2339 (1958)
13.29 R. Arad-Yellin, B.S. Green, M. Knossow, G. Tsoucaris: Tetrahedron Lett. *21*, 387 (1980)
13.30 K. Harata: Bull. Chem. Soc. Jpn. *50*, 1416 (1977)
13.31 A. Allemand, R. Gerdil: Acta Crystallogr. *A31*, 5130 (1975)
13.32 A.P. Dianin: Zh. Russ. Fiz. Khim. Ova. *46*, 1310 (1914)
13.33 J. Gall, A. Hardy, J. Mc Kendrik, D. Mac Nicol: Chem. Soc. Perkin Trans. *2*, 376 (1979)
13.34 A. Hardy, J. Mc Kendrik, D. Mac Nicol: J. Chem. Soc. Perkin Trans. *2*, 1072 (1979)
13.35 A. Hardy, D. Mac Nicol, D. Wilson: J. Chem. Soc. Perkin Trans. *2*, 1011 (1979)
13.36 A. Hardy, D. Mac Nicol, S. Swanson, D. Wilson: J. Chem. Soc. Perkin Trans. *2*, 999 (1980)
13.37 A. Freer, C. Gilmore, D. Mac Nicol, D. Wilson: Tetrahedron Lett. *21*
13.38 H.R. Allcock, R.W. Allen, E.C. Bissel: J. Am. Chem. Soc. 5120 (1976)
13.39 Communications of conference on clathrates and inclusion compounds: J. Mol. Structure *75*, 1-138 (1981)
13.40 A.I. Kitaigorodsky: *Organic Chemical Crystallography* (Consultantas Bureau, New York 1961)
13.41 L. Leichester, H. Bradley: Chem. Ind. (London) 1449 (1955)

13.42 T. Iwamoto, T. Miyoshi, Y. Sasaki: Acta Crystallogr. *B30*, 292 (1974)
13.43 A.R. Ubbelohde, F.A. Lewis: *Graphite and Its Crystal Compounds* (Clarendon, Oxford 1960)
13.44 G.R. Hennig: In *Progress in Inorganic Chemistry*, Vol.1, ed. by F.A. Cotton (Interscience, New York 1959) p.127
13.45 Yu.N. Novikov, M.E. Volpin: Usp. Khim. *40* (9), 733 (1971)
13.46 W. Metz, D. Hohlwein: Carbon *13*, 87 (1975)
13.47 D.D.L. Chung: J. Electron. Mater. *7*, 189 (1978)
13.48 E. Buscarlet, Ph. Touzain, L. Bonnetain: Carbon *14*, 75 (1976)

Chapter 14

14.1 B. Wunderlich: *Macromolecular Physics* (Academic, New York 1973)
14.2 A.A. Tager: *Fizsikokhimia polimerov* [Physical chemistry of polymers] (Khimia, Moscow 1978)
14.3 P.J. Flory: *Statistical Mechanics of Chain Molecules* (Wiley Interscience, New York 1969)
14.4 H.A. Stuart: *Die Physik der Hochpolymeren*, Vols.1-4 (Springer, Berlin 1952-1956)
14.5 R. Howard (ed.): *The Physics of Glassy Polymers* (Applied Science, London 1973)
14.6 P.H. Geil: *Polymer Monocrystals* (Wiley, New York 1958)
14.7 A.I. Kitaigorodsky: *Molecular Crystals and Molecules* (Academic, New York 1973)
14.8 A. Dobry, F. Boyer: J. Polym. Sci. *2*, 90 (1947)
14.9 R.L. Scott: J. Chem. Phys. *17*, 279 (1949)
14.10 G.L. Slonimsky: J. Polym. Sci. *30*, 625 (1958)
14.11 T.G. Fox: Bull. Am. Chem. Soc. *2*, 123 (1956)
14.12 P. Debye, H.R. Anderson, H. Brumberger: J. Appl. Phys. *28*, 679 (1957)
14.13 R.S. Stein, F.B. Khambatta, F.P. Warner, T. Russel, A. Escala, E. Balizer: J. Polym. Sci. *63*, 313 (1978)
14.14 D. Ya. Tsvankin: Vysokomol Soedin *6*, 2078, 2083 (1964)
14.15 D. Ya. Tsvankin, Yu.A. Zubov, A.I. Kitaigorodsky: J. Polym. Sci. (C) *16*, 4081 (1969)
14.16 F.P. Warner, W. Mac Knigt, R.S. Stein: J. Polym. Sci. *15*, 2113 (1977)
14.17 H.D. Keith, F.J. Padden: J. Appl. Phys. *35*, 1270 (1975)
14.18 B. Gallot: Pure Appl. Chem. *38*, 1 (1974)
14.19 S.S. Tarasov, D.Ya. Tsvankin, Yu.B. Godovsky: Vysokomol. Soedin. *20A*, 1534 (1978)
14.20 A. Keller, E. Pedemonte, F. Willmouth: Kolloid Z. *238*, 385 (1970)
14.21 J.M. Thomas: *Review of Pure and Applied Chemistry Proceedings* (Upac Meeting, York, September 1978)
14.22 J.M. Thomas: Phil. Trans. Roy. Soc. (London) *A277*, 251 (1974)
14.23 G. Wegner: Pure Appl. Chem. *49*, 433 (1977)
14.24 G. Gerasimov, M. Bazilevsky, V.A. Tikhomirov, A. Abkin: Dokl. Akad. Nauk USSR *244*, 1379 (1979)
14.25 R.H. Baughman: J. Polym. Sci., Polym. Phys. Ed. *11*, 603 (1973)
14.26 A.Sh. Cherdabayev, D.Ya. Tsvankin: Vysokomol. Soyed. *19*, 1237 (1977)
14.27 R.M. Myasnikova: Vysokomol. Soyed. *19*, 564 (1977)
14.28 A.Sh. Cherdabayev, D.Ya. Tsvankin: Vysokomol. Soedin. *18*, 2523 (1976)
14.29 R.M. Myasnikova, E.F. Titova, E.S. Obolonkova: Polymer *21*, 403 (1980)
14.30 A.A. Askadsky: Vysokomol. Soedin *19*, 1004 (1977)

Chapter 15

15.1   M.O. Dayhoff (ed.): *Atlas of Protein Sequence and Structure* (National Biomedical Research Foundation, Washington 1972)
15.2   A.V. Efimov: J. Mol. Biol. *134*, 23 (1979)
15.3   V.I. Lim, O.B. Ptitsyn: Biofizika *17*, 21 (1972)
15.4   V.G. Dashevsky, G.S. Sarkisov: Mol. Phys. *27*, 1271 (1974)
15.5   A.I. Kitaigorodsky: *Organic Chemical Crystallography* (Consultantas Bureau, New York 1961)
15.6   A.I. Kitaigorodsky: *Molecular Crystals and Molecules* (Academic, New York 1973)
15.7   A.A. Askadsky: Vysokomol. Soedin *19*, 1004 (1977)
15.8   A.A. Makarov, D.R. Monaselidze, N.G. Esipova: Int. J. Quantum Chem. *16*, 437 (1979)
15.9   N.G. Esipova: Mol. Biol. (Moscow) *12*, 1152 (1978)
15.10 M. King, J. Bello, E. Pignataro, D. Harker: Acta Crystallogr. *15*, 144 (1962)
15.11 V.I. Poltev: Int. J. Quantum Chem. *16*, 153 (1979)
15.12 K. Tatchell, K. van Holde: Proc. Nat. Acad. Sci. (USA) *75*, 3583 (1978)
15.13 B.K. Vainstein: *Modern Crystallography I*, Springer Ser. Solid-State Sci., Vol. 15 (Springer, Berlin, Heidelberg, New York 1981)
15.14 R. Kilkson, M. Maestre: Nature *195*, 494 (1962)

Subject Index

Additional intermolecular bond 281
Amorphous polymer 323
Amplitude of scattering 128
Anion radii 70
Asymmetric molecules 206
Atom-atom interaction 102
Atomic displacements in the lattice 107
- factor 130
- radii 61
- scattering 128
Attractive forces 68
Auxiliary gas 294
Average atom 134
- molecule 235
- radii 65
- unit cell 104,133
Azimuthal disorder 323

Bacterial cell 365
Bacteriophage 368
Biological macromolecules 354
Block copolymer 328,330
Body-Centered Cubic Structure 67
Boltzmann's formula 88
Bridgman technique 46

Cation radii 70
Chains of atoms 64
Channel compounds of tri-o-thimotide 302
- in thiourea 300
Channel-shaped cavities 286
Characteristic temperature 109
Charge exchange 283
- transfer complex 277
Chromosomes 365
Clathrate 285,290
- compounds formed with the aid of auxiliary gas 294
- hydrates 286,289
Close-Packed Spheres 60
Close packing 73
- - principle 157
- - principle for intermetallic compound 162
Cluster 93,158,159,160
Coefficient of correlation 137
- of geometric similarity 201
- of packing 62
- of increase in density 82,147
- of volume expansion 63
Coherent interfaces 122
- scattering 127
Complex 33,275,303
- based on tri-o-thimotide molecules 302
- formed by cyclodextrins 303
- of a polymer and a low-molecular-weight compound 350
- of nucleic acid and proteins 365
- with charge transfer 318

Component   19
Composite materials   20
- of polymers   327
Compton scattering   138
Concentration disorder   136
- waves   172
Condition of equilibrium   21
Configurational entropy   88
Conformation of molecules   89
Conformational disorder   218
- transformation   18
- transitions of molecules   218
Conformers   18
Connode   26,27
Continuous loss of symmetry   206,278
Coordination number   67,74,150
- polyhedra   70
- spheres   90
Correlation function   80
- coefficient   137
- parameter   261
Critical temperature   93
Crystal chemistry   49
Crystalline the spherulite   325
Crystallization front movement   122
Crystallographically independent positions   74

Daltonides   144
Debye-Waller factor   195
Decomposition   93
- of solid solutions   36,39,91,123,185
Deformation of block copolymer   332
- of polymer eutectic mixture   349
Degree of crystallinity   19,322
- of isomorphism   83
- of long-range order   96,176
- of spirality of a polymer chain   323
Dendritic growth   122

Dense packing in biological system   355
Density of packing   162
Differential thermal analysis   40
Diffusion X-ray scattering   89,268
Dilute solutions   93
Dipole-dipole interaction energy   241
Dipole moment   83
Disorder due to admixture molecules   223
Dissolution of a polymer in a low-molecular solvent   344
Distortion of lattice   179
DNA crystals   362
DNA-water system   364
Double clathrate hydrate   294
Double helix of DNA molecule   355,362

Effective atomic charge   194
Elastic stretching   123
Elastomer   327
Elements of symmetry   51
Electrical charges of atoms and ions   49
Electron component of free energy   88
- compound   145,162
- exchange   142
Electrostatic energy of interaction   71
- potential   68,191
Energy of atomization   194
- of mixing   92,138,192
- - for dilute solutions   97
- substitution   90
Equilibrium curves   19
- equations   25
- shape of crystals   124
Eutectic colony   112,119
- crystallite   115
- crystallization   112,114,115,116
- grain   112,115

Eutectic (continued)
- phase diagram 47,96
- point 27
- structure 122
Exchange of electron 194
Expansion coefficient 142
Experimental determination of the mixing energy 273

Flat layers 65
Flexibility 320
Folded chains 322
Formation of chemical compounds 85
Framework built of hydrogen bonds 285
- built of water molecules 291
- formed by rigid molecules 311
- with large voids 309
Free energy 20,21,101,103,175
- - of the crystal 94
- - of mixing 88
- - of a solid solution 89
Free rotation inside the cavity 314

Gas hydrate 275
General position 52,57
Geometric analysis 83,200
- - of intermetallic compound 147
- - of substitution 201
Geometric factor role in the formation of complexes 304
Geometrical model 60,63
Glide planes 53
Goldschmidt rule 188
Growth of mixed single crystals 43

Heat of formation of solid solution 193
Heat of mixture 188
Heterophase mixture 18

Heterophase system 111
"Hexa host" inclusion compounds 310
Hexahydrates of alcohols 297
Homopolymer 324
Hume-Rothery rule 82,146,188
Hydrates 286
Hydrophobic interaction 357
Hydroquinone-based clathrates 287

Icosahedron 158,161
Inclusion compounds 283,290
- - of graphite 316
Inclusion of molecules into intermolecular voids 305
- - into intramolecular cavities 301
Incompatibility of polymers 325
Increase in density parameter 82
Induction effect 280
Industrial purification 46
Inorganic solid solutions 181
Insolubility of organic systems 201
Intensity absolute scale 269
- of diffuse scattering 177,269
Interaction energy 98,106
- - calculation 255
- - of molecules 200
- potential 143
Interblock solubility 180,211
Intermediate phase 34,76,145,169
Intermetallic compound 17,142
Intermolecular hydrogen bonds breakdown 203
- radii 63,71
- vibrations 110
Internal energy 21
Interphase surface energy 122
Interstitial alloys 101
Interstitial phase 146
Interstitial solid solution 195,200,286,287

Ionic radii 61
Isomorphism 83
Isomorphous substitution 181,189
Isotherms of free energy 28

Khachaturyan's method 103
Kinetics of decomposition of solid solution 186

Lattice distortion 104,132,167
Lattice dynamics 88
Layer complexes 275,314
Limited solubility 31,93,96
Linear block copolymer 332
Liquidus curve 29,44
Long-range order 77,101,176

Macromolecular conformation 72
Macromolecule 319
Madelung constant 193
Mean square amplitudes of thermal vibration 110
Method of atom-atom potentials 105, 106,200
- of contact specimen 46
Microscopic investigation 126
Microstructure of heterophase systems 85
Mixing energies 268
Mixtures of alkali halide salts 181
- of enantiomers 40
- of low-molecular substance and polymer 350
Model of cut sphere 150
Model of solid continuum 108
Molecular compounds 17,76
Molecular crystals 75
Molecular engineering 340
Molecule of DNA 362
Molecule packing 72

Muscle tissues 354
Mutual solubility of substances 85

Natural composites 119
Noncontrolled crystallization 112
Nonequivalent positions 58
Nonstoichiometric chemical compounds 196
Normal lattice vibration 87
Nuclear quadrupole resonance 107
Nucleation 123
Nucleosome 365
Nucleotide 364

Octahedral void 61,69,101,146
Optical enantiomers 39
Order-disorder transition 172
Ordering 85,171
Organic solid solution 211
Orientational disorder 230
- - imitating the center of symmetry 245
- - in binary systems 203
- - one-component crystal with 219
Oriented short-range order correlation 260

Packets of molecules 321
Packing coefficient 62,65,74,149,283
Packing coefficient of clathrate 293
Packing complexes 275,283,308,312
Packing density 66,280
Packing of block copolymer 328
- of DNA inside the viral particles 368
- of particles 50
- of particles in disordered solid solution 84
- of polymer molecules 321
- of spherical atoms 62

Pairwise interaction 101,200,257,259
Pairwise potentials 96
Parameter of correlation 81
- of geometric similarity 219
- of long-range order 77
- of short-range order 179
Parasitic scattering 138
Partial ordering 89
Particle displacements 132
Pentagonal dodecahedra clathrate hydrates 291
Periodic domain microstructure 125
Peritectic diagram 47
- point 32
- reaction 168
Permutations of particles 101
Phase equilibrium 17,19
Phase diagram calculation 209
- transition of the second kind 171
Physico-chemical analysis 40
Plasticizer 344
Plastics 327
Point group 51,52
Polymer blend 324
Polymeric molecule 319
Polymorphic modification 17,36,63,142
- transformation 37
Polyphase equilibrium 22
Positional disorder 230
- ordering 217
Potential energy of interaction 91,162
Primitive cell 50
Principle of close packing 73
Probability calculation of different guest molecule orientations 257
Protein 19
- crystal 20,358
- crystallization 360
- globules 356

Protein-water system 358
Pseudo-atom 147
Pseudo-potential 143,172
Pseudoracemate 40,206

Quasi-harmonic approximation 209
Quasi-valence bond 275
Quasi-valence bonds between component molecules 275
Quenching from melt 42

Racemate without inversion center 249
Reciprocal lattice 127
Regular solution 89
Reliability factor 140
Repulsive forces 63
Ribonuclease molecules 361
Rotation axes 51,57
Rules of Vegard 167

Scattering 126
- due to lattice distortions 270
- caused by short-range order 270
Screw axis 57
Segment copolymer 331
Self-consistent field theory 102
Selective reflection 127
Semiclathrate 295
Series of electron density 140
Short range coefficient 177
- - component of potential 199
- - order 80,91,108,109,136,167,177,178,179
- - order of the segregation type 271
- - parameters 81,268
- - part of the energy of interaction 284
Single particle probability 257

Solid solution   164,343
- - in clathrates   293
- - of optically active substance   248
- - of polymer   339
- - of polymer in a monomer   340
- - single crystal   227
Solubility   89,98,231
- of low-molecular substances   347
Solvate   286
Special position   48
Spherical viruses   367
Spherulites   323,443
Spiral growth   321
Static distortion   105,132
Stoichiometric composition   28
Structure factor   128,139
Structure factor of solid solution   132
Substitution energy calculation   255, 267
Substitutional solid solution   46,74, 82,104,200
Superatom   147,156,157,160
Superstructure   75,76,135,157,172
Symmetry transformation   51

Technology of polymer blends   327
Temperature factor   132
- of transformation of the second kind   173
Terminal solid solution   167
Tetrahedral radii   61
- voids   61,69,101,146
Texture of polymer crystallites   349
Thermal motion   91
- scattering   138
- vibrations   109,130
Thermodynamic potential   20
Tobacco mosaic virus   366
Topochemical reaction   340

Transformation in solid state   36
Transitions of the first kind   171
Triple point   27
Two-phase mixtures of antipodes   40

Unidirectional crystallization   119,121
Unit cell   50
Urea channels   298
- complexes   298

Vacancy   108,196,197
Valence forces   64
Vibration frequencies in molecular crystals   88
Vibrational spectra of solid solution   88
Virus   366
Voids between the protein molecules   358
- in close packing   61

Waller formula   110

X-ray diffuse scattering   172
- - - by single crystals of disordered alloys   268
- - - distribution   268
- difraction   153
- intensity of diffuse scattering   177
- parasitic scattering   138
- scattering   126
- - caused by short-range order   270
- - due to statistical distortions   270
- - selective reflection   127
- structure analysis   139
- thermal scattering   138

Zeolite   287
Zone melting   46

Light Scattering in Solids I
Introductory Concepts

Editor: M. Cardona

2nd corrected and updated edition. 1983.
111 figures. XV, 363 pages
(Topics in Applied Physics, Volume 8)
ISBN 3-540-11913-2

Contents: *M. Cardona:* Introduction. – *A. Pinczuk, E. Burstein:* Fundamentals of Inelastic Light Scattering in Semiconductors and Insulators. – *R. M. Martin, L. M. Falicov:* Resonant Raman Scattering. – *M. V. Klein:* Electronic Raman Scattering. – *M. H. Brodsky:* Raman Scattering in Amorphous Semiconductors. – *A. S. Pine:* Brillouin Scattering in Semiconductors. – *Y.-R. Shen:* Stimulated Raman Scattering. – Overview. – Additional References with Titles. – Subject Index. – Contents of Light Scattering in Solids II, III and IV.

Light Scattering in Solids III
Recent Results

Editor: M. Cardona, G. Güntherodt

1982. 128 figures. XI, 281 pages
(Topics in Applied Physics, Volume 51)
ISBN 3-540-11513-7

Contents: *M. Cardona, G. Güntherodt:* Introduction. – *M. S. Dresselhaus, G. Dresselhaus:* Light Scattering in Graphite Intercalation Compounds. – *D. J. Lockwood:* Light Scattering From Electronic and Magnetic Excitations in Transition-Metal Halides. – *W. Hayes:* Light Scattering by Superionic Conductors. – *M. V. Klein:* Raman Studies of Phonon Anomalies in Transition-Metal Compounds. – *J. R. Sandercock:* Trends in Brillouin Scattering: Studies of Opaque Materials, Supported Films, and Central Modes. – *C. Weisbuch, G. R. Ulbrich:* Resonant Light Scattering Mediated by Excitonic Polaritons in Semiconductors. – Subject Index.

Light Scattering in Solids II
Basic Concepts and Instrumentation

Editors: M. Cardona, G. Güntherodt

1982. 88 figures. XIII, 251 pages
(Topics in Applied Physics, Volume 50)
ISBN 3-540-11380-0

Contents: *M. Cardona, G. Güntherodt:* Introduction. – *M. Cardona:* Resonance Phenomena. – *R. K. Chang, M. B. Long:* Optical Multichannal Detection. – *H. Vogt:* Coherent and Hyper-Raman Techniques. – Subject Index.

Light Scattering in Solids IV
Electronic Scattering, Spin Effects, SERS and Morphic Effects

Editors: M. Cardona, G. Güntherodt

1984. 322 figures. Approx. 560 pages
(Topics in Applied Physics, Volume 54)
ISBN 3-540-11942-6

Contents: *M. Cardona, G. Güntherodt:* Introduction. – *A. Pinczuk, G. Abstreiter, M. Cardona:* Light Scattering by Free Carrier Excitations in Semiconductors. – *S. Geschwind, R. Romestain:* High Resolution Spin-Flip Raman Scattering in CdS. – *G. Güntherodt, R. Zeyher:* Spin-Dependent Raman Scattering in Magnetic Semiconductors. – *G. Güntherodt, R. Merlin:* Raman Scattering in Rare-Earth Chalcogenides. – *A. Otto:* Surface Enhanced Raman Scattering: "Classical" and "Chemical" Origins. – *K. Arya, R. Zeyher:* Theory of Surface-Enhanced Raman Scattering. – *B. A. Weinstein, R. Zallen:* Pressure-Raman Effects in Covalent and Molecular Solids. – Errata for Light Scattering in Solids II (TAP 50). – Subject Index.

Springer-Verlag Berlin Heidelberg New York Tokyo

Secondary Ion Mass Spectrometry SIMS-II

Proceedings of the Second International Conference on Secondary Ion Mass Spectrometry (SIMS II) Stanford University, Stanford, California, USA, August 27–31, 1979

Editors: **A. Benninghoven, C.A. Evans, Jr., R.A. Powell, R. Shimizu, H.A. Storms**

1979. 234 figures, 21 tables. XIII, 298 pages (Springer Series in Chemical Physics, Volume 9)
ISBN 3-540-09843-7

Contents: Fundamentals. – Quantitation. – Semiconductors. – Static SIMS. – Metallurgy. – Instrumentation. – Geology. – Panel Discussion. – Biology. – Combined Techniques. – Postdeadline Papers.

Secondary Ion Mass Spectrometry, SIMS IV

Proceedings of the Fourth International Conference on Secondary Ion Mass Spectrometry (SIMSIV), Osaka, Japan, November 13–19, 1983

Editors: **A. Bennighoven, J. Okano, R. Schimizu**

1984. 392 figures. Approx. 480 pages (Springer Series in Chemical Physics, Volume 36)
ISBN 3-540-13316-X

Contents: Fundamentals. – Quantification. – Instrumentation. – Combined and Static SIMS. – Application to Semiconductor and Depth Profiling. – Organic SIMS. – Applications: Metallic and Inorganic Materials. – Geology. – Biology. – Index of Contributors.

Ion Formation from Organic Solids

Proceedings of the Second International Conference, Münster, Federal Republic of Germany, September 7–9, 1982

Editor: **A. Benninghoven**

1983. 170 figures. IX, 269 pages (Springer Series in Chemical Physics, Volume 25)
ISBN 3-540-12244-3

Contents: Field Desorption. – ^{252}Cf-Plasma Desorption. – Secondary Ion Mass Sepctrometry (SIMS) Including FAB. – Laser Induced Ion Formation. – Other Ion Formation Processes. – Index of Contributors.

Secondary Ion Mass Spectrometry SIMS III

Proceedings of the Third International Conference, Technical University, Budapest, Hungary, August 30–September 5, 1981

Editors: **A. Benninghoven, J. Giber, J. László, M. Riedel, H.W. Werner**

1982. 289 figures. XI, 444 pages (Springer Series in Chemical Physics, Volume 19)
ISBN 3-540-11372-X

Contents: Instrumentation. – Fundamentals I. Ion Formation. – Fundamentals II. Depth Profiling. – Quantification. – Application I. Depth Profiling. – Application II. Surface Studies, Ion Microscopy. – Index of Contributors.

Springer-Verlag Berlin Heidelberg New York Tokyo